THE AUSTRALIAN INDIAN SANDALWOOD PLANTATIONS

A Review and Summary of Research and Plantation Silviculture

KEN ROBSON

The Australian Indian Sandalwood Plantations
A Review and Summary of Research and Plantation Silviculture

ISBN: 978-1-923631-77-9 (paperback)

Published in Australia by Ken Robson and InHouse Publishing.
www.inhousepublishing.com.au

A catalogue record for this book is available from the National Library of Australia

THE AUTHOR

Ken Robson began his professional career in North Queensland, specialising in the cultivation and management of tropical pine plantations. Early in his career, he played a key role in rehabilitating degraded tropical landscapes across northern Queensland. This work became particularly significant following the introduction of World Heritage legislation for the North Queensland rainforests, which intensified efforts to restore and protect these globally important ecosystems.

Over the next 25 years, Ken dedicated himself to tropical forestry research across northern Australia and the western Pacific, spending much of this time with the Queensland Department of Primary Industries and Fisheries. His research and applied work were instrumental in advancing forestry practices in diverse and often challenging environments. He contributed to the development and implementation of improved silvicultural methods, strengthening both environmental outcomes and industry standards.

Following his extensive research career, Ken transitioned into a senior role with the two largest Sandalwood plantation companies operating in tropical Australia. For 17 years, he focused on improving operational efficiency and plantation performance. Drawing on his deep expertise in tropical forestry, he optimised management systems and enhanced productivity across large-scale Sandalwood operations.

Ken is widely recognised as one of the few professionals with comprehensive expertise in Sandalwood silviculture and breeding. He has worked extensively with all commercially grown Santalum species, establishing himself as a leading authority in the field. His contributions have supported the growth, sustainability, and commercial development of the Sandalwood industry throughout the region.

Now based in Brisbane, Ken operates his consultancy, Tropic Plantation Solutions. Through this practice, he continues to provide strategic advice and technical expertise in tropical forestry and plantation management.

PREFACE

This book is not intended as a practical manual for the establishment or day-to-day management of Sandalwood plantations. Rather, it presents the technical principles and scientific reasoning that underpin the range of silvicultural approaches developed to achieve reliable establishment, sustained growth, and optimal productivity. A proposed management framework is, however, outlined to provide context and practical direction.

While this account traces the evolution of Sandalwood plantation development and research over the past 45 years, it does not seek to be exhaustive. Instead, it highlights the most influential areas of investigation and acknowledges the contributions of those whose work has shaped the industry. Over this period, substantial and measurable advances have been made in seedling survival, growth performance, plantation productivity, and management practices.

For much of this time, senior managers and corporate executives often regarded Sandalwood plantations as modified horticultural enterprises rather than true forestry operations. In my view, this perspective constrained progress by limiting the application of established forestry principles and practices. Encouragingly, this mindset has shifted, and future research and management are increasingly being guided by sound silvicultural principles.

The book also draws on the outcomes of research field trials conducted over a 16-year period by my team and me. Much of this work was not previously published due to commercial confidentiality agreements with private organisations; with those entities no longer operating, this information might otherwise have been lost to the broader industry.

Despite these advances, much remains to be understood about the complex and dynamic relationship between sandalwood and its host species under plantation conditions. Although considerable research has examined the physiological interactions between sandalwood and its hosts, much of this work has been undertaken using potted seedlings. The extent to which these findings apply to mature trees growing under field conditions is not yet fully resolved.

The foundations of the modern sandalwood industry can be attributed to the pioneering foresters and early researchers of Western Australia, who recognised the potential for cultivating Santalum album in plantations across northern Australia. As many of these early stands now approach the end of their rotation, responsibility increasingly shifts to the marketplace, to demonstrate to global consumers that plantation-grown sandalwood can provide a consistent, high-quality, and sustainable alternative to material historically sourced from overharvested wild populations

Ken Robson (ITC/ Elders – TFS/ Quintis)

CONTENTS

Chapter 1. REVIEW OF SANDALWOOD R&D

This compendium presents a comprehensive review of research pertaining to the development of sustainable plantations of Indian Sandalwood (*Santalum album*), hereafter referred to as "Sandalwood," in Australia. It further describes current operational methodologies and examines the research underpinning contemporary knowledge regarding Sandalwood establishment and growth. The publication draws chiefly upon the insights and findings of Mr. Ken Robson, accumulated since 1991, and incorporates peer-reviewed studies relevant to the subject.

There are seven chapters. Chapter 1 highlights the key areas of research into Sandalwood in Australia and acknowledges the researchers and forest managers involved. It summarises this chronologically starting in the 1980's and concludes with the most recent aspects of research.

Chapter 2 describes the main species utilised in the Australian Sandalwood plantation system. Research undertaken by Quintis into other potential host species is discussed.

Chapter 3 describes the attributes of the sites, soils and climate suitable for successful establishment of Sandalwood plantations and correlates these attributes with the performance across these sites.

Chapter 4 examines the components for plantation development, which includes methodology for pre-plantation soil due diligence, nursery requirements, the various planting layout configurations, planting systems and infill planting.

Chapter 5 outlines key components in plantation management, including silvicultural stand management, weed control, pests and disease, plantation nutrition, and the range of associated and potential risks.

The chapter 6 provides a detailed outline of the Quintis Sandalwood breeding program, including development of seed production facilities and DNA analysis results.

The final chapter discusses topics of Sandalwood research where is there are still knowledge gaps and presents a plantation plan utilising the latest information described in the book.

1. INTRODUCTION

A paper titled "Sandalwood Research: A Global Perspective" by Subasinghe (2013) stated that scientific research on Sandalwood (*Santalum album*) started sometime in the 1940's. At that time, the species was starting to become of interest in countries outside of India. It can be assumed that this research primarily was undertaken within India, as Subasinghe (2013) does not provide any evidence, or examples of researchers or research undertaken in Australia (or any other country) in this period. However, the empirical evidence in Australia points to research specifically on *S. album* starting in Australia in the 1980's.

2. FOREST RESEARCH HISTORY

A brief review of the research into other key forestry species within Australia will put the period spent on research of *S. album* into perspective. The exotic pines and southern eucalypts planted in large areas across southern Australia have been the subject of research for over 120 years and there continues to be extensive work undertaken today.

In the initial research on plantation forestry species, the forestry departm*ents of Western* Australia (WA) and South Australia (SA) pursued similar strategies. WA's first trials with Pinus pinaster took place in the sand dunes near Bunbury in 1897. Additional trial plots *followed, an*d after extensive study and testing of more than 25 species, two pine*s emerged as* preferred: Pinus radiata was best suited to richer soils and areas of higher rainfall, while Pinus pinaster thrived on poorer, often sandy soils in both high and medium rainfall regions. Between 1900 and 1935, significant research focused on these two species (Williamson, Moore, & Warren, 2005). WA established its first large-scale eucalypt plantations in the Dryandra Forest, beginning planting in 1923 (Underwood, Early forestry in WA, 2023)

Figure 1: Hoop Pine research trial plot circa 1930's. (Original photo source; QLD Forestry Dept, Atherton).

South Australia's first tree nurseries began operating in 1876, followed by field trials in the mid-north and at Mount Gambier in the southeast to identify suitable tree species and plantation locations. Bundaleer Forest Reserve near Jamestown became Australia's inaugural commercial plantation forest. The Woods and Forests Department, now known as Forestry SA and considered one of the world's oldest plantation-based forestry authorities, was founded in 1882 (Johnston and Robinson 2015).

Edward Swain was appointed Queensland Director of Forests in 1918 and his directive to investigate the matching of exotic species around the world, using similar climatic patterns. This was highly successful in identifying slash (*Pinus elliottii*) and loblolly (*Pinus taeda*) pines as being suitable for Queensland. Trials in these species started in the late 1920's. At the same time there was extensive research into the exotic pines, they also started to look at the high value native species like hoop pine *(Araucaria cunninghamii,* see Figure 1), Kauri (*Agathis robusta*) and Red Cedar (*Toona ciliata*). These were identified as potential high value plantation species and trials were established in the 1910's.

In New South Wales (NSW), the first attempt at a commercial pine plantation was made at Tuncurry State Forest on the mid-north coast in 1912. Eucalypt plantations followed in 1939 at Walaroo and Whian Whian State forests on the north coast (Forest Corporation website).

Outside Australia, the high value tree Teak (*Tectona grandis*) has been researched since at least the 1600's with plantations being established in Sri Lanka as early as 1680 (FAO unpublished data).

3. SANDALWOOD RESEARCH

a) Sandalwood spicatum

Australian research into the native Sandalwood *(Santalum spicatum)* began in 1895. The research was focused on regeneration methods (McKinnell F. A., 1990). Lonergan (1990) published, "Historical review of Sandalwood (*Santalum spicatum*) Research in Western Australia" in 1990. That documents extensive details into the research undertaken. In summary, he shows the research focused on; regeneration trials, including some germination trials, plantings across a range of soils, seed storage, host trials, effects of browsing and fire on native generation, and growth rates across sites. Much of this work was undertaken in the 1960's and 1970's but studies into native Sandalwood continues today by the Western Australia government.

Figure 2: Wild growing Western Australian Sandalwood (Santalum spicatum) near Kalgoorlie (Plonk. 2022).

b) Sandalwood album

As indicated by Subasinghe (2013), the introduction of *S. album* into Australia led to an increase in research undertaken into the species. Credit must be given to the senior foresters in the Forests Department and the Department of Conservation and Land Management (CALM) in Western Australia for foreseeing the potential of Indian Sandalwood (*S. album)*.

1980's

The concept of growing Sandalwood in plantations in Australia first arose in the early 1980s. The Forest Department of Western Australia and the Australian Sandalwood Company sent a party to India to observe their Sandalwood industry and investigate the opportunities for development of Sandalwood forestry in WA. This trip led to the first Sandalwood seed being imported into Australia and thirty trees were established in Kununurra.

In the early 1980's Chris Done established Sandalwood seedlings in the Kununurra arboretum off Ivanhoe Road. He also planted some onto his property on Research Station Road. Although some of them survived they never thrived (Underwood, Early forestry in WA, 2023).

In 1987, a series of plantings evaluating various host combinations was initiated at the Frank Wise Institute within the Ord River Irrigation Area (Radomiljac A. M., 1998). During the same period, Murdoch University conducted trials on different soil types with a range of hosts, utilising newly imported genetic material from India. (Barbour, Plummer, & Norris, 2011)

1990's

The mid 1990's generated a lot of research interest in Sandalwood. Western Australia has had a large Australian Sandalwood (*S. spicatum*) industry for over 120 years, and researchers such as Dr John Fox, Dr Diane Barrett, Prof. Jen McComb, Owen Loneregan, Dr Frank McKinnell and operational foresters such as Peter Richmond, Roger Underwood, Peter Kimber and Ian Kealley were involved with developing sustainable management systems. This work led them to an interest in the Indian species (*S. album*).

It was well-known that Indian Sandalwood was becoming over harvested where it occurred naturally; and that demand for Sandalwood products was much higher than supply (Rashkow, 2022). CALM had commenced plantation establishment of Australian Sandalwood in southwest Western Australia, and the much higher value of the Indian Sandalwood made a research program looking at how to establish this species seem logical.

The earliest work on *S. album* was by Prof John Fox and Diane Barrett from Curtin University. Fox was initially working on *S. spicatum* but did look at *S. album* in the early 1990's. Their research focused on nutrition and shading of young *S. album* seedlings.

Prof. Jen McComb was a key researcher and was one of the first to look at the physiology and clonal propagation of the parasitic plant during the 1990's. Like most of the Western Australia researchers she was primarily working with the native species. A grove of *S. album* growing within the Murdoch University grounds enabled researchers like Fox and Barrett to observe various growth characteristics of S. album. In 1993, McComb utilised this grove to look at potential hybridisation between *S. album* and *S. spicatum* (Rugkhia & McComb, 1993).

A turning point came when Mr Andrew Radomiljac, funded by the department of Conservation and Land Management, commenced a PhD investigating the growing of *S. album* in plantations at Kununurra in 1995.

At the same time in Queensland, there was also an interest in a Sandalwood research program. The technical program was run by Mr Ken Robson and supervised initially by Dr Grahame Applegate, in the late 1990's and then Dr Rod Keenan, in the early 2000's. Like Western Australia, the Queensland interest in Sandalwood grew from the management of the native Sandalwood (*S. lanceolatum*). The trial work with *S. album* also focused on seedling survival. This initially examined germination and pot hosting trials.

Radomiljac completed his PhD thesis in 1998 (Radomiljac A. M., 1998) and this provided a boost to the interest in a plantation industry based in the Ord River Irrigation area (ORIA) in WA. His thesis provided solutions to Sandalwood survival after planting with the introduction of the pot host *Alternanthera* at about 15 weeks before planting out. He also examined the effect on attached Sandalwood to a variety of hosts. Nitrogen fixing hosts proved to improve Sandalwood growth more than non-N-fixers (this has been confirmed by several other researchers since). Additionally, Radomiljac investigated the effect of the diurnal profile of photosynthesis and transpiration of various hosts on Sandalwood, the root to shoot ratio and the relationship of sap amino acids when paired to different hosts. In almost all his experiments, Sandalwood performed best when paired with Sesbania formosa. (This is discussed in more depth in Chapter 2).

Finally, Radomiljac researched ways to influence heartwood initiation with the injection of paraquat (a contact herbicide) and ethrel (a growth regulator, a.i Ethephon).

First commercial plantations

During this period, Managed Investment Scheme (MIS) were starting to invest in forestry projects across Australia. An MIS is a managed fund where investors' money is pooled. However, the first attempt by MIS to grow Sandalwood failed. This was blamed on several reasons including the rate of expansion in relation to system knowledge, use of the wrong hosts, poor seedling quality and the many aspects of management that still needed to be investigated and understood for reliable tree growth and oil production (Barbour, Plummer, & Norris, 2011). Nevertheless, MIS would be the main driver for Sandalwood plantation growth for the next ten years. MIS investment in forestry peaked in about 2007 but would continue to fund Sandalwood projects for another nine years.

An example of a failed attempt was the 75 hectares planted at Arawadi Road in Kununurra and flood irrigated. This was established by Kimberley Timber Co, an African mahogany grower, in 1997. The planting was effectively a Mahogany (*Kaya senegalensis*) plantation with Sandalwood inter-planted amongst the mahogany. The Sandalwood under the mahogany were few and far between due to poor survival of seedlings and high mortality over the years. The trees that remained at harvest had never developed a long bole, with all surviving trees tending to be small and multi stemmed.

Figure 3: Within the Arawadi plantation at harvest. White arrows are pointing out Sandalwood trees (Robson, 2012).

In the late 1990's, a new company was formed, Tropical Forestry Services Corporation Ltd, specifically to take advantage of the MIS program and develop Sandalwood plantations. The East Kimberley Sandalwood Project No. 1 (EKS) was TFS's first project. This was 109 hectares established on lease land (Stock Route Road, ORIA) in 1999. Although this plantation was moderately successful, compared to the Arawadi plantation, the basis of the host configuration was still African mahogany (*Khaya senagalensis*). However, there was also *Senna siamea, Dalbergia lanceolaria and some Cathormion umbellatum* in the mix. The EKS plantation was harvested in 2013.

Tropical Forestry Services Limited

Figure 4: The EKS plantation did produce some very good quality Sandalwood logs even though Mahogany was the main host. White arrows pointing out the Sandalwood (Robson, 2012).

2000's

In 2001, a second private Sandalwood company started developing plantations in ORIA using MIS funds. Integrated Tree Cropping (ITC) was a major competitor for TFS for land and resources for new plantations. As each year passed, the amount of land being planted with Sandalwood increased in the valley.

As the plantation size ramped up, the focus of field research was on configurations and hosting, nursery and propagation. The early trials at Frank Wise Research Station (FWRS) demonstrated that using the agricultural machinery available for the site preparation and cultivation in the flood irrigated valley could be suitable for Sandalwood. For no reason other than this equipment was available, the standard distance between rows for Sandalwood plantations became 3.6m.

The term "configurations" generally refers to the species mix, the distance between trees along the row and the layout of those species within a cycle. Configurations will be discussed further in Chapter 4.

At the time, the investigations into configurations were not led by research but by the senior operations managers in each company at the time. As the Sandalwood plantation enterprise was still new, Mr Mal Baker and Chris Done (TFS) and Mr Bruce Matherson (ITC) had their own ideas on the best layouts and during the early 2000's they put their ideas into their annual planting programs.

At this time there was considerable interest in irrigation practices. Concern about root pathogens (Phytophthora spp) led to a reduction in irrigation frequency and volumes, and generally to drier plantations.

Interestingly there was limited further research into new potential host species. The species trials undertaken at FWRS in 1987, which indicated that *Cathormion umbellatum*, *Dalbergia lanceolaria* and *Senna siamea*, were adequate long-term hosts and *Sesbania formosa* and *Acacia trachycarpa* performed well as intermediate host, seem to be sufficient for the companies as they developed new plantations.

Initially, it was anticipated that the long-term host could also provide a commercial product either at Sandalwood rotation or soon after. Hence the use of species such as Khaya and Tectona in the original plantings. However, it became clear that a mixed species plantation of these high value trees as hosts and Sandalwood could not work. Optimal management had to be for one or the other. For this reason, it became clear early in the rotation that those high value species would not be suited and were dropped from all plantings from 2000 onwards. Although there was not great interest in looking for new hosts, TFS was doing some work investigating new provenances of *Senna siamea*.

Early Sandalwood plantations experienced low seedling survival rates, often reaching only 50%. This challenge prompted extensive research into propagation and nursery practices conducted by companies, universities, and State government agencies. Areas of investigation included optimal potting media, seed management, enhancement of germination rates, nursery pest and disease control, as well as mechanisation of both Sandalwood and host propagation and establishment. As a result of these efforts, survival rates exceeding 90% are now commonly achieved.

In this period there a growing interest in the genetics of Sandalwood in Australia. An interesting study undertaken by Ms Danica Harbaugh (University of California, Berkeley) for her PhD was to investigate the Phenology of the Sandalwoods (Santalaceae). This study provided insights into the biogeography, taxonomy and ethnobotany (Harbough & Baldwin, 2007) of the Sandalwoods.

Both the West Australian and Queensland government agencies were dabbling in Sandalwood research. In Western Australia, Dr Liz Barber and Mr Jon Brand continued the trial work started during Radomiljac PhD at Frank Wise Research Station. This work mostly analysed the performance of Sandalwood with various hosts and spacing regimens as reported in Barbour, Plummer, & Norris (2011). During a harvest of some trees in these trials significant heartwood rot was observed. Barbour, Norris and Dr Treena Burgess from Murdoch University undertook a study investigating the various causes of the rot (Barbour, Norris, & Burgess, 2010).

In Queensland there was growing interest in rehabilitation of degraded tropical lands and Researchers thought that Sandalwood establishment could play a role in getting trees back into weed and grassed areas. In 2003, the Queensland Department of Primary Industries (QDPI) established a genetic resource stand of four of the Santalum species at the Walkamin research station in north Queensland. Additionally they established test plantings of Sandalwood on sites at Babinda, Mareeba, Charters Towers and Ayr.

Figure 5: QDPI researcher with a Sandalwood host and species trial at Charters Towers in 2002 (Robson, 2002).

By about 2006 Bruce Matherson had left ITC and was one of the largest private growers of Sandalwood in the Ord valley. He embarked on an experimental establishment program growing Sandalwood without hosts. He provided his Sandalwood nutrition with regular and constant application of aerial liquid foliar fertiliser.

Before 2005, research activities for Australian Sandalwood companies were primarily conducted by operational personnel. The majority of these "trials" conducted between 2000 and 2005 and did not adhere to the principles of true experiments. These initiatives were typically the result of someone proposing a novel idea without incorporating essential experimental practices like replication or randomisation. Documentation and recording of the implemented strategies were, in most cases, contingent on luck rather than a systematic approach.

In 2007 TFS employed Dannielle Wiseman, then in 2009 Michael Patterson, and ITC employed Ken Robson in 2008 as dedicated researchers to undertake research trials. With these researchers working their respective R&D programs, several field trials were established across ORIA. This work included the Sandalwood breeding work, investigations into the effect of soil type and irrigation on growth, pest control and the use of chemicals. A new era had begun where research projects were being properly designed, enabling statistical analysis of results.

By 2008, the forestry section of CALM became the Forest Products Commission (FPC) and Liz Barbour and John Brand were the key Sandalwood researchers. As the FPC had the largest genetic resource of

Sandalwood they undertook a diversity study with Dr Marget Byrne, Dr Julie Plummer and Dr Chris Jones from UWA. This was really the first time a study using DNA molecular markers was utilised to identify genetic variation. As no genomic library existed for *S. album*, RFLP markers developed for *S. spicatum* where used. As well as *S. album,* this study also looked at and compared other Sandalwood species, *S. austrocaledonicum* and *S. spicatum*. This work formed the baseline for the genetic analysis that Quintis would first undertake in 2012.

At the same time, Dr Tony Page at James Cook University was looking at inter species hybridisation. He made several successful hybrids with *S. album* and *S. austrocaledonicum* and *S. lanceolatum* (Page, Hanington, & Clement, 2012).

Both TFS and ITC (which was acquired by Elders Forestry by 2010) started a Sandalwood improvement program. Although it was well understood that the genetic diversity of the gene pool held by both companies was quite small, they started on a basic improvement program with the resources they had. The TFS/Quintis breeding program will be discussed in detail in Chapter 6. It should be noted that the key players in this initial breeding work were Chris Done from TFS and Andy Wright and Ken Robson from ITC.

2010's

The MIS scheme was still providing funding for new plantations with nearly 300ha planted by TFS in the Ord Valley in 2010. During this period many of the other MIS schemes with Eucalypt species were finishing due to poor financial management (Locke, 2012) and TFS was looking for other means of funding. A further 350 ha was planted at Kingston Rest funded by an Institutional investment, which would be the main funding behind TFS plantations for the next seven years.

From 2012 to 2014, Andrew Brown was the Quintis Forestry research team leader with Ken Robson and Dr Dan Manson working as researchers. In 2014, Ken was appointed head of Forestry Research while Andrew focused on oil production research. In 2014, Emma Mende was employed to assist with the research activities.

The research discussed in this section was undertaken by TFS/Quintis. Due to the secrecy and lack of openness and collaboration with other Sandalwood growers, the research work by Santanol (ex-Elders) is difficult to define. However, it was well known that Santanol's Charmain Matherson was working on tissue culture of Sandalwood. Her lab was located at their nursery complex on Packsaddle Road.

At that time, TFS was expanding their plantation estate into the Northern Territory and Queensland. These sites had never been planted with Sandalwood before, there were a lot of unknowns. The same establishment system and species were used across all the new sites regardless of climate or soil type. The other big difference was the application of water by drip irrigation rather than flood application. Until 2010 all the plantations had been irrigated with a flood system. All the subsequent plantations established in Kingston Rest and the Northen Territory would be irrigated by a drip system.

Consequently, in the early 2010's the research focus on establishment with the key areas being, weed control, testing various herbicides and pre-emergents, establishment nutrition, pest control (particularly giant termite and new problem species such as fig leaf beetle) and development of an irrigation scheduling system.

Figure 6: Experiments testing pre-emergent herbicides were established (Robson, 2012).

Before 2010, weed management within newly established plantations relied primarily on manual methods, such as hand pulling and chipping, performed by teams of workers. As annual plantation establishment increased to approximately 1,500 hectares, this approach proved unsustainable. Consequently, research was conducted to evaluate various systemic and contact herbicides, along with different application techniques. Additional studies also examined alternative weed control strategies, including the use of weed matting and tree guards. The outcomes of these trials will be analysed comprehensively in Chapter 5.

The majority of the plantations in the Northern Territory are established on Kandosol and Tenosol sandy loams soils. It is well documented these soils have poor nutritional value (Moody & Bolland, 1999), and trials were established to investigate the benefit of fertiliser application at establishment. As the plantations aged this research continued with the development of a foliar baseline for key species, elemental deficiencies and optimal methodologies for applications to trees, and potential benefit of bio-stimulants and inoculants. The results of this nutrition work will be discussed in depth in Chapter 5.

Initially the purpose of growing the trees was to produce oil. The oil is obtained by removing the heartwood, chipping and distillation. However, as the markets evolved for TFS/Quintis, it was realised there was significant potential markets for the timber as single logs. This realisation led to a greater focus on research to improve stem form, quality and size.

Figure 7: Host overcrowding was becoming obvious by 2014 (Robson, 2014).

Up until the late 2000's only minimal stem form pruning was undertaken, and a high number of the final crop trees were double leaders. When harvesting commenced in 2013 it was discovered that a single stem produced more area of heartwood than two stems on the same tree. The development of a pruning regime to ensure single stem development was implemented on research advice in 2014. Although thinning is a standard silvicultural practise to improve stem form and size in almost all forestry organisations, despite advice from Kimber and Underwood it was not considered an option for TFS/Quintis. Plantations. This was partially due to a failure of not understanding forest stand dynamics and a false belief that more trees equated to more oil. However, it was clear by 2014 that the high numbers of hosts were creating overcrowding and a shading effect on the Sandalwood reducing their size and increasing mortality. An initial host thinning trial and a study on thinning lines was undertaken in 2013. The importance of this work will be discussed more in Chapter 5.

Other areas of research included development on new grafted seed orchards located in Northern Territory and Queensland. The work undertaken with this will be discussed in Chapter 6.

As mentioned, the move into new areas asked the question, are current hosts suited to these new sites? Research into potential new hosts was undertaken and will be discussed further in Chapter 2.

The change in water distribution from flood to drip also presented several challenges for plantation establishment and tree growth. The water for all the Northern Territory plantations is sourced from underground aquifers and requires a complicated design of pumps and motors, underground mains, sub mains, risers and drip lines. The design for this was contracted out to specialist experts. It is important to note that the plantation row orientation for all Quintis plantations is based solely on the optimal irrigation design without any environmental or growth benefit considerations. The R&D undertaken to assist with irrigation scheduling, determining how much water is required for optimal growth will be discussed further in Chapter 5.

The move to the NT and Qld also meant there were new risks, particularly in regard to pests and diseases. The risk from the giant termite (Mastotermies) was known but was not fully understood. A system of control and prevention needed to be developed. This was undertaken with the assistance of Dr Brian Thistleton from the Northern Territory DPI. Bob Sandery was employed to manage and improve the program. The termite was just one of the P&D risks and has been found in the Kingston rest, Northern Territory and Queensland sites. During the initial years in the Northern Territory, several important pests and disease challenges emerged, which will be examined in more detail in Chapter 5.

Although there was significant research into nursery productivity during the 2000 period, TFS continued to improve their seedling quality with research in the 2010's. A trial looking at seedling size and that effect on seedlings was undertaken by Robson in 2012, which showed seed size does carry through to seedling vigour in the field (Robson K. , 2012). Dr Dan Manson investigated the inoculation of Dalbergia with Rhizobium to improve growth and quality. Successful completion of the trial led to a recommendation from this research and was implemented into the nursery manual. In 2016 Mende looked at the effect of Sandalwood development with varied alternanthera introduction.

2020's

The main focus of R&D since 2020 has been in mid rotation nutrition and stand management.

In 2019, experiments looking at the application of phosphorus to deficient sites began. These initial trials investigated two rates and two methods of application across three sites. Results for this work have been implemented with almost all the Northern Territory estate receiving a broadcast application of 60 kilograms/ha P. These initial trials have led to looking at application of P by fertigation and are ongoing. The results of this work will be discussed more in Chapter 5.

Mostly due to a change in leadership and the appointment of trained foresters in late 2010's, there was renewed interest in stand management. Particularly in the thinning of both hosts and Sandalwood to suit site quality. In 2021 the first trial thinning Sandalwood was established. Consultation with investors saw a further four thinning trials added in 2022 and 2023. This topic will be discussed in more depth in Chapter 5.

With the addition of Dr Precila Gonzales in 2020 Quintis started a project looking at the use for the excess biomass (hosts) after harvest was completed. Dr Gonzales undertook trials with production and use of biochar from host species.

The University of Sunshine Coast appointed a PhD student in 2022 to look more deeply into Sandalwood/host relationships. Although much of this work has been done previously, particularly in potted trials, this work brings some advantages for forest managers.

Two new Sandalwood planation management companies entered the management scene in 2022/23, Apical Forestry Management (AFM) and Great Northern Plantations (GNP).

CHAPTER 2.SPECIES CHARACTERISTICS

It was established by Radomiljac (1998) that Sandalwood performed better when parasitising a nitrogen fixing host. All the tree hosts utilised within the plantation are N-fixing.

The Sandalwood plantation system uses a three host arrangement. A pot host while in the nursery, then an intermediate or short-term host (STH) planted next to the Sandalwood, and lastly a long-term host (LTH) planted a row across or distant from the Sandalwood.

A component of this arrangement is that the earliest host self-thin. After 12 months the pot host dies, and the Sandalwood is reliant on the intermediate or short-term host for a period. This period gives the slower growing long-term host time to become large enough to sustain Sandalwood hosting. At about three to four years after establishment, the intermediate or short-term host starts to die off and the Sandalwood moves across to the now established long-term host.

The development of this system is critical for the success of current plantations. It provides the Sandalwood with hosting adequate for good growth as well as provide an optimal stand arrangement. This will be discussed in greater detail in Chapter4. Plantation development.

Potted sandalwood seedlings in the nursery with alternanthera pot host.

Sandalwood seedling approx. 6 months after planting with pot host

2 year old Sandalwood next to shorth term host

4 year old sandalwood plantation recently moved to long term hosts

Mature 8 year old sandalwood tree

Full rotation (15 years) sandalwood tree

Figure 8: The three host stages are an important component to the success of Sandalwood plantations.

This chapter discusses the principal species involved in this host system. It provides an in-depth look at S. album, as well as the primary host species currently in use and the potential hosts that have been studied.

1. SANTALUM GENUS

The Santalum genus currently comprises 16 species, with *S. album* being the most renowned, the species often referenced when discussing Sandalwood. This genus is categorised into five genetic sections: Eucarya, Santalum, Solenantha, Hawaiiensia, and Polynesia. While they all belong to the Santalum genus, the Eucarya species exhibit more primitive botanical features compared to the modern Santalum group. The Eucarya group includes *S. spicatum, S. acuminatum*, and *S. murrayanum*.

Biogeographic reconstructions point to Santalum originating in Australia approximately 50 million years ago. The genus split into two groups around 20 million years ago (Harbough & Baldwin, 2007). The Eucarya species can be distinguished from the rest of the Santalum by their primitive reproductive characteristics. They feature leathery calyxes with minimal colouring and large globular fruit attractive to larger mammals and marsupials. Their thick leaves are suited for the arid Australian landscape.

Approximately ten million years ago, the Santalum group diverged into the other four sections, as noted by Harbough and Baldwin (2007). These sections exhibit modern botanical features, with petite flowers and fleshy drupes appealing to birds. Their thinner leaves are characteristic of monsoonal tropical species. In turn approximately 1 million years ago the Santalum genus diverged into the various species found today that are spread out across the Pacific.

The Solenantha, Hawaiiensia, and Polynesia species have notably small populations, with many facing threats of extinction due to overexploitation, fire, and habitat loss from agriculture and urban expansion. In fact, a known species *S. fernandezianum* from the Juan Fernandez islands (Chile) was declared extinct in 1978 due to over exploitation (IUCN , 1998).

Table 1: Sections, species and native locations for Sandalwood species. The brackets next to some species indicate there are sub species.

Section	Species	Native location
Santalum	*S. album*	Australia, Timor, Indonesia (India)
Santalum	*S. yasi*	Fiji, Tonga, Niue
Santalum	*S. austrocaledonicum (3)*	New Caledonia, Vanuatu
Santalum	*S. macgregorii*	Papua New Guinea
Santalum	*S.lanceolatum*	Australia
Santalum	*S.leptocladum*	Australia
Santalum	*S. boninense*	Bonin Islands
Santalum	*S. obtusifolium*	Australia
Eucarya	*S. acuminatum*	Australia
Eucarya	*S. spicatum*	Australia
Eucarya	*S. murrayanum*	Australia
Solenantha	*S. freycinetianum (3)*	Hawaiian Islands
Solenantha	*S. haleakalea*	Hawaiian Islands
Hawaiiensia	*S. ellipticum (2)*	Hawaiian Islands
Hawaiiensia	*s. paniculatum*	Hawaiian Islands
Polynesia	*S. insulare (9)*	Society Is, Marquesas Is, Pitcairn Is, Cook Is, Austral Is.

Aside from the phenotypic and geographic variation between the species, it is the oil characteristics that has determined the value and in some cases the vulnerability of the species in their natural habitat. All of the species discussed have commercial quantities of desirable oil traits. This is specifically alpha (*α*) and beta (β) santalol. The higher the percentage of these components the more desirable the oil.

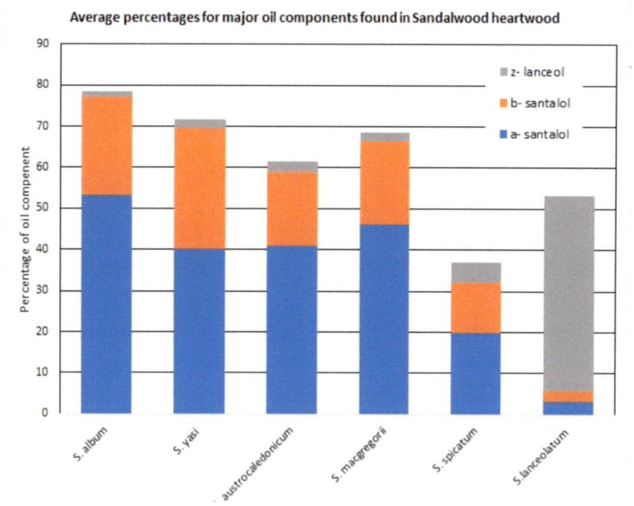

Figure 9: The Average key oil constituents for 6 Santalum species (Doran, Brophy, & Thompson, 2002).

Figure 9 shows the average amounts of alpha (*α*) and beta (β) santalols and lanceol found in these Australian and Pacific Sandalwood species.

a) Pacific Species

While all Santalum species contain some quantity of aromatic heartwood oils, those with the highest levels of desirable santalols include *S. album, S. yasi,* and *S. austrocaledonicum* (Figure 9). These species are closely related and no reproductive barriers that inhibit hybridisation (Page, Hannington, & Clement, 2012). These species thrive in similar environments, typically hot, humid and moist monsoonal climates. They are often found growing on dry, well-drained western slopes shielded from natural fires.

Figure 10: S. yasi growing under Pinus plantation in Tonga (Robson 2004).

S. yasi, an attractive upright tree closely resembling *S. album*, and can be distinguished by its longer thin leaves, especially in seedling stage. Indigenous peoples of Fiji and Tonga have traditionally valued *S. yasi*, propagating it mainly for agroforestry and garden planting. Both countries regulate commercial harvesting of native Sandalwood.

Figure 11: S. austrocaledonicum plantation in Vanuatu (Robson, 2001).

S. album, introduced to Fiji, is favoured by the Fijian Indian population for planting, it can be found in wildling outbreaks in Suva, raising concerns about potential hybridisation and dilution of native *S. yasi.*

S. austrocaledonicum, smaller than *S. album* and *S. yasi,* is distributed across Vanuatu and New Caledonia. While it can develop a long stem bole in native environments, this is uncommon in garden settings. Extensive population studies have been conducted, leading to the development of clonal seed orchards in Vanuatu. Notably, *S. austrocaledonicum* seeds are black, unlike the light brown seeds of *S. album* and *S. yasi*.

While *S. austrocaledonicum* is primarily planted for agroforestry and garden purposes, some efforts have been made towards commercial-scale plantations. Both Vanuatu and New Caledonia regulate native harvesting, with New Caledonia implementing a reforestation program to replace harvested trees.

b) Australian Species

S. spicatum is one of the species in the older Eucarya group of Santalum species. It occurs across a wide range of environmental conditions and has a broad distribution throughout the southern semiarid, and into the northern arid areas of Western Australia. There are distinct morphological variations that have been observed across this range suggesting the presence of two ecotypes within the species (Brand, 1993).

This species has been exploited in Australia for its aromatic wood since the mid 1800's and managed harvesting from native stands still occurs. Additionally there have been several planting programs undertaken to support and rehabilitate harvested areas. Establishment of this species is by direct seeding, burying seed near a potential host, rather than planting seedlings. Plantations generally use *Acacia acuminata* as the main host species.

Figure 12: S. spicatum plantation with A. acuminata hosts, near Beverly WA (Robson, 2009).

S. lanceolatum is northern Australian Sandalwood and has the broadest natural distribution of any Sandalwood species occurring in Australia. This species typically presents as a shrub, but can also develop into a small tree, reaching heights of up to eight metres, as documented by Applegate and McKinnell (1993). The distribution of S. lanceolatum is patchy, with occurrences recorded from the tropical north of Australia, across the arid central regions, and extending into the southern temperate zones (Applegate & McKinnell, 1993). Across its geographic range, considerable variation has been observed in morphological characters such as leaf shape, colour and size (Gardner, 1929) (Applegate, Davis, & Annable, 1990). *Santalum lanceolatum* grows in alluvial and colluvial soils, around drainage areas, and in sandy clay soils with high levels of calcium carbonate (Applegate, Davis, & Annable, 1990)

Since the mid-19th century, *S. lanceolatum* has been harvested from New South Wales, Victoria, and from northern Queensland (Taylor , Swift , & Collins, 2000) (Warburton , James, Fripp, Trueman, & Wall, 2000). In 2007, *S. lanceolatum* was geographically split into two species, S. *lanceolatum* north of 20° latitude and *S.leptocladum* south of that line (Harbaugh, 2007).

2. SANTALUM ALBUM (INDIAN SANDALWOOD)

Sandalwood is a highly valued aromatic hardwood and has significant cultural, medicinal and commercial importance in Asian, Middle Eastern and Pacific Island countries. The oil has been used in many of these countries since the 7th century BC (Srinivasan, Sivaramakrisnan, Rangaswam, Ananthapadmanabha, & Shankarnarayana, 1992). The oil is distilled from heartwood and has many uses. This includes, an important fixing agent used in perfumery, aromatherapy, the powder is used for incense and joss stick manufacture. Sandalwood logs are sought after for furniture and carving. Sandalwood oil content from *S. album* is the most commercially important of all the Sandalwood species.

Species Description

The natural range of Indian Sandalwood is from the islands of Flores, Sumba and Timor and the top end of Northern Territory in Australia. However, the largest population for this species is distributed widely throughout India, with the larger occurrences in the central Kerala region.

It is a medium size upright hemi root parasitic tree with a typically bushy crown. It can parasitise to a wide range of species from grasses to other Sandalwood plants. Under gregarious growing conditions, such as growing in plantations, self-parasitism is common. Sandalwood establishes haustorial connections with the host plants and depends on them for its requirement for a range of nutritional elements and amino acids (Karmakar. A, 2020) (Radomiljac A. M., 1998). It can obtain some nutrients on its own. Seedlings can survive without a host for three years but thereafter they tend to die (Tewari, Naik, Veerapur, & Chandras, 2014). More information on Sandalwood nutritional requirements can be found in Chapter 5, Plantation Management.

The average height in plantations from age ten, is between eight metres and nine metres, while maximum tree height within Quintis plantations is no more than 12 metres (Quintis, PSP Data, 2019).

In its natural environment Sandalwood can be found growing on most types of soils, at elevations from sea level to 1800 metres and with mean annual rainfalls between 500-3000 mm (Tewari, Naik, Veerapur, & Chandras, 2014). This is true in plantations as well, with successful Quintis plantations on heavy clay soil through to light sandy soils.

When it is young, it grows well under partial shade, but at the middle and late stages of its life, it shows intolerance for heavy over-head shade. It adapts well in different climatic conditions except for regularly waterlogged soils. Sandalwood can tolerate occasional waterlogging, as seen in the irrigation practices of the Ord Valley, where flood irrigation covers the rows with water for about three to five days at a time, depending on the slope of the land. This flooding happens roughly every six weeks.

Sandalwood tolerates occasional frosts and has been cultivated as far south as Albany in Western Australia. It has also been observed growing in Brisbane, Queensland. The Quintis Dalbeg plantations experience winter temperatures close to zero Celsius, and frosts have occurred without affecting Sandalwood trees as young as four years old.

Although Sandalwood can handle some waterlogging and occasional frosts, it cannot tolerate fire. Even a mild fire will cause mortality from young to mid rotation (seven years) trees.

Leaf size and shape can vary considerably but tend to be ovate-elliptic (Figure 13a), acute at the base and acute or obtuse at the apex. Leaf margin can vary with the different provenances. Indian provenance trees have smooth entire leaf margin, while Timor provenance trees are also entire, but the margin is wavy (Figure 13b).

Figure 13: Variation in leaf margins between provenances a: Indian & b: Timor (Robson, 2023).

Seed

Trees generally flower and set fruit twice a year. There is a fruiting set in February to April and then again from August to October. However, this can vary considerably dependent on the site and yearly climate.

Flower panicles are terminal and lateral, reddish inside with simple white hairs at the base of anthers. Fruit is a drupe globose about the size of a small cherry, green to red becoming black before dropping from the tree. The mesocarp is pale yellow and the endocarp spherical, between six to nine mm diameter.

Seed is collected manually with selection of ripe drupes, deep red to black. Fruit is usually collected from the bottom portion of the tree but can be collected higher with the use of an elevating platform. Tarps can be laid on the ground to collect fallen seed and then collected into buckets. Note that seed that has previously fallen and has been on the ground usually has much poorer germination and should not be collected.

After a period of soaking in water (up to 12 hours) the seed is cleaned by having the mesocarp removed and the endocarp preferably dried, within 48 hours of being collected. Cleaning is either by hand, physically rubbing the mesocarp off, or mechanically use of a rotating drum that agitates the mesocarp off. It is important that all the mesocarp is removed as even small amounts remaining can inhibit germination. The cleaned seed is allowed to dry on a cool shaded bench for up to 48 hours. Once the seed is bagged and labelled, the seed is stored in a cool room in an airtight container at four to eight degrees Celsius and 40 to 60% humidity. In these conditions seed can store for up to two years without loss of viability. There are approx. 5000 cleaned seed per kilogram.

Young sandalwood flower buds (stage 2)

Sandalwood flowers stage 3 (green and just emerged from bud) and stage 5 (pink). With stage 6 the flowers turn red.

Young fully formed fruit (stage 8), these are approx. 30 to 40 days from being stage 10 and ready to pick

Fully sized green fruit (stage 9) and ripe dark purple/back fruit (stage 10)

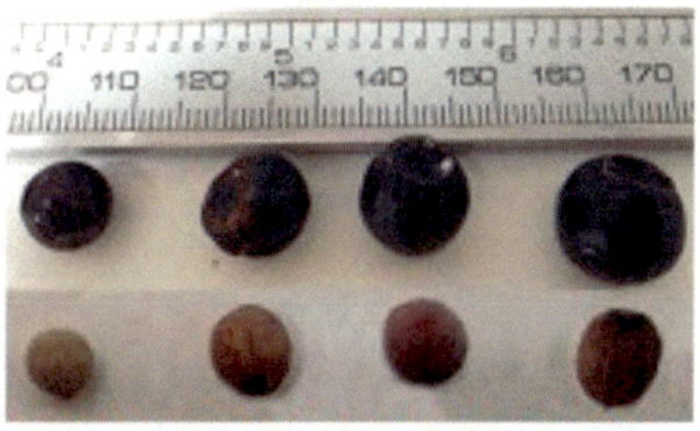

Fruit size does have an influence on seed size. Seed sizes range from 7 to 9mm in diameter.

Cleaned ready for storing S. album seed

Figure 14: Flowering and fruiting stages of S. album.

Physiology

Sandalwood utilise an organ called a root haustoria to obtain a range of nutrients and amino acids. The below ground physiological characteristics of Sandalwood with various hosts is not fully understood. For example, how much does Sandalwood utilise and source water, either by itself or from hosts, is not clear. There are many papers that indicate that Sandalwood sources water from the host (Tennakoon & Cameron, 2006), (Doddabasawa, Chittapur, & Lokesh, 2020) but none that indicate at what proportions? It is well known and understood that hemi-parasitic plants acquire solutes through Xylem transfer from the host, but the makeup of tat transfer is less understood. Sandalwood acquires some water, organic solutes and amino acids by tapping into the xylem of the host. Research suggests that Sandalwood connects to its host through haustoria, which link to xylem pits rather than forming direct lumen-to-lumen xylem connections. As a result, water or nutrients are unlikely to flow directly from the host's xylem to the Sandalwood through tracheary connections (Tennakoon & Cameron, 2006). This implies that although Sandalwood may absorb some water from its host, this contribution is relatively minor compared to its overall needs.

The requirement for acquisition of nitrogen is critical for above ground crown develop and photosynthesis. It has been suggested that a host association is essential for Sandalwood as a source of N and P (Iyengar A. , 1960) and our own research suggests that as well. Noting that Sandalwood may also rely on the hosts for other elements such as K, Mg (Rangaswamy, Jain, & Parthasarthi, 1986), Ca and Fe (Rao, 1942), Cu and Zn (Ramaiah, Parthasarathi, & Rao, 1962). Nitrogen availability can decide host suitability in Sandalwood, and there are substantial benefits in terms of nitrogen gain accruing from a well nodulated legume host (Radomiljac, McComb, Pate, & Tennakoon, 1998). The benefits of increased foliar N concentration, leaf chlorophyll contents and leaf photosynthesis, lead to larger increases in leaf area and therefore biomass accumulation.

There is a relationship between water and nitrogen levels in Sandalwood that affects the Sandalwood water use efficiency (WUE). Sandalwood matches the transpiration of the host so that they can maintain a lower leaf vapour pressure (ψ). A study with mistletoe (Schulze & Ehleringer, 1984) found that increased transpiration was an adaptive response to situations where N levels in a host transpiration stream were low. Radomiljac, McComb, et al. (1998) also discovered that *S. album* transpiration was found to be significantly lower when attached to the low N yielding host rather than when attached to hosts with low N, suggesting essentially little capacity to adapt to hosts with low N concentrations in their xylem stream or for that matter poorly hosting N fixing hosts.

Research has indicated at the ψ only showed slight differences when on different hosts or in the absence of a host. These differences proved to be insignificant, implying relatively little effect of a host on directing osmotic adjustment.

This physiological relationship between hosts and Sandalwood is an area that need further study. Much of the reported research is from pot or lab trials. These findings need to be confirmed in field scenarios.

3. HOSTS

For a tree to be a good host they need an expansive shallow feeder root system as well as a deeper structural and water sourcing roots. The host needs to have an aggressive root system that can continue to grow well as soils gradually deplete of nutrients and Sandalwood haustoria numbers increase. Associations with soil microbes such as Rhizobium are a significant advantage,

The host must not have an aggressive crown which will quickly compete for space and light. It must be able to establish quickly and develop an expansive root system before transition and therefore, it must be good at sourcing recycled nutrients. Many N-fixers particularly those with soil biota associations fit this criterion, and it must be reasonably drought tolerant, as Sandalwood will take an unknown proportion of the sourced water.

The effect of parasitisation

This discussion is included in this section as the effect of parasitisation can eventually lead to later age transition. *S. album* does not seem to treat all hosts as equal. Pot trials have shown that *S. album* is selective with its hosts, however, this does not seem to be the case with other Sandalwood species. Some species are better hosts than others, generally these are N-fixers but not always and it is unclear why this is the case. Some species are so good at providing to the Sandalwood that the hosting can kill them in a short period of time. Within the plantation we clearly see the effect of Sandalwood parasitisation on the hosts (host pressure). This is most obvious with *Senna siamea*. Prior to hosting, the Senna crown growth if not controlled will quickly take over the site and out compete all the other species. However, after transition there is significant slowdown of the Senna above ground growth. On many sites this has reached the point of mortality. Many Senna within the plantation have died due to hosting pressure exerted by the Sandalwood. As mentioned previously, this event is documented with other species. For example, Citrus, although not a N-fixer, is considered a good host for Sandalwood but not recommended due hosting by Sandalwood has shown to cause host death. Evidence that parasitisation leads to mortality is apparent when observing Senna rows on the edges of the plantation or in places where Sandalwood density is low. Senna in these areas typically appear significantly healthier compared to those growing in the middle of the plantation, surrounded on both sides by Sandalwood.

Current long-term Hosts

When the two of the key host species used by Quintis are considered, Cathormion, and Senna neither satisfy all the criteria above.

Cathormion is a riparian species, occurring as a sub dominant species along riverbanks from eastern Kimberly to Cape York. Unlike some other species, Cathormion is recognised as a slow starter. During its early growth stages, the plant prioritises the development of deep structural roots, rather than producing shallow feeder roots. This strategy enables the plant to secure itself and access resources from deeper soil layers, although it may limit rapid initial growth. In almost all cases, Cathormion does not establish a supportive root system for Sandalwood by the time of transition, typically between ages two and three. This lack of available support may impact the successful co-cultivation of Sandalwood with Cathormion during these early years. However, Cathormion is a N-fixer and good nutrient recycler. It also has a mildly aggressive above ground biomass (AGB) after year four and is reasonably drought tolerant.

Cathormion does not handle competition well, either by weeds or other Cathormion. For quick early development of this species weed and grasses must be controlled. There are many areas with very high stocking of Cathormion (2 metre spacing). This is reducing individual tree development and is slowing the species ability to become an effective host.

Senna is an exotic wet tropics species for southeast Asia. It is a rapid aggressive grower above and below ground, quickly forming a broad feeder root system that supports Sandalwood's fast transition. However, due to the aggressive AGB without crown management, it can quickly out compete other species and is detrimental to their growth. It is not drought tolerant and suffers to the point of mortality due to parasitisation by Sandalwood once full transition has occurred.

The Dalbergia species used as hosts fit the criteria better. Although they are also wet tropics species, they grow best in monsoon regions with a dry spell of six months or longer. They can establish both AGB and root systems rapidly. The root structure has expansive shallow feeder roots as well as deeper structural roots. They are N-fixers and the associations with soil bacteria are similar too and advantageous to Sandalwood (known). They have been traditionally used and favoured as a host for Sandalwood across India.

Current Short-term hosts

Another important consideration is the strategic selection of short-term host species. Historically, *Sesbania formosa* has been identified as the most effective host for Sandalwood, significantly enhancing above-ground biomass (AGB) during early development. Young Sandalwood stands exhibit rapid growth when paired with Sesbania; however, this often results in an excessive root-to-shoot ratio imbalance. Sandalwood can develop a strong dependency on Sesbania, leading to a marked decline in health upon the host's mortality. Introducing *Acacia trachycarpa* alongside Sesbania can help mitigate this issue. When only Acacia remains, there is reduced AGB growth, and the transition effect diminishes as the Acacia population declines.

a) Potential host species

The research clearly indicates that *S. album* prefers a host that can fix nitrogen (Radomiljac, McComb, Pate, & Tennakoon, 1998). The majority of these fall into the Fabaceae or Leguminosae family. However there are some other groups that use associations with microorganism to fix nitrogen and have shown potential for use as a host.

In determining which species to investigate, Quintis primarily focused on the Fabaceae family. Within this family there are 104 genera of which many have dozens of species. Of the 104 genera, 25 are small to medium trees, 23 are medium to large trees and 41 are shrubs or ground covers. There are 14 vines and one is a water plant.

When the zone they naturally occur in was considered, five genera occur in the dry tropics, 12 large tree genera occur in the tropics and ten genera of small trees in the tropics. This gives 27 genera with potential species as possible hosts.

The removal of the hard-to-get genera, specifically those from South America and Africa, and then removed genera with species only from the wet tropics, this left 20 genera with species that could provide species suitable for use as hosts. These 20 genera are listed in Table 2

Desirable traits

Sandalwood hosts can be divided into 2 groups, short-term hosts and long-term hosts. Generally, other than longevity, the desirable traits are the same.

1. N fixing: all must be N fixers
2. STH – live for no longer than four years, LTH must live actively for more than 15 years
3. Quick to moderately quick establishers
4. Extensive shallow feeder root system
5. Wind resistant
6. Drought tolerant
7. Stress resistant
8. Can grow well on a range of soils and sites
9. Reasonable straight single bole form.
10. Grows similar rate to Sandalwood

By considering these desirable traits, 12 of the potential genera has species that fit.

Table 2 Fabaceae genera with species potentially suitable as Sandalwood hosts. Those shaded green are species with desirable traits.

Indigofera	Acrocarpus	Pterocarpus	Cathormion
Erythrina	Caesalpinia	Tipuana	Cercidium
Tamarindus	Cassia	Senna	Sesbania
Erythrophleum	Albizia	Bauhinia	Calliandra
Delonix	Dalbergia	Acacia	Ceratonia

b) Testing potential hosts

The research undertaken in the 1990s at Frank Wise research station implemented the current species that we primarily use now (see Chapter1. Review of Sandalwood R&D). Quintis has undertaken research to find other potential hosts. Table 2 indicates the genera with potential species to investigate. Potential species within these genera were selected and if seed could be purchased, they went into the testing program. There were many species from South America within these genera that could not be tested due to no seed being available or allowed to be imported into Australia. The methodology for testing selected species is outlined below.

Research methodology

The testing of potential hosts is a 2-stage process.

1. Large potted trial, and then testing those that performed well in a,
2. Field trial

Potted trial

Sandalwood and potential hosts are grown together for 18 months in 75 to 100 litre pots. After the growth period Sandalwood growth is compared across all host species. A control is usually one of the existing long-term host species. For the trials conducted, *Senna siamea* was used as the control. Sandalwood trees were assessed for above ground height, below ground root biomass and a leaf Chlorophyll measure. An example of data is shown in the Table below.

Table 3: Results for host potted trial conducted in Kununurra.

Rank	Species	Sandalwood Height (cm)	Sandalwood Total Dry Weight	Chlorophyll
1	*Acacia dunni*	113.0	231	26.00
2	*Albizia julibrissin*	108.0	265.5	23.33
3	*Acacia glaucocarpa*	97.8	202	6.00
4	*Erythrina*	87.3	199.5	15.27
5	*Casuarina glauca*	86.7	51.5	5.20
6	*Acacia shirleyi*	86.3	177	21.93
7	*Lawsonia inermis*	85.0	213	4.00
8	*Senna siamea*	85.5	150	5.70
9	*Melaleuca linearis*	81.8	38.5	6.23
10	*Acacia multisiliqua*	77.0	128	13.98
11	*Acacia auriculiformis*	63.3	61.5	24.28
12	No host	54.5	10	5.65

Overall Quintis has tested 43 species in pots.

Table 4: List of species tested in pot trials by Quintis.

Acacia aneura	*Acacia trachycarpa*	*Cathormion umbellatum*	*Melaleuca linearis*
Acacia auriculiformis	*Acacia victoriae*	*Chamaecrista rotundifolia*	*Micromelum minutum*
Acacia cinncinata	*Adenanthera pavonina*	*Dalbergia lanceolaria*	*Millettia pinnata*
Acacia deanei	*Albizia julibrissin*	*Dalbergia latifolia*	*Pterocarpus indicus*
Acacia difficilis	*Albizia lebbeck*	*Desmanthus virgatus*	*Pterocarpus santolinus*
Acacia dunni	*Arachis*	*Erythrina vespertilio*	*Senna siamea*
Acacia glaucocarpa	*Cajanus*	*Erythrophloem chlorostachis*	*Sesbania formosa*
Acacia multisiliqua	*Cassia didymobotrya*	*Hibiscus tilliaceus*	*Sesbania sesban*
Acacia platycarpa	*Casuarina cunninghamiana*	*Lawsonia inermis*	*Vachella valida*
Acacia shirleyi	*Casuarina equisetifolia*	*Leucaena leucocephala*	*Vachellia suberosa*
Acacia simsii	*Casuarina glauca*	*Lysiphyllum cunninghamii*	

Field trial

If a species shows potential in the pot trial, it is then planted in the field to assess its suitability, not only as a host but also as a plantation species. *Acacia dunni* proved to be an excellent host for Sandalwood in the pot trial. However, when it was planted out in the field, it failed to meet the desired characteristics for forestry purposes. It exhibited high susceptibility to wind and easily broke off, making it unsuitable for plantation forestry.

Another species, *Albizia julibrissin*, also performed well as a pot host. However, it was highly vulnerable to constant insect attacks, resulting in a significant mortality rate during the first two months. These challenges make it less feasible as a field species.

In the pot trial, *Erythrina vespertilio* showed promise as a host species. However, due to its large thorns, it was deemed unsuitable for a field trial as the thorns could pose foreseeable management issues.

Table 5: List of species tested in field trial by Quintis.

Acacia aneura	*Acacia plectocarpa*	*Cajanus*	*Erythrophloem chlorostachis*
Acacia auriculiformis	*Acacia shirleyi*	*Cassia didymobotrya*	*Hibiscus tilliaceus*
Acacia cinncinata	*Acacia simsii*	*Casuarina cunninghamiana*	*Leucaena leucocephala*
Acacia difficilis	*Acacia trachycarpa*	*Casuarina equisetifolia*	*Micromelum minutum*
Acacia dunni	*Acacia tumida*	*Casuarina glauca*	*Millettia pinnata*
Acacia glaucocarpa	*Acrocarpus fraxinifolius*	*Cathormion umbellatum*	*Pterocarpus indicus*
Acacia lamprocarpa	*Adenanthera pavonina*	*Chamaecrista rotundifolia*	*Pterocarpus santolinus*
Acacia leptocarpa	*Albizia julibrissin*	*Dalbergia lanceolaria*	*Senna siamea*
Acacia multisiliqua	*Albizia lebbeck*	*Dalbergia latifolia*	*Sesbania formosa*
Acacia platycarpa	*Arachis*	*Desmanthus virgatus*	*Tamarindus indica*

Quintis tested 40 species for host potential. The pot and field trials are part of this process. From this a matrix of desirable traits was developed.

c) Host potential Matrix

A host matrix was created to evaluate potential hosts based on desirable traits. If a species possesses a trait, it receives 10 points; if the trait status is unknown, it gets 5 points; and if it lacks the trait, it earns no points. Scores are presented as percentages, with the highest possible score being 100%. The matrix does not judge how effective a host is but rather assesses its potential suitability.

Current suite of hosts and matrix scores

*Acacia trachy*carpa (83%) and *Sesbania formosa* (83%); these are the current and preferred short-term hosts (STH). *A. trachy*carpa lost points due to no commercial use and no bole. *S. formosa* lost points due to no drought tolerance and no commercial use.

Acacia simsii (83%) was previously utilised as a STH, but its use was discontinued because it did not decline after four years and demonstrated inconsistency as a host. Consequently, Sandalwood planted adjacent to the host did not consistently exhibit healthy growth.

Cathormion umbellatum (82%), *Dalbergia lanceolaria* (95%) and *Dalbergia latifolia* (100%) are the three main long-term hosts (LTH) species used in Quintis plantations from 2016 to 2022.

Senna siamea (Cassia siamea) (45%) was used extensively from 2000 as LTH but was removed from the program after 2014 due to mass die off issues.

Albizia lebbeck (79%) and *Acacia difficilis* (71%) has been used across several plantations as infill species to replace missing LTH's. They have performed well in this role.

In the host matrix, all the current hosts score over 80%. Cathormion is the lowest of the current host due to its slow start, no commercial use and poor form (Table 6).

Table 6: Host suitability matrix showing currently utilised host species.

Species	N fixing	STH – live for no longer than 4 years	LTH must live actively for more than 15 years	Quick to moderately quick establishers	Extensive shallow feeder root system	Wind resistant	Drought tolerant	Stress resistant	Can grow well on a range of soils and sites	Reasonable straight single bole form.	Grows similar rate to Sandal	Potential commercial use	Proven host	Matrix SCORE %
Dalbergia laterfolia	10		10	10	10	10	10	10	10	10	10	10	10	100%
Dalbergia lanceolaria	10		10	10	10	10	10	5	10	5	10	10	10	92%
Acacia trachycarpa	10	10		10	10	10	10	10	10	No	10	no	10	83%
Acacia simsii	10	5		10	10	10	10	10	10	10	10	no	5	83%
Cathormium umbellatum	10		10	5	10	10	10	10	10	5	10	no	10	83%
Sesbania formosa	10	10		10	10	10	No	10	10	10	10	no	10	83%
Albizia lebbeck	10		10	10	10	10	10	10	10	No	No	5	10	79%
Acacia difficilis	10		10	10	10	No	10	10	10	No	10	no	5	71%
Senna siamea	10		10	10	10	No	No	No	No	No	No	no	10	42%

Potential new hosts

There are three potential new LTH that could be established now.

Casuarina cunninghamiana (96%) is frequently used in plantations in India and China. In our matrix it has scored better than *D. lanceolaria*. It does lose points because there is no commercial use for the timber.

Based on trials conducted at a Quintis site in Katherine, it has been observed that *Acacia shirleyi* (92%) has exhibited exceptional performance. This species displays a pine tree-like crown shape when planted with a spacing of three metres and typically grows with a single trunk. However, it is important to note that *Acacia shirleyi* has a site restriction, as it does not thrive well in areas with high humidity, such as Daly, Dalbeg, or the Kununurra region. Despite this drawback, its overall growth and performance at the Quintis Katherine site have been exceptional.

Millettia pinnata (92%) is proven as a valuable host internationally. It is cultivated in Australia and various other countries for its fruits, which can be used to produce biofuel. There are two recognised stem forms of this tree: a compact, smaller variety that grows up to eight metres in height and has a branching structure, and a taller form that can reach heights exceeding 15 metres. This species exhibits vigorous growth when planted in wet areas and is also known for its ability to withstand drought conditions. The common name for this species is Pongamia.

Other strong host possibilities

Acacia leptocarpa (71%) is a long-lived Acacia usually found in western QLD on a range of soils all in dryland zones. It can be branchy and yet has not been trialled in the field as a host, but it belongs to the same Acacia clade (Juliflorae) as other well performing Acacia's. It would have a similar crown structure to *M. pinnata*.

Acacia platycarpa is also an untested STH but has many of the desirable traits. It belongs to the Plurinerves acacia clade like *A simsii*. However, it should be recognised that known Acacia host like A. *trachycarpa, A. shirleyi, A. difficilis* belong to the Juliflorae clade. It is unknown if the clade relationship affects hosting ability.

Table 7: Host suitability matrix showing best potential host species.

Species	N fixing	STH – live for no longer than 4 years	LTH must live actively for more than 15 years	Quick to moderately quick establishers	Extensive shallow feeder root system	Wind resistant	Drought tolerant	Stress resistant	Can grow well on a range of soils and sites	Reasonable straight single bole form.	Grows similar rate to Sandal	Potential commercial use	Proven host	Matrix SCORE %
Casuarina cunninghamiana	10		10	10	10	10	10	10	10	10	10	5	10	96%
Acacia shirleyi	10		10	10	10	10	10	10	No	10	10	10	10	92%
Millettia pinnata	10		10	10	10	10	10	10	10	No	10	10	10	92%
Acacia multisiliqua	10	5		10	10	10	10	10	10	10	10	no	5	83%
Acacia leptocarpa	10		10	10	10	10	10	10	5	No	10	no	No	71%
Acacia platycarpa	10	5		10	10	5	10	10	10	No	10	no	5	71%

d) Site specific host selection

Currently there are only a few options to modify host species makeup dependant on the site, but observation has shown that there is growth and performance variation between hosts due to site differences. The graphic below attempts to show the current host site preference specifically for soil moisture / soil texture / site humidity/ plantation estate, based on observation, growth and mortality.

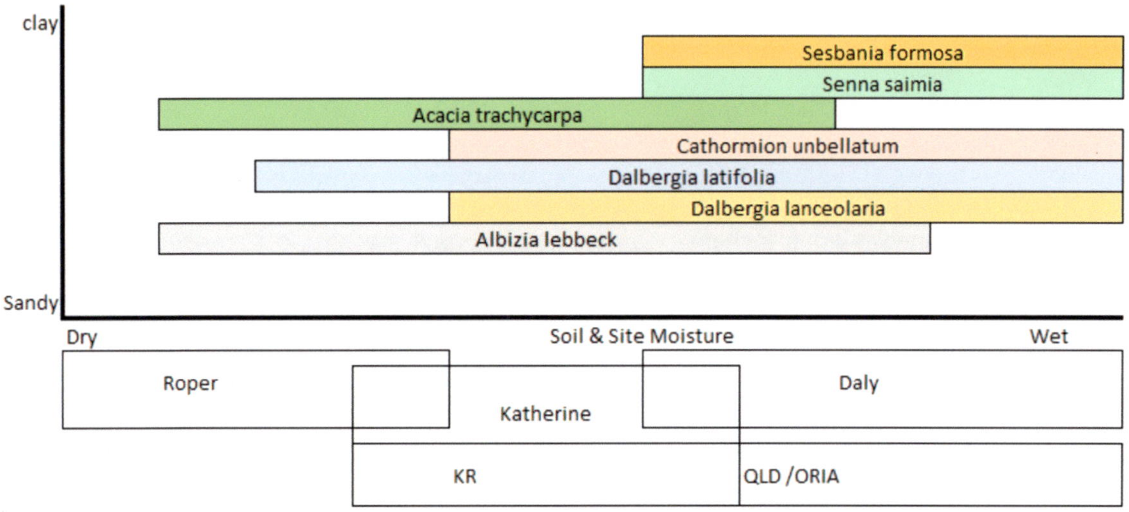

Figure 15: Graphic that indicates the current suite of species tolerance to climatic and soil moisture.

This graphic has been developed with knowledge of site preference based on growth performance and survival of each species and is an approximate (best guess) for the site plasticity of those species (Figure 15).

The graphic shows that most hosts thrive in moist locations. Planting species in drier areas leads to poor growth, stress, and higher mortality.

Most of the potential new host prefer dry sites and tend to be more drought tolerant. The graphic below attempts to show how they fit within the sites planted by Quintis (Figure 16).

The addition of these species will provide the grower with a greater section to ensure the best hosts are selected for a particular site.

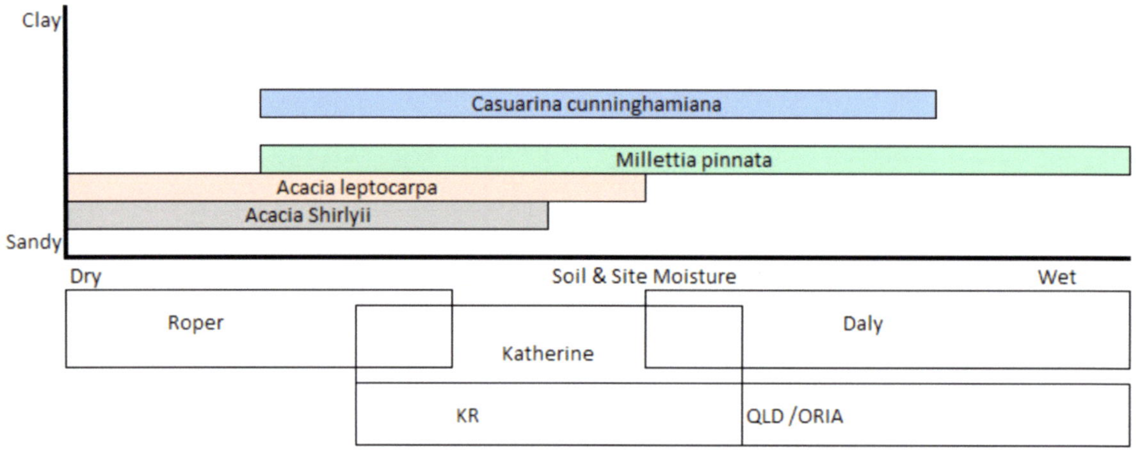

Figure 16: Graphic that indicates for potential new host species their tolerance to climatic and soil moisture.

Site selection potential with new species

With the addition of three new LTH species, the ability to select for individual sites is significantly increase.

For instance, in establishing a site with environmental conditions comparable to those of Roper Plains, characterised by a dripper irrigation system, sandy soils, cool winters, and extremely hot summers, the recommended LTH species would include *A. shirleyi, M. pinnata, A. leptocarpa, C. cunninghamiana,* and *D. latifolia*. Established STH selections remain appropriate for such conditions, with *A. trachycarpa* serving as the principal host species.

A site like KR or Taylors Park, drip irrigation and sandy loam soils, would see potentially *S. formosa* and *A. trachycarpa, A. shirleyi, A. leptocarpa, M. pinnata, D. latifolia*, and C. *cunninghamiana* in the mix.

The moister sites at the Daly, Dalbeg and ORIA region would see *S. formosa, D. latifolia, Cathormion, D. lanceolaria,* C. *cunninghamiana* and potentially *M. pinnata* available as key species.

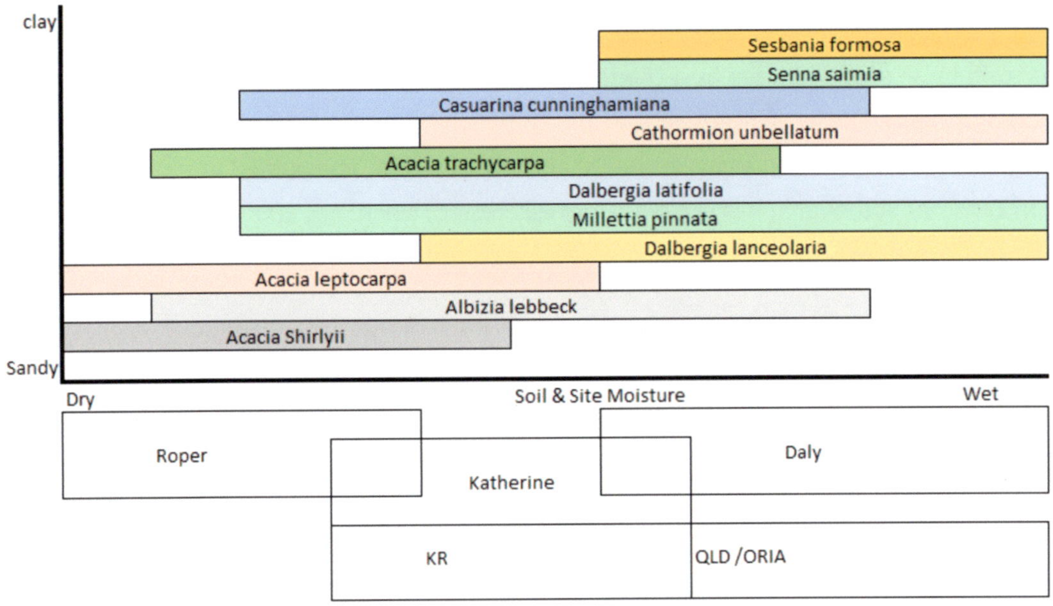

Figure 17: Graphic that indicates for current and potential new host species their tolerance to climatic and soil moisture.

Figure 18: Casuarina cunninghamiana in trials has shown to be a good LTH (Dalbeg, Robson, 2023).

Figure 19: Millettia pinnata located between two Sandalwood trees in a MIX configuration. Note that the shape and size of the host is similar to the Sandalwood (Eagle Park, Robson, 2020).

e) Host utilised in other Countries.

Sandalwood album is cultivated commercially in both India and China. Many host species, especially those in use within India, have a long history of application. Some of those species evaluated in the trials were deemed unsuitable for forestry purposes (as indicated by green cells in Table 8). Species marked 'Yes' in the 'possible for Aust' column have undergone testing and demonstrate potential for utilisation, including examples such as *Albizia lebbek*. Conversely, species denoted 'No' in yellow have not been tested and there is insufficient information regarding their viability as hosts. Those indicated as 'No' in red are excluded due to their potential as weed species.

Research conducted by Quintis identified that several species considered as effective hosts in India did not perform well in field trials in Australia. For example, Eucalyptus camaldulensis, commonly used in India, has proven to be an inadequate host (Radomiljac A. M., 1998). It does not fix nitrogen and fails to supply the necessary levels of nitrogen and phosphorus required of a good host. Additionally, *Azadirachta indica* (Neem), while present in the list and used in India, is classified as a Class B weed in the Northern Territory. Although it has not been formally tested, Neem has appeared invasively within Sandalwood plantations and has demonstrated poor efficacy as a host.

Table 8: List of species that are utilised for Sandalwood host in India and China *(Teixeira da Silva, , et al., 2016)* and their possibility as a host in Australia.

Species	Family	Country	tested in Aust	Possible host for Aust
Albizia lebbek	Fabaceae	China	Yes	Yes
Cajanus cajan	Fabaceae	China	Yes	Yes
Cajanus cajan	Fabaceae	India	Yes	Yes
Calliandra haematocephala	Fabaceae	China	Yes	Yes
Casuarina equisetifolia	Casuarinaceae	India	Yes	Yes
Dalbergia latifolia	Fabaceae	India	Yes	Yes
Dalbergia odorifera	Fabaceae	China	Yes	Yes
Dalbergia sissoo	Fabaceae	India	Yes	Yes
Pongamia pinnata	Fabaceae	India	Yes	Yes
Cassia siamea	Fabaceae	India	Yes	Yes
Casuarina junghuhniana	Casuarinaceae	China	No	Yes
Acacia auriculaeformis	Fabaceae	China	Yes	No
Acacia confusa	Fabaceae	China	Yes	No
Acacia nilotica	Fabaceae	India	Yes	No
Erythrina corallodendron	Fabaceae	China	Yes	No
Eucalyptus camaldulensis	Myrtaceae	India	Yes	No
Tectona grandis	Moraceae	India	Yes	No
Azadirachta indica	Fabaceae	India	No	No

Species	Family	Country	tested in Aust	Possible host for Aust
Bauhinia biloba	Fabaceae	India	No	No
Caesalpinia pulcherrima	Fabaceae	China	No	No
Caesalpinia sappan	Fabaceae	China	No	No
Cassia fistula	Fabaceae	India	No	No
Ficus bengalensis	Moraceae	India	No	No
Gardenia jasminoides	Rubiaceae	China	No	No
Grevillea robusta	Proteaceae	India	No	No
Hibiscus rosa-sinensis	Malvaceae	China	No	No
Kuhnia rosmarnifolia	Asteraceae	China	No	No
Lantana camara	Verbenaceae	China	No	No
Melia dubia	Meliaceae	India	No	No
Phyllanthus reticulatus	Euphorbiaceae	China	No	No
Pithecellobium dulce	Fabaceae	India	No	No
Syzygium cumini	Myrtaceae	India	No	No
Terminalia arjuna	Combretaceae	India	No	No
Wrightia tinctoria	Apocynaceae	India	No	No
Zizyphus mauritiana	Rhamnaceae	India	No	No

These examples show that because a species is indicated as a host in other countries, a prospective Sandalwood grower needs to do their own testing before settling on a species to utilise in their plantations.

4. CURRENT HOST DESCRIPTIONS AND SILVICULTURE

This next section will describe the characteristics of the current suite of host, *Cathormion umbellatum*, *Dalbergia lanceolaria* and *Dalbergia latifolia*, *Senna siamea*, *Sesbania formosa*, *Acacia trachycarpa* and Alternanthera.

a) Cathormion umbellatum

Cathormion umbellatum subsp. *moniliforme* (DC.) Brummitt, is a native Australian species that also occurs across southeast Asia (Thailand, Sumatra, Java, Maluku, Lesser Sunda Islands, and New Guinea) (ALA., 2023). It occurs across northern Australia, less than 200 kilometres from the coastline and north of 19° latitude. Its altitudinal range is quite small, only being found from sea level to 100 metres. It is found growing in monsoon forest and other closed forests, usually on heavy soils which are periodically flooded, such as closed forests close to the sea often near mangroves or salt pans. The provenance planted within Quintis plantations was sourced from the Keep River area in Northern Territory (Figure 20).

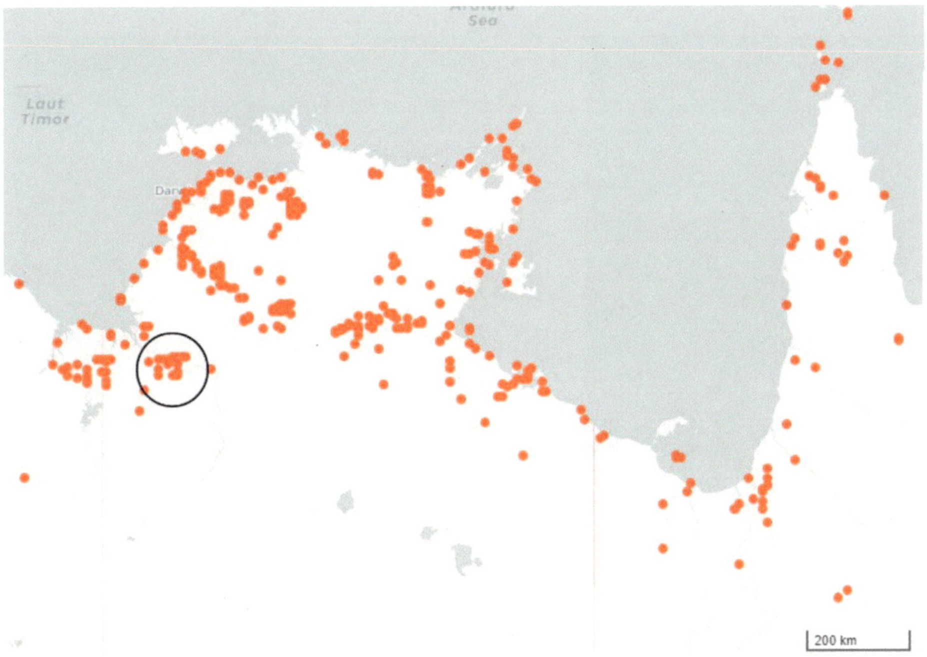

Figure 20: Distribution of C. umbellatum (Atlas of Living Australia). Circle indicates location of Keep River Provenance.

It is a reasonably drought tolerant plant once it establishes a root system, but it also likes plenty of water to grow well during this establishment period. On most Quintis sites in NT there is adequate water supply to enable Cathormion to survive until it can access more ground water. However it cannot survive on some sandy soil sites where the water supply is inadequate to sustain good Cathormion growth. Due to this slow growth trait, it can make it difficult to determine if the tree is suited to the site or not. There are examples of some sites where Cathormion struggled for the first two to three years and later, at five to six years of age, is equal to any of the good sites elsewhere. In general, on Kandosol (red/white sandy soils) with deep sandy A horizons, Cathormion will struggle. It very rarely dies due to droughting, but with the physiological stress that droughting brings, there can be mortality associated with disease or insect pressure or competition from grasses. In many cases it can take two to three years before it is obvious that Cathormion will not grow on that site.

It is typically a small tree (five to seven metres), although it can grow to 20 metres in natural stands. In plantation at the end of rotation (15 years) it is usually similar size to Sandalwood (seven to eight metres). The larger trees have very coarsely flaky bark with thorny stems and branches. The form tends to be quite branchy (Figure 21) rarely forming a bole longer than one metre. However, occasionally there are examples of single stemmed individuals (Figure 22).

While small, it is susceptible to leaf rollers (Pyralidae). It can also be attacked by Crusader Bugs (*Mictis profana*) and their nymphs which feed on young shoots.

Figure 21:Typical eight year old plantation Cathormion. Wide crown and multi stem form (Midway, Robson, 2022).

Figure 22: Occasionally single stem phenotypes can be found within the plantation (Sandy Ck, Robson, 2021).

Leaves are pinnae with individual creamy coloured flowers on distinct pedicels. Fruits are flat pods, about 10-20 cm long, constricted between the seeds, breaking into one seeded segment at maturity. Flowering occurs September to October. Collection of pods is from October to November. Seed will hang on the trees for significant periods of time. The pods are dark brown/black when ripe and they are conspicuous on the tree (Figure 24). There is approximately 3,500 cleaned per kg. Seed can be stored without much loss in viability for several years.

Silviculture

Weed control is critical for good establishment of Cathormion seedlings. This species does not like within row competition and uncontrolled weeds and grasses while they are young will slow growth further. When planted closer than three metre spacing there is higher variability in seedling height and vigour. If planted closer than three metres spacing, a thinning would be beneficial.

The tree naturally has a spreading crown. Although a single stem form pruning would be beneficial at about age six months, as it will give the tree a short bole, however, even with stem form pruning the tree will not develop a single apical leader and develop a wide crown.

Figure 23: C. umbellatum flowers, (Dalbeg, Robson, 2023).

Figure 24: C. umbellatum seed pods, (Dalbeg, Robson, 2023).

b) Dalbergia lanceolaria

Dalbergia lanceolaria is one of the key long-term host species utilised by Quintis and has been planted very successfully across a wide range of sites. Since 2015, it has been very reliable at establishment. Prior to that there were issues with small nursery seedlings and high mortality. Since 2014, research into better nutrition, seed quality and addition of specific rhizobia through the nursery stage has significantly improve seedling quality and from that establishment survival.

D. lanceolaria is an exotic tree native to mixed deciduous, evergreen and open forests, in India, Sri Lanka, Pakistan, Nepal, and across southeast Asia (Myanmar, Thailand, Cambodia, Laos and Vietnam (Dhola Riya., Sukhadiya, Behera, & Mehta, 2019).

Due to the unimproved seed source that Quintis has been planting, there is a high degree of growth and form variation across the estate. The original seed source used is unknown. Although upright, it tends to be a multi stemmed tree with a wide spreading dome-shaped crown, and able to grow between 15 to 25 metres. If the crown size is not controlled, it will overtop the Sandalwood and inhibit Sandalwood grow.

It is deciduous (drops its leaves) during the cooler drier months with little to no transpiration occurring during this time (Figure 26). This does raise question about its efficiency as a consistent Sandalwood host.

It has smooth pale grey bark which can have a greenish yellow tinge. The leaves are compound leaflets, ovate, obovate, or elliptic, often emarginate, glaucous. Flowers are large axillary or terminal panicles unilaterally arranged. The fruit is thin, flat pods, with tapering ends. The pods are lanceolate, faintly nerved, dark green with one to three seeds. Flowering in most areas commences in May and can extend into July but is usually finished by the end of September. Flowers quickly develop into seed pods and start to yellow and brown off by late August. Pods are ready to pick from August, although pods will hold on to the branch for significant periods. Approximately 5000 seeds in a kilogram of seed pods, and approx. 17,000 seeds per clean kilogram of threshed. Seed does not store well, and viable seed needs to be collected each year.

Figure 25: Although a good host, it tends to be multi-stemmed spreading crown tree, (Robson, 2021).

Figure 26: The tree is deciduous in the cooler months of the year, (Midway, Robson, 2023).

The root system is well developed, consisting of deep tap roots and long lateral roots. When near the soil surface, roots can produce suckers. This long lateral rooting structure is beneficial for quickly connecting to Sandalwood.

If trees are not developed enough by the time Sandalwood hosting occurs, and due to hosting pressure, it can be susceptible *to Lasiodiplodia theobromae*. This can cause significant mortality to small and weak trees. Lasiodiplodia will be discussed in more details in Chapter 5: Plantation management.

Silviculture

Close within row spacing will improve stem form as individuals will compete and there will be increase bole height and reduced stems. However, this tree is a quite a vigorous grower and individuals will quickly compete leading to variable growth within rows. Closer than four metre spacing is not recommended unless a non-commercial thin is intended by age three.

If growing for a commercial timber crop, Dalbergia is not a short-term rotation species, a rotation length of at least 25 to 30 years is required. Although the Sandalwood rotation may not be long enough to produce a commercial timber product, a stem form prune early after establishment will assist with silvicultural management. Removal of competing leaders within the first 12 months after establishment will encourage a single apical form. A second stem form prune early in the second year will assist with this desired form.

To control the crown, the tree can withstand regular siding and topping by a mango pruner. However, low topping should be avoided as this will encourage a wider crown, with more stems and branches. The hedging of Dalbergia will be discussed further in Chapter 5: Plantation management.

As an exotic species it can show signs of Phosphorus deficiency. This usually is not obvious until Sandalwood start hosting. This will be discussed more in Chapter 5: Plantation management.

c) Dalbergia latifolia

Dalbergia latifolia was a relatively new long-term host species for Quintis, only being planted operationally since 2016. However, it has been tested as a host species since 2012. Interestingly, *D. lanceolaria* was incorrectly identified as *D. latifolia* from 1999 to 2012 and until that point all the Sandalwood plantation companies thought they were planting *D. latifolia* but were in fact planting *D. lanceolaria*.

D. latifolia is an exotic tree native to low-elevation tropical monsoon forests of southeast India. Commonly known as Indian Rosewood. It tends to be a single stemmed tree with a wide spreading dome-shaped crown, and able to grow up to between 15 to 40 metres. It remains evergreen throughout the year although on drier sites it can be semi-deciduous.

Quintis did plant *D. latifolia* across most of their estate. New plantations in Queensland, Mataranka and Douglas Daly, and as an infill species for failed long-term hosts in Kingston Rest and Katherine. It has performed very well on all these sites.

Leaves are alternate, odd-pinnate with unequal-sized leaflets originating from the same rachis. Leaflets are broadly obtuse, dark green above and pale below while the flowers are white in axillary panicles. The brown pods are oblong-lanceolate and pointed at both ends. They contain one to four smooth brown seeds and do not open at maturity. The bark is grey, thin with irregular short cracks, exfoliating in fibrous longitudinal flakes. Flowering in most areas commences in August and can extend into October but is usually finished by the end of September. Flowers quickly develop into seed pods and start to yellow and brown off by late September. Pods are ready to pick late September / early October, although pods will hold on the for significant periods. Approximately 5000 seeds in a kilogram of seed pods or approximately 17,000 seeds per clean kilogram.

The root system is well developed, consisting of deep tap roots and long lateral roots. When near the soil surface can produce root suckers. The root structure close to the surface is a good attribute to have for a Sandalwood host as the Sandalwood feeder roots that connect to host roots are all in the top 20 cm of soil profile.

Figure 27: Quintis plantation D. latifolia, (Dalbeg, Robson, 2018).

Figure 28: D. latifolia flowers and leaves, (Mustang Hill, Robson, 2020).

Silviculture

Although primarily characterised as a single-stem tree, it may develop multiple leaders in certain environments. Similar to D. lanceolaria, close within-row spacing enhances stem form due to increased competition among individuals, resulting in greater bole height and fewer stems. However, this species exhibits vigorous growth, leading to rapid competition and variability among individuals within rows. Spacing closer than four meters is not advisable unless a non-commercial thinning is planned by age three..

Dalbergia is not a short-term rotation species, a rotation length of at least 20 to 30 years is required for a commercial timber. *D. latifolia* is grown in Sri Lanka and southern India in a rotation of this length (Bhandari, 2023). Although the Sandalwood rotation may not be long enough to produce a commercial timber product, a stem form prune early after establishment will assist with silvicultural management. Removal of competing leaders within the first 12 months after establishment will encourage a single apical form. A second stem form prune early in the second year will assist with this desired form.

As an exotic species it can show signs of Phosphorus deficiency. This usually is not obvious until Sandalwood start hosting. This will be discussed more in Chapter 5: Plantation management.

d) Senna Siamea

Senna siamea, also known as *Cassia siamea*, is a medium-size, evergreen tree growing up to 14 to 18 metres tall, with a multi stemmed truck of up to 50 cm in diameter. It is an exotic species, native to Asia, (Brunei, Cambodia, China, India, Laos, Malaysia, Myanmar, Nepal, Philippines, Sri Lanka, Thailand, and Vietnam). *S. siamea* will grow in a range of climatic conditions but is particularly suited to lowland tropics with a monsoon climate. It grows best when its roots have access to groundwater. It is susceptible to cold and frost and does not do well at altitudes above 1300 metres. Its light requirements are high. Unlike the other long-term hosts and although included in the Leguminosae family, there is speculation as to whether this species can fix atmospheric nitrogen. Many references indicate that this species is a non-nitrogen fixing, leguminous tree (Echo Community, 2024) (Winrock, 2024), while others indicate it has the ability to fix nitrogen (Plantin, 2024).

The bole is short, usually multi-stemming from the ground, the crown is usually dense and rounded at first, later becoming irregular and spreading with drooping branches. The bark is a grey or light brown, smooth but becoming slightly fissured with age. The root system consists of a few thick roots, growing to considerable depth, and a dense mat of rootlets in the top 10-20 cm of soil, it may reach a distance of seven metres from the stem in 1 year and eventually a distance up to 15 metres (Orwa, Mutua , Kind, Jamnadass, & Imons, 2009). The root structure close to the surface is a good attribute to have for a Sandalwood host as the Sandalwood feeder roots that connect to host roots are all in the top 20 cm of soil profile (Robson K. , 2019).

Leaves are compound with about eight-20 leaflets. Crushed leaves emit an unpleasant odour. Flowers about 25 mm diam., but only a few open in each inflorescence. Pedicels between 30 to 35 mm long. Fruits between 16 to 20 cm long. Seeds between 10 to 20 per pod, each seed between 8 to 9 x 5 to 6 mm, flat and thin (Irwin & Barneby, 1982).

In Quintis plantations this species establishes vigorously achieving over three metres height within 12 months. It has been planted on a range of soil types and provided that enough water is supplied it has grown exceptionally well. It is a heavy user of water and grows better on flood sites rather than drip sites. It does not tolerate long periods of droughting or low temperatures. Although there was no mortality, trees located in Dalbeg that went through a frost at age four were severely burnt and died back from the top.

Silviculture

Once Sandalwood hosting occurs from about age three years, growth is stalled and on sites with vigorous Sandalwood growth Senna mortality can occur. This is predominantly on sandy sites irrigated by drip system but there is evidence of it also occurring on flood irrigated sites in ORIA and Dalbeg. The physiological stress of hosting large Sandalwood trees subjects Senna to *Lasiodiplodia theobromae* vulnerability. Prior to Sandalwood hosting, the tree crown can be controlled with annual hedging. After hosting occurs, care must be taken to ensure that any hedging is not undertaken in stressed sites.

Due to the pressure of hosting, Senna is also susceptible to pest attack, particularly caterpillar defoliation by Lemon Emigrant (*Catopsilia pomona*). Prior to hosting, a level of caterpillar defoliation is acceptable to control crown size and overall growth.

It very rarely will have a single stem but tends to be multi stemmed with a spreading crown. A form prune in the first six month with a follow up form prune at 12 months can produce trees with single boles of two metres. This assists with tractor access in rows for weed control activities.

Figure 29: Typical three year old Senna unmanaged crown, (Robson, 2017).

Figure 30: Row of multi-stemmed Senna in a mature plantation, (Smiths, Robson, 2017).

On NT sites it can show symptoms of phosphorous deficiency from age five. A broad acre application of granular phosphorous can assist with reduction of these symptoms. Due to high crown management cost in the first few years, the lack of drought tolerance and high mortality in plantations, *S. siamea* has not been planted by Quintis since 2014.

Figure 31: Senna form pruned to a single stem in plantation, (Florina Rd, Robson, 2023).

Figure 32: Row of two year old Senna side hedged to control crown size, (Dalbeg, Robson, 2016).

e) Sesbania formosa

Sesbania formosa is a rapidly growing, nitrogen-fixing tree that typically reaches less than ten meters in height, though it can occasionally grow as tall as 20 metres. Endemic to Australia (Burbidge, 1965),it is found in the Pilbara and Kimberley regions of Western Australia, extending eastward to Arnhem Land in the Northern Territory. This species thrives from sea level up to 550 metres in elevation, inhabiting closed forests or swampy areas, especially on alluvial plains where soils range from deep sands to dense, black, alkaline clays.

The bark is pale grey, featuring furrows and a corky texture. Its leaves are petiolate and have compound axes that are grooved or channelled on the upper surface. The tree produces large, creamy flowers with calyces measuring between 20 to 25 mm and lobes roughly 5 mm long. Petals are variable and distinctly clawed, with the largest petals reaching approximately 90–100 mm in length.

Figure 33: Sesbania plantation edge trees in flower at Millaroo QLD, (Robson, 2022).

Figure 34: White flower of Sesbania, (unknown).

Seed pods are narrow and curved, 40 to 60 cm long and pointed at both ends. The seeds are bean-like, smooth, shining and dark brown. Flowering occurs from August to September; seed develops quickly and is ready to pick when pods turn brown but before they dry out. Seed can be ready to pick in late November to early January. There are 40,000 seed per kilogram. The seed stores well in a cool room with good viability after 5 years.

Sesbania formosa is the optimal short-term host for the successful establishment of Sandalwood. In an experiment conducted by Radomiljac (1998) and Bui et al. (2026) using potted plants, Sandalwood performance was compared with several host species, including Sesbania. A range of growth assessments, including total height, leaf, stem and root biomass, and root-to-shoot ratio were recorded over a 38 week period in the Radomiljac trial and 72 week period for the Bui experiment. The results indicated that Sandalwood exhibited superior growth when paired with Sesbania compared to the other host species. This trend has been consistently observed in both controlled observations and field testing, highlighting Sesbania's significant role in promoting faster Sandalwood growth.

Furthermore, the research demonstrated that sesbania growth rates were significantly influenced when hosting Sandalwood, particularly when compared to another short-term host (STH*), A. trachycarpa.* The findings suggested that more assimilates were extracted from sesbania when paired with Sandalwood than from other hosts.

When planted in the field, a key dynamic between Sesbania and Sandalwood is Sesbania's rapid height growth within the initial six months of establishment creates a shaded microclimate beneficial for the coexisting Sandalwood. The roots of Sesbania quickly extended toward the Sandalwood, forming haustoria on the sesbania root (see Figure 36). This interplay of root systems contributed to the establishment of a dappled shade, leading to increased competition for light. This competition, in turn, tends to prompt the Sandalwood to grow upward, initiating a longer bole.

In summary, *Sesbania f*ormosa not only facilitates accelerated Sandalwood growth but also tends to produce predominantly longer bole Sandalwood plants. This underscores the importance of Sesbania as an effective short-term host in fostering optimal conditions for Sandalwood cultivation.

Silviculture

Sesbania is a very rapid establisher, reaching over three metres within six months after planting on most sites. It does tend to have a single stem but can get large lower branches that will impede Sandalwood growth if not pruned off. It is a heavy user of water and does not tolerate droughting when young. It prefers heavier soils over sandy soils. On sandy sites, without applications of extra water it can start to die out after 12 months. On clay sites with plenty of water, it can last for four to five years and will need to be managed out.

On the 'wetter' sites, it can withstand side and top hedging. On the sandy sites heavy hedging is not recommended. If Sesbania is still present after four years, it can inhibit long-term Sandalwood viability and should be managed out. This is usually done with heavy side and top hedging and then droughting.

During the period when most sesbania are starting to die off, it is common to see wildlings germinating along the drip line or on mound in flood stands. The wildling seedlings must be removed if they are in the Sandalwood row as they can inhibit Sandalwood growth. The wildling seedlings usually do not appear to be acting as efficient hosts when in the Sandalwood rows, however, if present they compete for other resources such as water and space.

Figure 35: Young plantation with stem pruned Sesbania as a STH, (Roper Plains, Robson, 2017).

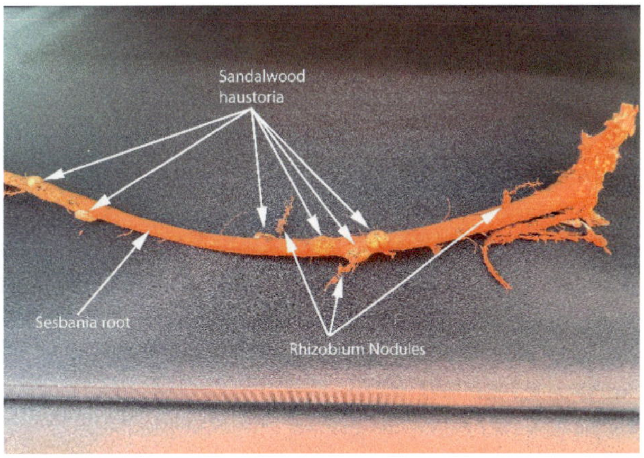

Figure 36: Large Sandalwood haustoria along a sesbania root, (Robson, 2023),

f) Acacia trachycarpa

Acacia trachycarpa is a native tree mostly confined to the Pilbara region of northwest W.A. where it is widespread and common. Typically *A. trachycarpa* is a spreading and sometimes flat-topped crowned tree (Figure 37) about two to five metres high with phyllodes mostly four to ten cm long, one to two mm wide and not rigid.

A reasonably fast-growing species that is reported to be frost sensitive and not very drought tolerant. Plants are killed by intense fires but can resprout after cool fires or mechanical clearing.

The mature bark is known as 'Minni Ritchi' and is a burgundy red to reddish-brown colour which peels off the trunk and branches in thin strips that curl backward (Figure 41). Branchlets sparsely appressed-hairy, sometimes glabrous. The new shoots slightly viscid, light green.

The yellow flowers are a simple inflorescence as a flower-spike that is 9 to 30 cm in length (Figure 38). The seed pods that form later have a narrowly oblong to linear shape. The pods are flat with straight sides with little constriction between seeds (figure 39). The resinous, sticky, shiny brown pods have a length of seven to 11 cm and a width of seven to 12 mm and are strongly irregularly curved. The seeds within are arranged obliquely and have a spherical to obolid shape. The shiny dark brown to black seeds are five to eight mm in length and four to seven mm wide with a dull yellow to brown depression at the centre.

Figure 37: Typical A. trachycarpa shape, with wide crown, (Robson, 2012).

Flowering occurs June to July but can be as early as May on some sites. Collection of pods is from September to October. Seed will hang on the trees for significant periods of time but will eventually dry out and drop off if left too long on the tree. The pods are light brown when ripe and they are conspicuous on the tree hanging in bunches. There are 17,000 seed per kg. Seed stores well in a cool room with good viability after five years.

Figure 38: A. trachycarpa flowers, (Robson 2020).

Figure 39: A. trachycarpa Seed pods, (Robson 2023).

Figure 40: Spreading crown of A. trachycarpa over top of Sandalwood, (Roper Plains, Robson, 2021).

Figure 41: 'Minni richi' bark of A. trachycarpa, (Robson, 2021).

Silviculture

Acacia trachycarpa is utilised as a short-term host and planted only 1.2 metres away from the Sandalwood at establishment. On sites that are preferred by this species, the wide crown can be an issue as it can over top the Sandalwood (Figure 40). Branches that are over the top of small Sandalwood can be removed with clipping, cutting back the branch. This should only be done while the branch stem is still green to light brown. Any branches cut once the bark has turned 'minni richi' will die back. For this reason, side hedging is not recommended.

As *Acacia trachycarpa* is a native species, it is a good scavenger of soil nutrients and very rarely shows any signs of nutrient deficiencies. This will be discussed more in Chapter 4: Plantation establishment.

g) Pot host

What is the purpose of the pot host?

When researchers initially started investigating how to develop Sandalwood plantations in the early 1990's, one of the big issues was very poor initial survival when planted into the field. The reason behind the poor survival was mostly due to the Sandalwood root systems not having consolidated sufficiently within the pot. When removed the potting media would fall away and the seedling would effectively be planted 'bare rooted'. A Sandalwood seedling is in the nursery for up to six months (25 weeks), which is a similar time to most host species. However, unlike the Sandalwood, the hosts are well consolidated and can be removed from the nursery tube without significant disturbance to the roots system.

The use of a pot host was proposed as a possible solution for this poor consolidation issue. It was also thought at the time that the addition of a host would be beneficial. A range of species were tested. These were primarily annuals with a root structure of fine roots. Species tested were grasses, capsicums/chili, tomato, various annual flowering exotic species, pinto peanut (*Arachis pinto*).

This issue was one of the objectives of Radomiljac PhD project (1998). He discovered that the use of a native sprawling low growing perennial herbaceous plant, *Alternanthera nana*, significantly increased survival. This plant was easy to propagate from stem cuttings that quickly produced masses of fine roots that quickly consolidated the space within the pot. This enables the Sandalwood seedling to be removed from the pot without all the potting media falling away and ensure minimal root disturbance at planting.

It should be noted that although the native *Alternanthera nana* was used, this is not the species currently utilised by plantation companies. There is a more readily available and site variable species *Alternanthera ficoidea*. *A. ficoidea* propagates easily from stem cuttings and tends to have green leaves with yellow variegation.

Figure 42: Sandalwood seedlings ready for dispatch from the nursery showing the pot host *Alternanthera* (Unknown).

Figure 43: Growth of 2 species of *Alternanthera* surrounding the Sandalwood seedling in the field, (Dalbeg, Robson 2016).

Although Sandalwood does host to the pot host during this nursery phase, Alternanthera does not fix atmospheric nitrogen and the numbers of haustoria connected during the nursery phase is minimal. It has been clearly shown that the presence of or lack of alternanthera in nursery seedling does not have any effect on seedling growth and development whilst in the nursery (Robson & Mende, 2016). In fact, establishing alternanthera too early can have a detrimental effect on Sandalwood growth, particularly if regular pruning of the pot host is not undertaken.

In the Robson and Mende (2016) study, it did not appear that the alternanthera was performing as a host for the Sandalwood. However, alternanthera does increase planting survival. Anecdotal evidence suggests that Sandalwood seedlings without alternanthera surrounding them have higher mortality in the first three months after planting. It can be theorised that although not a significant host while in the pot, the alternanthera does become an important early host when the Sandalwood is starting to send roots out into the soil. It also provides a cool moist microclimate for the surface feeder roots of the young Sandalwood.

ALTERNANTHERA SPECIES

There are many Alternanthera species, some native and many exotics. Alternanthera belong to the Amaranthaceae family of which many are known as Joyweed. All are herbs, some are aquatic, but most are terrestrial. Some species are known to be toxic to plants around them (allelopathic). Many species have been reported as noxious weeds, including *A. pungens, A. philoxeroides*, and *A. sessilis*. They range in colour from yellow to green to red to purple.

Interestingly, some Alternanthera species undergo C3 carbon fixation, and others have the C4 pathway. Certain other species, like *A. ficoidea*, have an intermediate C3-C4 pathway (Sage, Sage, Pearcy, & Borsch, 2007). C4 plants are more efficient at growing in high temperature and high light environments.

The initial species used in trials as a pot host was *Alternanthera nana.* This was the original pot host used in Radomiljac thesis. (1998). It is commonly known as Hairy Joyweed. It is native in Western

Australia, the northern part of the Northern Territory and northeastern Queensland where it can grow in vine thickets.

The preferred species utilised by the commercial Sandalwood companies is *Alternanthera ficoidea*. Commonly known as Parrot leaf or Tricolour Joyweed. It is an exotic and is native to Mexico and South America. Although it grows well in most soil types, in does best a well-drained loam. It does not like heavy shading and once the Sandalwood seedling becomes established the Sandalwood crown shade it out and it dies.

The third species that has been used as a pot host is *Alternanthera dentata* (Ruby Leafed Alternanthera). It is also an exotic to Mexico and South America. It prefers a more humid environment than A. *ficoidea* and tolerates frost better than the other two species. This species has successfully been tested by Quintis and could be used as a replacement species.

Figure 44: A. nana leaves and flowers, (© G. Sankowsky).

Figure 45: A. ficoidea is the species utilise by Sandalwood companies as a pot host, (unknown source).

Figure 46: A. dentata has been trialled for use as a pot host, (unknown source).

5. TRANSITION FROM HOSTS

Sandalwood is categorised as a hemi-parasitic species, indicating that while it can perform some life-sustaining processes such as photosynthesis for energy production, it relies on a host plant to fulfill most of its nutritional and some of its water requirements. Once attached to a host, Sandalwood generally thrives. However, complications arise when the host undergoes changes or dies off, prompting the Sandalwood to seek another host. This transition period, from the onset of host decline until the Sandalwood establishes enough haustoria connections with a new host to sustain growth, is termed "Transition."

During transition, several above-ground visual symptoms manifest, starting with pronounced crown yellowing, followed by leaf drop and sparse crowns. These symptoms likely indicate the Sandalwood terminating existing haustoria from the dying host while accelerating feeder root growth. Consequently, nitrogen levels decrease, leading to reduced transpiration, which in turn diminishes carbon uptake and halts or reduces basal growth.

With reduced photosynthesis, the Sandalwood depletes stored carbon reserves to develop new root systems, diverting most of its energy into this process rather than above-ground growth. Anecdotal evidence suggests that a healthy two to three year-old Sandalwood plantation can survive for four to eight months without adequate haustoria. In monsoonal regions, this period typically allows the tree to find a new host before exhausting its carbon reserves, albeit this is speculative.

The timing of transition is crucial, particularly in monsoonal climates, as it is coinciding with the end of the wet season. This may lead to mortality if a new host is not found during the high evapotranspiration months of September and October.

Haustorial proportion and root to shoot ratio.

Understanding the relationship between haustoria numbers and tree growth is imperative. As the above-ground biomass increases, so does the requirement for more haustoria, necessitating an increase in root size.

Young Sandalwood trees exhibit a large root to shoot ratio (1:5+), primarily concentrated around the host roots within the top 40 to 50cm of soil, with the majority in the top ten cm (Robson K. , 2021)

As Sandalwood grows and hosts, the root mass increases, albeit not at the same rate as above-ground biomass. While there are enough host roots available in its current vicinity, the Sandalwood gets bigger and the root to shoot ratio increases. The root mass will increase but not at the same rate as AGB. So, although there is an increase in haustoria the actual root to shoot ratio significantly increases to 1:8+ (figures are mostly speculative).

Sandalwood's root system lacks a prominent tap root or large structural roots. Instead, it features several main roots along with numerous fine feeder roots that terminate in haustoria. In its native habitat, Sandalwood grows as an understory species within the mid canopy, so it does not require deep structural roots for survival. Rather, it invests energy in expanding a web of smaller feeder roots. As the tree matures, it needs an exponentially increasing number of haustoria to support its growth above ground.

Length of transition

As previously mentioned, the typical duration for a Sandalwood to transition from a short-term host to a long-term host is approximately six to eight months. Following this period, an overall improvement in Sandalwood health is often observed. Factors influencing this transition phase include Sandalwood size, host size, time of year, water availability, soil type, and levels of insect predation.

The duration of the transition phase appears to be largely influenced by root size and the number of haustorial connections. For instance, a one year old tree may possess a root system spanning one m^2, with aboveground biomass (AGB) of ten kg or a diameter at breast height (DBH) of five cm. To maintain continued development, it might require approximately 100 haustoria per kilogram of AGB, equating to 1,000 haustoria per square meter of root area. When the tree's DBH increases to ten cm, effectively doubling its diameter, the aboveground biomass volume increases eightfold, necessitating roughly 8,000 haustoria to sustain growth at the same rate.

Considering this haustorial ratio, it could be interpreted that transition is not strictly a six to eight month process. Instead, although there is a general improvement in Sandalwood health after this period, there may not be a significant increase in AGB. This theory suggests that it takes considerably longer to regain the necessary proportion of haustoria to resume increased growth, highlighting the need for increased nutritional inputs during this time, especially to aid in root development for both host and parasite.

Transition's role in heartwood.

Transition also plays a role in heartwood development. While existing literature suggests that Sandalwood does not begin developing heartwood until it is ten to 13 years old (Kumar, Srinivasa, Joshi, & Seetharam., 2010). Australian plantation Sandalwood start producing heartwood around five years of age. This heartwood initiation, starting in the root ball approximately 10cm below ground and ascending up the tree over time, is suspected to be triggered by a stress event such as transition from short-term to long-term hosts. See a detailed description of this in Chapter 5: Environmental Stress

Reducing the effect of transition

Although transition is an important phase in the lifecycle of a plantation Sandalwood, it would still be beneficial to minimise the effect of transition. To mitigate the effects of transition, the MIX stand configuration was developed to minimise the changeover from one host to the next. It has mostly achieved this, with little evidence in existing stands of Sandalwood going through a major transition stage.

In standard (STD) plantation configurations, successful host transition can be greatly improved through careful host selection. Dalbergia latifolia has proven particularly well suited to this role, owing to its ability to rapidly develop an extensive root system and tolerate the high level of parasitisation imposed by Sandalwood. The extensive use of *D. latifolia* as the key host since 2016 will assist with reducing transition. This species generally has good single bole stem form and although establishes quickly has not shown a general tendency to over grow Sandalwood. Although there is possibly some evidence of a slowdown of tree AGB there is no evidence of mortally due to hosting.

The encouragement of host root development particularly in the first two years after establishment can reduce transition time. Potentially the average root to shoot ratio of trees is 1 to 5 /6, then by age three the host trees need to be minimum three metres tall (speculative and would vary with species) to have a large enough root system to reach across the inter-row to the Sandalwood. We know that for connection to occur, the host roots need to be in near vicinity to the Sandalwood row as the Sandalwood root will be centred along the irrigated row and current hosts.

Promoting Sandalwood root development from the initial establishment stage is essential. While it is recognised that Sandalwood roots generally remain proximate to the short-term host during this period, strategic management may be required to minimise the transition phase. Techniques may include early removal of Sesbania through crown management and maintaining *A. trachycarpa* for as long as feasible. Additionally, minimising interrow compaction and increasing mulching can enhance soil carbon, water retention, and microbial activity. The introduction of interrow crops, such as millet or non-climbing nitrogen-fixers like Wynn Cassia, may also be beneficial.

Chapter 3. SITE CHARACTERISTICS

Indian Sandalwood is a tropical species that demonstrates successful establishment in various regions of northern Australia (Done , Kimber, & Underwood, 2005). The tree itself will grow well on a range of soil types but prefers well-drained soils. Suitable soils range from light-medium clays through to the lighter sandy. Given Indian Sandalwood requires the presence of suitable host tree species which must also be compatible with the site conditions and soil types preferred by Sandalwood. Commercial plantations may incorporate up to five different species, presenting challenges in identifying locations that accommodate all requirements.

There are many factors which need to be considered when selecting a site for an Indian Sandalwood plantation. The two most important are:

- Soil type (texture and nutrition),
- Climate (rainfall, temperature, humidity and extreme weather events).

The inability to meet all key factors has rendered much of northern Australia unsuitable for plantation establishment. Identifying sufficiently large areas for plantations has posed a significant challenge for Quintis.

The following will discuss site classification processes and their impact on operational management.

1. AUSTRALIAN PLANTATION LOCATIONS

In Australia, Sandalwood plantations occur in three Australian states, Western Australia, Northern Territory and Queensland. These states cover the tropics in Australia, the northern areas are within the "goldilocks" zone (a zone which is most likely to contain suitable sites) for Sandalwood and have an environment that is suited to Sandalwood growth. The climate is warm and humid, and the soils are of the desired quality. Current plantations are generally located more than 70 kilometres from the coastline. This is to minimise the risk from the high category cyclones that cross the coast in the tropics. [more in Chapter 5].

Figure 47 shows the locations of Northern Territory and West Australian plantations that are ten hectares and larger. The ownership of these plantations varies.

Western Australia

There are two plantation sites in Western Australia, the Ord River Irrigation area (ORIA) and Kingston Rest. These sites are discrete and distinct from each other. Although the climate is similar, the soils and the irrigation system are different. Flood irrigation is used in the ORIA and dripper irrigation at Kingston Rest. The first plantations were established in ORIA in 1999 and the last in 2018. Kingston Rest was planted between 2010 and 2012. At the time of writing this book, harvesting had commenced only in ORIA.

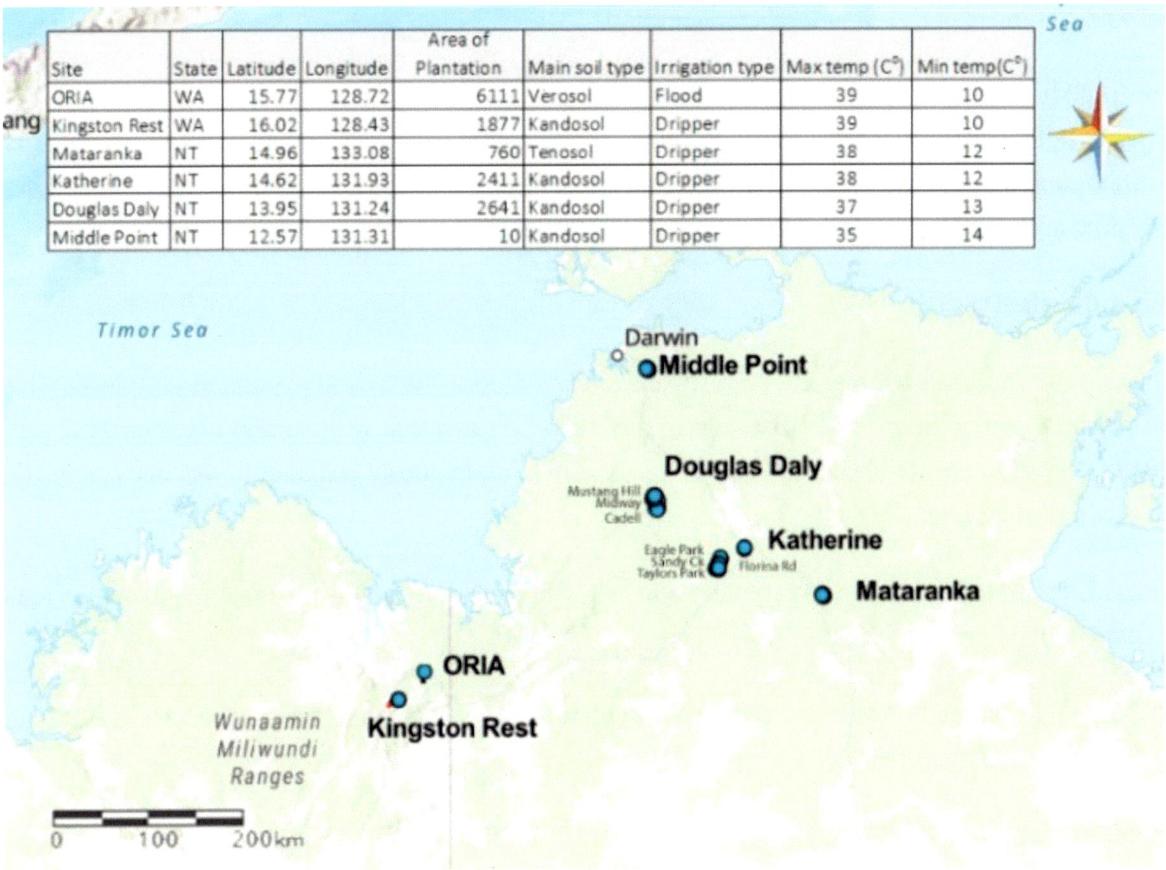

Site	State	Latitude	Longitude	Area of Plantation	Main soil type	Irrigation type	Max temp (C°)	Min temp(C°)
ORIA	WA	15.77	128.72	6111	Verosol	Flood	39	10
Kingston Rest	WA	16.02	128.43	1877	Kandosol	Dripper	39	10
Mataranka	NT	14.96	133.08	760	Tenosol	Dripper	38	12
Katherine	NT	14.62	131.93	2411	Kandosol	Dripper	38	12
Douglas Daly	NT	13.95	131.24	2641	Kandosol	Dripper	37	13
Middle Point	NT	12.57	131.31	10	Kandosol	Dripper	35	14

Figure 47: Map showing location of commercial Sandalwood plantations in WA and NT (circa 2021).

Northern Territory

There have been three sites for plantation establishment in Northern Territory. All are irrigated with a dripper system, pumping water from deep aquafers. Additionally, Middle Point, on the map is a privately owned small plantation of ten hectares.

Mataranka

The plantation at Mataranka, known as Ropers Plains, was established in 2016 and 2017. The soils at this site are a mix of Tenosols and Kandosols. Although this plantation is 100 km south of Katherine, the climatic conditions are similar. Unfortunately, this plantation was sold in 2021 to an agricultural group, and the Sandalwood plantation has been converted to farmland.

Katherine

The Katherine region consist of four separate plantations: Taylors Park, Sandy Creek, Eagle Park and Florina Road. Eagle Park is 107 hectares, and Florina Road is 178 hectares and were planted in 2012. Although similar in size and planted in the same year, they are also quite different sites. Eagle Park is a drier Tenosol soil and planted in the ' mix' configuration, while Florina Road is a moister heavy Kandosol soil and planted with the 'Standard' configuration.

The 1930 hectares at Taylors Parks was established between 2013 and 2015. This plantation also has a mix of both loamy Kandosols and sandy Tenosols. A smaller plantation, Sandy Creek consists of 194 hectares and was planted in 2016. As the name suggests, in the northern section of Sandy Creek there are large areas with sandy Tenosol soils. However, the southern section is mostly Kandosols.

Douglas Daly

The Douglas Daly plantation estate consists of five plantations. The largest plantation is Midway. This was a cattle station prior to the plantation. It is 1290 hectares that was planted between 2013 and 2015. The site consists almost entirely of sandy clay loam Kandosol soil, and it was planted in the standard configuration.

In 2015, Cadell plantation was established and is 274 hectares on Kandosol soils but with some areas that are inundated when the Daly River floods.

Mustang Hill plantation was also a cattle station prior to Sandalwood and has 348 hectares planted in 2016 in a mix configuration.

Maneroo plantation is 500 hectares and was planted in 2016 and 2017. It has also been established on good well drained Kandosol soils.

The Early Storms plantation is 226 hectares established in 2017 on a former banana and melon farm.

Queensland

Dalbeg & Millaroo

Three plantations have been established in Queensland. The largest plantations are in the Burdekin River Irrigation Area (BRIA) and are flood irrigated. The plantations at Dalbeg and Millaroo were planted between 2012 and 2017. Predominantly the soils are Sodosols and Kurosols, but there are other small areas of other soil types.

Goldtyne

The third area at Goldtyne near Lakeland Downs, planted in 2008 was one of the first plantations in Queensland. The drip irrigated plantation is established on deep red volcanic Ferrosols which are highly sought after for horticultural cropping.

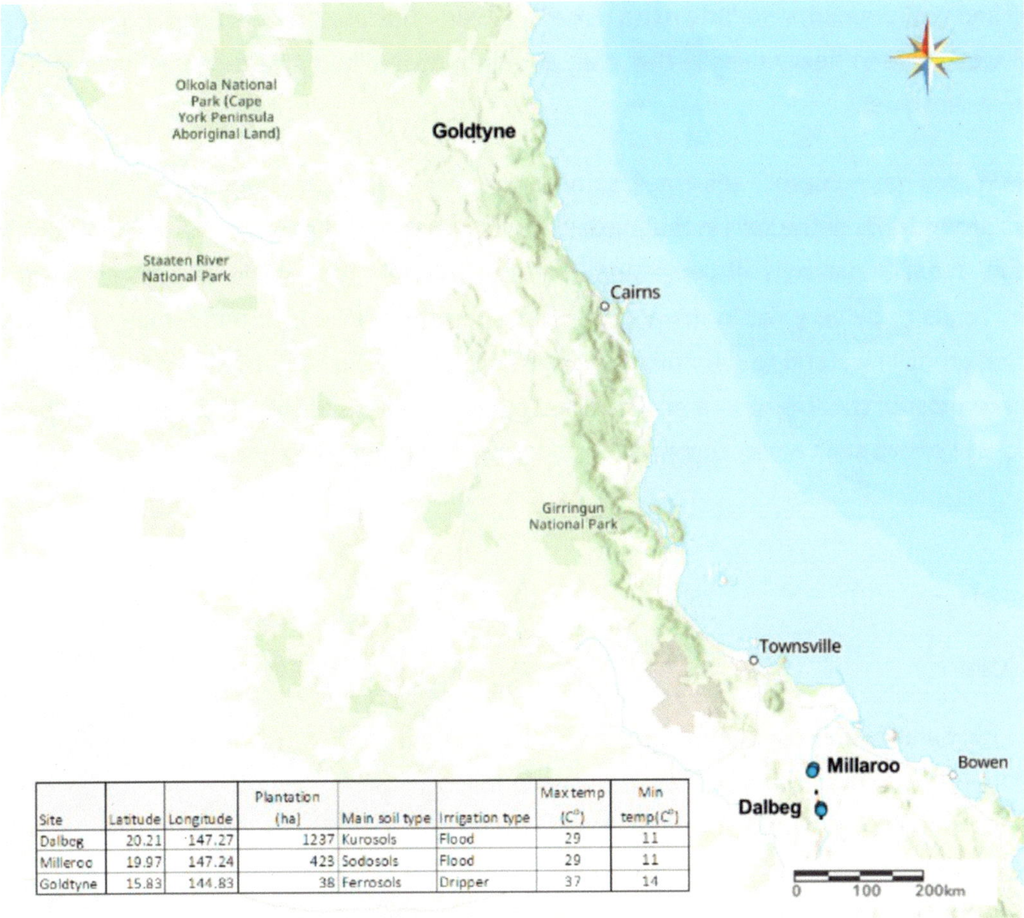

Site	Latitude	Longitude	Plantation (ha)	Main soil type	Irrigation type	Max temp (C°)	Min temp(C°)
Dalbeg	20.21	147.27	1237	Kurosols	Flood	29	11
Milleroo	19.97	147.24	423	Sodosols	Flood	29	11
Goldtyne	15.83	144.83	38	Ferrosols	Dripper	37	14

Figure 48: Location and site details of Queensland Plantations (circa 2021).

2. SOIL

Although Sandalwood has been shown to grow on a range of soil types, a major constraint for successful establishment and survival is drainage. The sites must drain well both from overland flow and through the soil profile. Quintis was very selective about soil types and spent considerable effort on drainage planning in the plantation design. The soil due diligence process will be discussed later in this chapter.

Although there is predominantly one soil type planted with Sandalwood plantation in the Ord Valley Irrigation Area (ORIA), there are two variants of this type. The black cracking clays which are Vertosols can be divided into Cununurra and Aquitaine clays. There are also some patches of alluvial sands, which are either Tenosols, Kandosols or Rudosols. These Vertosol soils are a heavy clay soil and do not drain well and this can cause survival and growth issues if the site is wet and not prepared to drain correctly.

The soils in the Northern Territory are generally good for plantation establishment. Most are sandy clay loams to well-drained light clays, Tenosols and Kandosols. For most sites, the land is flat or slightly undulating and well-drained, with little to no risk of flooding. The sites that do flood, do so because of backed up water due to heavy rains in the river, and then generally they are only inundated for less than five to seven days.

The Burdekin area has numerous soil types, some of them are heavy cracking clays (alluvial soils) and sodic soils. Around 40% of the soils in the Burdekin delta are classified as sodic, mostly duplexes. With high rainfall, these sodic soils disperse and drainage through the profile is compromised. The topography tends to be very flat in many places, so when the region is in flood, many parts of the Burdekin delta remain waterlogged for prolonged periods. This would be detrimental to a Sandalwood plantation, so most of the irrigation area was ruled out on this basis alone. The areas of Dalbeg and Millaroo consist of two main types, Kurosols and Sodosols. The topography is slightly sloping away from the river.

a) Main soil types

Vertosols (Ug)

Predominantly Sandalwood plantations in the Ord valley are planted on "Cununurra" clay soils. This is the main soil type found across the ORIA and Packsaddle Road. There are several variants of this soil type, but primarily there are three common phases, normal, eroded and flooded. These soils are uniform, dark brown to greyish brown, medium to heavy clay (40% to 60%) with pronounced swelling and shrinkage (self-mulching). It normally contains nodules of white carbonate and iron/manganese concretions. Soil pH ranges from 7.5 (mildly alkaline) to 9.5 (highly alkaline).

Figure 49: Typical Vertosol soil profile under Sandalwood plantations in ORIA, (Robson, 2013).

The other Vertosol soil type is the Aquitaine clay soils. Although not preferred, due to the waterlogging nature, Quintis had several plantations on this soil type in the northern section of the Ord irrigation area.

These Aquitaine clay areas are subject to more frequent, prolonged inundation and generally have higher clay content than Cununurra clays. Although these soils are also defined as cracking clays, they are generally found in flatter, lower lying areas than in the Cununurra clays. They also exhibit a weaker self-mulching behaviour and have a coarser underlying structure.

Tenosols/Kandosols (Um)

These soil types occur in ORIA, across most of the Northern Territory and some are also found in the Burdekin River irrigation area (BRIA). Although the Ord valley plantations are mostly on the cracking clay soils, there are some areas of red loamy soils that are fertile, permeable and well drained, these are mostly Kandosols. The plantations in Northern Territory are also predominantly planted on well drained red Kandosols but there is some Tenosols in the Katherine region.

Both soil types have a profile that transitions from sandy loam to medium clay soils and tend to be a reddish in colour. Tenosols tend to have a paler A horizon. While Kandosols have a topsoil A horizon that is sand loams to sandy clay loams while Tenosols top horizon ranges from pure sand to loamy sands. Neither soils have a strong texture contrast between the A and B horizons. They have a massive or weakly structured B horizon and are not calcareous. Parent material in the Northern Territory tends to be predominantly limestone. These soils are slightly acidic at depth to neutral pH in the upper horizons (5.5 to 6). When drainage is impeded the B horizon can be mottled and variable amounts of iron nodules can be found (McKenzie, Jacquie, Isbell, & Brown, 2004).

Figure 50: Kandosol profile from Midway plantation (Robson, 2012).

Due to their sandy texture they can be frequently eroded, especially in areas with significant slope (>1%). Due to this erosion over time the loamy surface horizons may be absent, exposing the clayey subsoil (eroded phase). The exposure of the subsoil in the more eroded variants can result in a hard setting surface and infiltration problems. Where these soils have been severely eroded a calcareous and sometimes saline subsoil occurs near or at the surface.

Generally, these soils have low to moderate agricultural potential with moderate chemical fertility and low water-holding capacity. The readily available water (RAW) of soil is a critical component for their eligibility to be suitable for Sandalwood plantation. Very sandy soils do not hold enough water to sustain a large crop of trees. Calculation of RAW is essential during the due diligence phase of plantation development.

Kurosols (Db) and Sodosols (Dy, Dg)

Most of the plantations in the Burdekin are grown on these two soil types. Kurosols have a strong texture contrast (also known as a duplex soil) with a strongly acid B horizon that may or may not be sodic. Kurosols are formed from parent materials that are highly siliceous, siliceous to intermediate in composition and where rainfall is from 50 to 1350 mm in poorly drained sites or 750 to 1300 mm on well-drained sites. The surface of Kurosol soils is often acidic. They generally have very low

Figure 51: Kurosol soil from Dalbeg, QLD, (Robson, 2013).

agricultural potential with high acidity (pH < 5.5) and low chemical fertility. Kurosols commonly have low water-holding capacity and are often sodic (Thompson, Cannon, Reid, & Baker, 1990).

Sodosols show strong texture contrast with highly sodic B horizon, but they are not highly acidic (pH > 5.5). The parent material of Sodosols ranges from highly siliceous, siliceous to intermediate in composition (Isbell, 1998). Sodosols are only found in poorly drained sites with rainfall between 50mm and 1100 mm. Generally, Sodosols have very low agricultural potential with high sodicity leading to high erodibility, poor structure and low permeability. These soils have low to moderate chemical fertility and can be associated with soil salinity (Thompson, Cannon, Reid, & Baker, 1990).

Chromosols (Dy)

Found in the BRIA, Chromosols have a strong contrasting texture. They are not strongly acidic or sodic in the upper B horizon. The parent material of Chromosols ranges from highly siliceous, siliceous to intermediate in composition. The Chromosols found in BRIA are brown and red in colour and are generally on well-drained sites with rainfall between 350 mm and 1400 mm. These soils have moderate agricultural potential with moderate chemical fertility and water-holding capacity. They can be susceptible to soil acidification and soil structure decline.

b) Mineral nutrition

Tropical Australian soils are known to be low in phosphorous and nitrogen and the soils on which Sandalwood plantations have been established are no exception (Wild, 1958). The total phosphorus content of most surface soils are low. There are many factors that affect this including soil parent material, degree of weathering, climatic conditions and previous use. Soluble phosphorus, either from fertiliser or natural weathering, reacts with clay, iron and aluminium compounds in the soil, and are bound up and converted readily to less available forms by the process of phosphorus fixation. This reactivity between applied phosphorus and the soil is called 'phosphorus buffering capacity' and is measured by the Phosphorus Buffering Index (PBI) soil test. Consequently, a soil with a high PBI value will require more phosphorus fertiliser than a soil with a low PBI.

Due to the fixation processes, phosphorus moves very little in most soils. Phosphorus stays close to its place of origin, crops seldom absorb more than 20 percent of fertiliser phosphorus during the first cropping season after application. As a result, little soil phosphorus is lost by leaching. This fixed, residual phosphorus remains in the rooting zone and will be slowly available to succeeding crops. Soil erosion and crop removal are the significant ways soil phosphorus is lost.

Phosphorous levels in soil can be tested using a range of chemical tests, Bray-P1, Olsen, Mehlich-3 and Colwell. The main differences are in the extraction process, for example Bray-P1 test extracts Phosphorous with acids and Olsen test extracts Phosphorous using sodium bicarbonate. The Colwell test used by Quintis is derived from the Olsen test and uses sodium bicarbonate to extract Phosphorous.

When utilising the Colwell test, the value is independent of the soil's ability to bind phosphorus. So, when interpreting the Colwell test, it is important to also consider the PBI.

The Colwell values needed for optimum tree production increase as PBI values increase. The reason for this is, in low PBI soils more Phosphorous is available for plant uptake than in soils with a high PBI. The levels of soil Phosphorous for healthy plant growth varies dependant of the soil type, specifically the amount of organic matter and clay, iron and aluminium compounds. However, for most soils with less than 20 mg/kg, fertiliser applications are recommended.

Table 10: Soil Colwell P and PBI required for good soil status (Summit Fertilisers, 2023),

Cassification	Colwell	PBI
Extremely low	20 to 27	<35
Low	27 to 36	36 to 140
Moderate	36 to 44	141 to 280
High	45 +	280 +

Table 9: Average Colwell P and PBI in top 60 cm of profile for pre-establishment soil in Quintis estate.

	Colwell P	Colwell PBI
Taylors Park	12.5	23.3
Midway	11.0	36.5
ORIA	33.5	98.9
Dalbeg	27.0	47.3

Although Australian soils are low in Phosphorous and Nitrogen, many agricultural soils contain adequate levels of Potassium (Gourley, 1999). However, overall the potassium balance across the country is negative (Reuter, Duncombe-Wall, & Judson, 1997). The soils with Sandalwood plantations tend to fall into this negative group.

Potassium availability in soils is due to a series of factors, including the parent material, organic matter content, extent of weathering, leaching of soil minerals and soil texture. The total amount of potassium available is correlated related to the amount of clay. As clay increases so too does potassium availability. Sandy soils also can lose potassium due to leaching.

Potassium concentrations are typically low in sandy loam Kandosols planted with Sandalwood, especially in the upper 60 cm of soil. Recommended levels for exchangeable potassium are between 0.3 and 0.6 cmol(+)/kg (Gourley, 1999). Almost all the soils in Northern Territory and Queensland are well below this level.

The Kandosols in Taylors Park on average have 0.11 cmol (+)/kg, the Kandosols in Midway have slightly, less averaging 0.09 cmol (+)/kg. The Queensland sites have more exchangeable Potassium averaging 0.2 cmol (+)/kg, whilst the soils in Kingston rest have on average 0.27 cmol (+)/kg. The heavy clay Vertosols in ORIA have good levels, averaging 0.54 cmol (+)/kg.

Generally, potassium availability declines down the soil profile but may be available in high levels due to leaching (Robson, Longnecker, & Osborne, 1992) and this may be the case on the Northern Territory Kandosols. Analysis of potassium totals from 60 to 120 cm down the soil profile at Taylors Park show that available potassium has increased to 0.19 cmol (+)/kg and at Kingson Rest to 0.32 cmol (+)/kg. However, this generally was not the case at Midway, where only plantation Midway A, the soils increased available potassium at depth. This is possibly because this compartment has a higher clay content at depth than the rest of the plantation.

pH

Soil pH serves as a metric indicating the acidity or alkalinity levels within the soil. Natural soils exhibit varying degrees of acidity or alkalinity, which can be quantified through pH testing. The pH value represents the concentration of hydrogen ions within the soil solution, utilising a logarithmic scale where a pH decreases of 1 corresponds to a tenfold increase in acidity.

The inherent pH of soil is influenced by its parent material, i.e., the rock from which it originated, and the subsequent weathering processes it underwent. Over time, these processes often lead to a gradual decrease in pH, resulting in increased acidity.

The pH of soil plays a crucial role in determining the solubility of nutrients and chemicals in soil water, thereby influencing the availability of essential nutrients for plant uptake. While certain nutrients are more accessible in acidic conditions, others exhibit greater availability in alkaline environments. Nevertheless, a soil pH close to neutral generally ensures optimal availability of most mineral nutrients for plant growth.

Strongly acidic soils, characterised by a pH below 5.5, may induce toxicities of aluminium and manganese, while also causing deficiencies in calcium and magnesium. Additionally, they hinder the accessibility of essential plant nutrients like phosphorus and molybdenum.

Conversely, plants growing on alkaline soils may experience deficiencies in nutrients such as zinc, copper, boron, and manganese. Extremely alkaline soils, with a pH exceeding 9, are prone to high sodium levels. An ideal balance is achieved when soil pH falls within the range of 5.5 to 7.5, ensuring optimal conditions for nutrient availability and plant growth.

The Kandosol and Tenosol soils tend to be quite weathered and are acidic at depth. However, the topsoil is inclined to be neutral to very slightly alkaline. Analysis in 2023 of the topsoil across the plantation indicated an increase in pH. Sampled data clearly showed there has been an increase in soil pH within Sandalwood plantations irrigated via drip from aquafer water.

In Midway A, B, and D, the average pH (measured using a 1:5 water ratio) beneath the drip tape rose from 6.1 in 2011 to 7.6 in 2023. Moving away from the irrigated line, the average pH in 2023 was 6.9. Increased alkalinity was observed only under the drip line. Research by Quintis found significant Sandalwood and host root activity across the row, where pH levels have increased only slightly and remain within the neutral range. The rise in pH under the drip line has also been observed at Taylors Park.

Irrigation water is drawn from the limestone aquifer beneath the plantation. The aquifer's water is slightly alkaline, with an average pH of 7.5. Evidence suggests that consistent application of this water is responsible for the observed increase in soil pH. Additionally, the soil pH under the drip line appears to have risen to a level comparable to that of the irrigated water, without further significant increases.

As mentioned, nutrient availability to plants decreases as pH increases. However, Figure 52 indicates that all macronutrients are readily accessible within the pH ranges observed in the Quintis plantations. Although within these levels the Figure 52 diagram shows potential deficiencies in the micro elements, zinc, boron, iron, manganese and copper. However, foliar analysis generally does not show this, and foliar levels are within a healthy range.

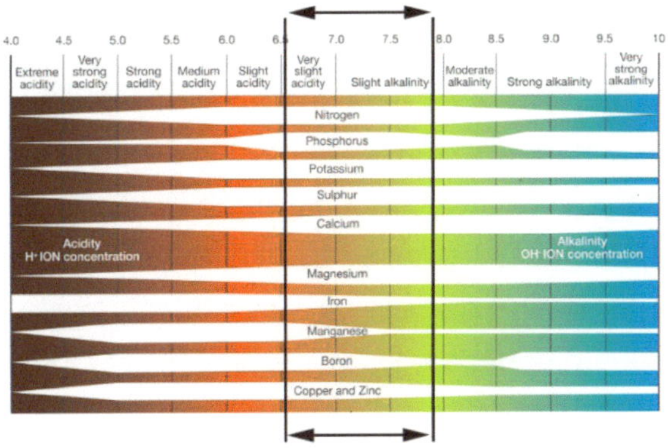

Figure 52: As pH increases, some elements become unavailable to plants. The vertical lines show the range of topsoil pH in Sandalwood plantations. This indicates that most nutrients are still readily available. Adapted from (Brady & Well, 2017).

CEC

Cation exchange capacity (CEC) is a useful indicator of soil fertility because it shows the soil's ability to supply three important plant nutrients: calcium, magnesium and potassium. What CEC measures is the soil's ability to hold cations by electrical attraction. Cations are positively charged elements, the positive charge indicated by a + sign after the element symbol. The number of + signs indicate the amount of charge the element possesses. Cations are held by negatively charged particles of clay and humus called colloids. Colloids consist of thin, flat plates, and for their size have a comparatively large surface area. For this reason, they can hold enormous quantities of cations. They act as a storehouse of nutrients for plant roots (Lines-Kelly, 1993).

There are several factors that affect the CEC, including pH, drainage and soil texture. As the pH becomes more alkaline the number of negative charges on the colloids increase, thereby increasing CEC. Soils with high clay content have the greatest capacity to attract and hold cations, whereas sand has no capacity. Consequently, the Vertosols in ORIA all have good CEC levels between 15 and 45 cmol(+)/kg). Whereas the Kandosols in Northern Territory all have CEC levels well below what is required (less the 1 cmol(+)/kg).

The CEC in these sandy soils can be enhanced by increasing pH levels, this is usually with the addition of lime. The addition of organic matter is also an effective way of improving the CEC. This can be done with inter row pasture, regular slashing and the addition of mulch.

As described in the section above, there has been an increase on topsoil pH under the drip line. It may be the case that instead of inhibiting nutrient up take, it is improving the availability due to an increase in CEC. Soil data within Quintis plantations at Midway confirmed that there has been an increase in CEC between 2011 and 2020. For example at Midway A, the pre establishment CEC was 2.8 cmol(+)/kg and in 2020 it had increase to 7.4 cmol(+)/kg. This is most likely due to additions of organic matter and increased soil pH.

c) Mineral toxicities

At depth, there are several mineral nutrients that are found at levels toxic to Sandalwood and host species, in particular Boron and Aluminium. As the Sandalwood roots are primarily in the top 40 cm, and usually the toxic elements are at levels deeper than that, this toxicity usually causes more host mortality than Sandalwood. Quintis testing indicated that some of the Aquitaine soils in the ORIA can have high mineral levels and when Boron levels are above 2mg/kg they are toxic to Sandalwood plantation species and soils with this amount of Boron will be detrimental to tree health.

Soil testing in the Kandosol soils in NT show that on average the topsoils have 0.15 mg/kg of Boron, and the lower profile has 0.18 mg/kg. These levels are considered low, and availability would be considered marginal.

Salinity

Increased salinity results in a decline in water quality, soil properties and plant growth. There are many indicators of potential salinity problems. While individual indicators are not conclusive proof that salinity exists at a site, the more indicators that are present, the more likely it is that an area is saline or is becoming saline.

The Sandalwood plantation soils most susceptible to saline issues are the Vertosol and the Sodosols. Saline soils are often bare or sparsely vegetated. Salt crystals may be visible on the surface and sheet, or gully erosion is often present. Saline soils may be puffy when dry or slippery when wet. The soil may be continuously or seasonally wet or waterlogged.

A convenient method of estimating the amount of salt in a soil or water sample is to measure the electrical conductivity (EC) of a solution. Dissolved salts enable the conduction of electrical current, with conductivity directly correlating to the concentration of salts present in the water. The measure Quintis uses to determine salinity of a site is ECe. ECe considers the soil texture and is determined by multiplying EC 1:5 by a soil texture factor.

The salinity tolerances for Sandalwood plantation species in this table have mostly come from observations, natural habitat and limited literature. The ECe for soils in the ORIA range from less than 0.3 Ece to over 10 decisiemens per metre (dS/m). Quintis testing did show that soils with ECe over 1.0 dS/m will affect healthy tree growth. This is not saying that plantation cannot survive on soils with ECe between 1and 2 dS/m, only that trees will perform much better on soils with ECe lower than 1 dS/m. It should be noted that generally ECe and high boron levels a commonly found together in Cununurra clays.

Table 11: Plant species tolerance to soil Salinity.

Salinity tolerance levels - crop species	**Level of soil salinity (ECe) at which plants begin to decline.**
Sensitive crops turnip, strawberry, beans, carrot, apple, avocado, *Dalbergia latifolia, Dalbergia lanceolaria*	0 - 1 dS/m
Moderately sensitive crops potato, grapes, corn, cowpea, linseed, Citrus, *Santalum album , Senna siamea , Sesbania formosa*	1 - 2 dS/m
Moderately tolerant crops grain sorghum, rice, *Cathormion umbellatum*	2 - 4 dS/m
Tolerant crops oats, sorghum, wheat, canola, safflower, soybean, *Casuarina cunninghamiana*	4 - 8 dS/m
Very tolerant crops barley, cotton	8 - 16 dS/m
Generally, too saline for all crops	> 16 dS/m

Sodic soils

Most of the ORIA and BRIA soils are slightly sodic. Due to the low infiltration capacity, water can stagnate in the soil easily and, in dry periods, can be difficult to wet up without copious amounts of irrigated water and good drainage. Generally, agriculture on these soils is limited to crops tolerant to surface waterlogging. While Sandalwood can withstand some dry periods, it does not like to be waterlogged.

Optimal exchangeable sodium percentage (ESP) levels of soils would preferably be less than 2, Cununurra clays tend to be low, less than 3 at the surface but increase with depth to values of 10 to 20. Aquitaine clay with native vegetation has higher ESPs of 4 to 6 at the surface, increasing to 27 at depths between 1.2 metres and 1.5 metres. These levels are quickly reduced under irrigation. ESP levels above 6 cause soil dispersion and a lowering of hydraulic conductivity. The other area with sodicity issues is in BRIA. Almost all the plantation sites have an ESP above 3 and some as high as 40 at depth.

To deal with these soils, gypsum was applied during site preparation. As these are flood irrigated, the tree rows are planted on mounds to provide good drainage. Increasing organic matter into the soils can also reduce sodicity.

d) Soil Erosion control

All the soils under plantation in NT have an erosion risk due to their sandy topsoils. Strategies must be in place to inhibit soil loss.

Erosion is driven by two processes, wind and water. Water and wind erosion interact. Both are influenced by climate, soil state and vegetation cover. The soil state is predominantly influenced by both human and animal activities. The soil state (exposure to erosion) across the Quintis estate was very much human driven.

The soils found in Taylors Park 3, 4, 15 and 16, Sandy Creek and some areas of Drovers are mostly Tenosols. This soil type has been reported to be 5 and 13 times more erodible than slightly heavier Kandosol soils like those found in Taylors Park 9 and Midway (Miles & McTainsh, 1994).

Without adequate strategies in place, water erosion (Malinda, Fawcett, Little, & Bligh, 1998) (Malinda, 1995) can remove a significant amount of topsoil. To prevent this, a permanent cover crop must be maintained in the sloping inter-rows, where erosion is most likely to occur.

In areas that receive high rainfall and wind, the requirement for high levels of soil surface cover is critical if erosion is to be reduced. The wet season is a concern if the soil is bare and mostly surface crusted as volunteer weeds will only give sporadic cover.

Soil organic matter in these sandy soils is very low, in most cases less than 1% and there was often some compaction by agricultural machinery, which lowered infiltration rates to less than ten mm per hour. Where crusting had occurred, infiltration rates have been measured less than 2mm per hour. Low infiltration means high surface run-off, and increased risk of erosion.

Soil erosion control strategies

Due to erodibility of the soils the following land management strategies should be implemented. In undulating topography, plantation rows should be designed to run across the slope and not down the slope. The design of the drip irrigation system and the plantation design must be correlated.

Application of ameliorants such as lime, gypsum or dolomite as recommended by soil scientists should be applied prior to ripping as it is preferable for them to be incorporated to depth.

In almost all cases deep ripping due to soil compaction or the hard setting characteristics of the soil is required. All deep ripping should be below the compaction depth identified by soil surveyors. The operation should be carried out with a winged tyne and tillage point that does not invert the soil. It should lift and crack the compaction layer. Ripping should be carried out when the soil is dry to avoid smearing. To avoid the loss of soil moisture to depth and maintain a manageable soil surface, the ripped line should be rolled with a heavy steel roller or cultivated as soon as possible after ripping. If very large soil clods are being left on the surface the soil may be too dry, or the ripper is performing the wrong action. If a blunt shanked ripper is being used, it would be advisable to weld a rounded or V shaped piece of steel on the shank to assist soil flow.

Ideally a program of sowing a cover crop should be implemented and maintained. This should be done prior to the wet season and rely on the initial storms to obtain a germination before the high rainfall events which can start in December. This preferably should be a N-fixing cover crop such as Wynn Cassia (*Chamaecrista rotundifolia*) but can be one of the tropical low growing (not N-fixing) pasture species such as Jarra grass (*Digitaria milanjiana*), Buffel grass (*Cenchrus ciliaris*) or Sarbi (*Urochloa mosambicensis*). The use of minimum disturbance sowing machinery to plant cover crop is recommended. Once established, the stands should be allowed to set seed for at least two years to increase future potential recruitment. Once set, the stands can be slashed but no lower than 200 mm high to ensure good groundcover.

Figure 53: Wynn Cassia ground cover at the plantation edge, (Midway. Robson, 2013).

Once the tree lines are established, a slashing program can be implemented to place excess organic material along the tree row (side throw slasher). If the rows have been mounded, the operator should try to get as much material along the side of the mounds as possible. As the canopy closes the leaf litter drop and mulching of any dead trees are a bonus to provide ground protection, organic matter and nutrients.

e) Compaction and porosity

Soil compaction refers to the deterioration of soil structure characterised by an increase in soil bulk density and a decrease in soil porosity. Compaction can adversely affect all the physical and biological properties of the soil. The overall effect of the soil compaction on the plant yield is detrimental.

Compaction leads to restricted root growth, reduced nutrient accessibility, and increased soil nutrient loss through leaching, runoff, and gaseous losses to the atmosphere, all of which hinder plant growth.

The ability of roots to penetrate soil is significantly hampered by compaction, a critical concern for Sandalwood root systems, which predominantly rely on feeder roots within the top 20 cm of the soil profile. Compacted soils can restrict Sandalwood ability to make haustorial connections. Generally, compaction results in a decrease in root length, root penetration, and rooting depth (Farrakh Nawaz, Bourrié, & Trolard , 2012).

Soil compaction commonly occurs when heavy machinery repeatedly traverses the surface, particularly in wet conditions, as wet soils are more prone to compaction. Overstocking of cattle and pathways leading to water points can also contribute to compaction.

The soil type and texture makeup also has an influence on compaction depending on their moisture level and the contents of the soil's organic matter. Silty loam soils are more susceptible than medium or fine textured loamy and clay soils at low water contents, while the sandy soils are slightly susceptible to the soil compaction (Farrakh Nawaz, Bourrié, & Trolard , 2012).

While compaction is often attributed to management practices, Kandosol soils under Sandalwood plantations naturally undergo hard setting during dry seasons due to specific physical, chemical, and mineralogical properties. (Fabiola, Giarola, Pires da Silva, Imhoff, & Dexter, 2003).

Bulk density (BD) and porosity are indicators of soil compaction and soil health. Bulk density is the weight of dry soil per unit of volume typically expressed in grams/cm^3. Bulk density over 1.58 grams/cm^3 in sandy clay loams and clay loams are considered to restrict root growth. While bulk density at 1.60 grams/cm^3 in these soils is considered to affect root growth (Moore, 1998).

Figure 54: A ring with known volume is tapped into the soil. This known volume is dried and weighed to calculate the bulk density in grams/cm^3 ,(Robson, 2021).

The soils in the Douglas Daly are predominantly sandy clay loams and clay loams, bulk density in the eight to ten year old plantations ranges between 1.43 grams/cm^{3}and 1.62 grams/cm^3. This variation in data is contingent upon sampling locations, namely, just off the planting row (50 cm), where tractor wheels consistently run (100 cm), or the centre of the rows (150 cm). It is anticipated that compaction may occur where tractor wheels consistently run, which indeed seems to be the case in some areas.

Across all sites, the average bulk density 100 cm from the planting line is 1.52 grams/cm^3, with some sites reaching as high as 1.62 grams/cm^3.

Compaction resulting in an increase in bulk density also leads to a decrease in porosity. For instance, if compaction raises bulk density from 1.3 g/cm^3 to 1.5 g/cm^3, porosity decreases from 50 percent to 43 percent. Porosity, defined as the pore space in soil between mineral particles (and solid organic matter) filled with either air or water, averages 42% in the Douglas Daly region but can reach as high as 46%. However, increased bulk density reduces porosity to as low as 37%.

Currently, the higher levels of compaction observed in the Douglas Daly region do not inhibit root growth. Nonetheless, considering the Sandalwood rooting system connects Sandalwood to the host within the top 20 cm of the soil profile, it is imperative to prevent further compaction.

f) Soil assessments

Santalum album is a species that is found on a wide range of soil type in its native environment in India. From laterites, loams, sand, clays and black cotton soils, and usually soils of poor nutritional status. In Timor, Sandalwood is found on rugged stony topography with heavy textured soils, often sodic and in some cases saline (Radomiljac & Vernes, 2019). It is well known that Sandalwood does not tolerate prolonged waterlogging.

Quintis had a rigorous procedure in place to ensure that only soils suited to effective growth of Sandalwood was planted. The soils consistent with the following are considered suitable for Sandalwood establishment:

- Non waterlogging, well drained or could be well drained with the use of various cultivation methods.
- Soils with a non-impeded depth of at least 1400 mm. That is soil without hard pans such as rock layers or compacted clay or sand.
- Adequate readily available water (RAW). A minimum of 25 mm of RAW, preferably 50 mm or above within a 700 mm profile.
- Soils with moderate to neutral subsoil acidity.
- Soils free of toxic levels of certain elements such as Boron and Aluminium

The soils work undertaken by Quintis was very important for gathering different information. In the first instance, it provides an insight into the suitability of land which Quintis was interested in acquiring. This is the soil assessment for land and due diligence (SALADD) but it can also be used for gathering data in conjunction with growth data and eventually harvest data to gain knowledge on the relationships between soils and plantation performance.

Soil science and soil chemistry are complex fields. Essentially there are a few major components which must be looked at:

- The physical attributes of the soil profile. This will determine how well the profile drains and how the soil needs to be treated from an operational point of view. The physical descriptions provide clues on how the soil will behave and a history of events and treatments.
- The topography and drainage. The slope of the land, soil texture and the drainage zones. This will determine how well water drains away from the site. This will play a big part in plantation and field design.
- Chemical properties. The soil chemistry provides information on how fertile the soil is, any toxicities that may be present and other chemical properties that influence the behaviour of the soil and performance of the plantation.

Physical characteristics

The physical attributes of the soils are assessed on site by digging soil pits and assessing the profile of the soil down to 1.4 metres. By utilising pits, the assessor can see the soils in a manner which represents how they would be in an undisturbed environment as opposed to using cores for example.

The attributes that are required to be assessed are:

<u>Soil texture</u>

Provides an indication of water holding capacity and specifies how different soils can be managed. The preparation and management of plantations will be vastly different between a sandy loam and a medium clay. Sandalwood will grow on a wide range of soil textures from sand through to the clays. Host selection may differ depending on soil texture.

<u>pH</u>

pH needs to be in the correct range to suit Sandalwood. The optimum pH is between 6.5 and 7.5. Occasionally there are soils with acidic topsoil that be rectified with the additions of lime.

<u>Colour</u>

Can give clues about the parent material and some history of drainage. Mottled colours are a sign of poor drainage and waterlogging.

<u>Dispersion/slaking</u>

This test shows how the soil reacts when it gets wet. The dispersion test is an indicator of sodicity. Sodic soils do not drain well through the profile but can be treated to improve them.

<u>Root depth of existing plants</u>

Existing root depth gives clues on how other plants have performed in the soil at that site. It can also indicate when there are problems for areas that lack root depth.

<u>Signs of water logging</u>

There are signs of water logging which can be present in the profile. These include mottled clays and manganese nodules.

Compaction

Compaction is particularly important to identify as part of the assessment. It is normally caused by machinery and farming practices but can be caused by water or livestock. Compaction can generally be managed but has a cost associated with repair. Some soils correct more easily than others and the depth of the compaction can influence the correction techniques.

Compaction is given a score in the field between 0 and 2 (soil pak score), where 0 is like concrete and 2 is non compacted or friable. Compaction can be rectified by physically breaking down the compaction by deep ripping or growing a cover crop or applying gypsum. Some of the Kandosol soils are naturally hard setting.

Coarse fragments or rock layers

Coarse fragments and rock areas influence drainage and can provide physical barriers for root growth and problems for machinery.

Soil horizons

The horizons are identified and give clues on how water and roots will move down the profile. The surface horizons may influence what operational methods are used to prepare the soil for the trees.

Samples are taken at each depth down to 1.4 metres and at two and three metres (usually six samples per pit) and sent away for chemical analysis.

Chemical analysis

Chemical analysis supplies extra information to back up the observations made in the pits. It shows nutritional information and elemental toxicities. The lab data provides pH, EC, major nutrient levels, micronutrient levels, sulphates, chlorides and organic matter.

Decisions on fertiliser application, addition of ameliorants such as gypsum/lime and other operations needed can be made once the pit analyses have been done. In the case of due diligence for land acquisition, the data and pit analyses are used to decide whether the land is suitable for Sandalwood plantations. An operational plan and a budget can be developed once the soil results are known.

Topography

Assessing the topography of the land is important to determine how well the land will drain and how the plantations should be designed. Things to consider are:

- The slope of the fields - this is important for runoff of rainfall but is also critical for flood irrigation.
- Drainage points - fields/farms must have points where water can exit and get away. The areas of tropical Australia where the plantations are grown experience high rainfall events and plantations must be designed to cope with these events.
- The topography will determine whether the fields need to be mounded and what type of mounds to use.

- The topography will often dictate the irrigation type. Some fields will have to be drip irrigated because they are too steep to flood irrigate.

Upon completion of all assessments, plantation design and operational maps can be developed. These provide foundational data essential for effective plantation management. Additionally, this data supports the evaluation of growth across various sectors and facilitates accurate yield estimation.

3. CLIMATE

Sandalwood occurs in the warm humid tropical regions of India, Indonesia and Australia. It can be found in areas with monsoonal rainfall of over 2000mm but tends to occur on western slopes where rainfall is much less, and the soils are free draining. Temperatures range from 13°C to 37°C but Sandalwood has been grown in extremes of 4°C to 46°C (Singh, 1995). However, for good plantations, temperature is not the only factor to consider, humidity and natural rainfall are important criteria as well. These add to the evapotranspiration which can determine the rate of physiological activity of both Sandalwood and host.

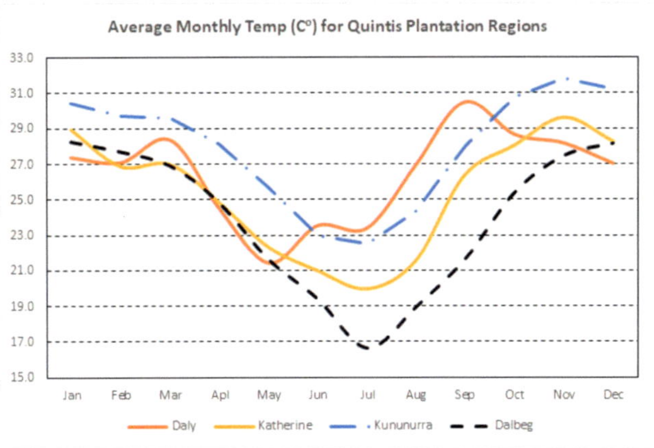

Figure 55: Mean annual temperature for the four Quintis plantation regions (Meteorology, 2023).

a) Temperatures

Summers at the plantation sites are typically characterised by high humidity and heat, with mean temperatures ranging between 27°C to 31.7°C, often reaching maximums in the low 40°C range. Winters tend to be relatively cooler, with mean temperatures approximately 10°C lower. However, the Queensland site has experienced at least three instances of light frosts since its establishment, resulting in observed damage to host trees, particularly *S. siamea* and *C. umbellatum*. Although the frost caused damage to the host trees, they quickly recovered, while the Sandalwood trees themselves showed no adverse effects from the cold weather.

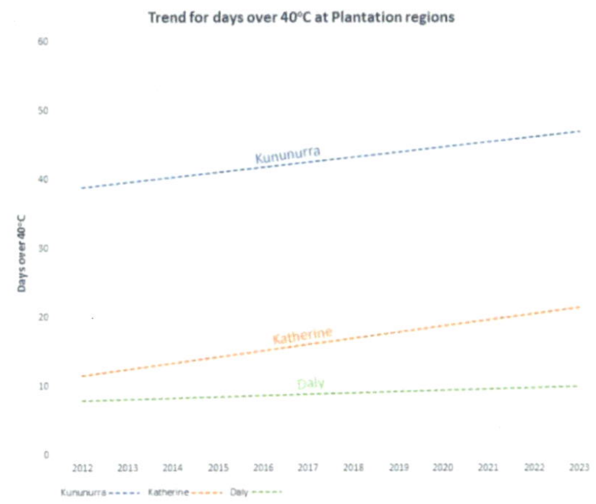

Figure 56: The linear trends for days over 40°C in each of the plantation districts (Meteorology, 2023).

There is a discernible trend towards increasing extreme temperature events. While some years have always had more days than average with temperatures over 40°C, this occurrence is on the rise. This trend appears somewhat linked to El Niño events, where the number of hot days significantly increases during those years.

A study conducted by CSIRO (Carter, et al., 2010) in Kununurra's Sandalwood plantation demonstrated that when temperatures increased during the afternoon, Sandalwood transpiration rates decreased, particularly in conditions where vapour pressure deficit increased (low humidity and high ETo).

b) Rainfall

All the plantations are in northern Australia, subject to a monsoonal rainfall season. Approximately ninety percent of the annual precipitation falls during the wet season between December and March. Although the average yearly rainfall at Kununurra Airport from 1971 to 2023 was 839mm per year, there is significant variation in wet season rainfall totals between years. Figure 57 illustrates the combined rainfall totals for each reported year since 2008 for Kununurra and since 2012 for the other sites. This figure indicates that since 2008, exceptionally wet years (greater than 20% to 45% above the long-term average) occur every three to five years. However, the data may not fully capture the extent of wet season rainfall as it is reported yearly, whereas the wet season spans from December to March. Nonetheless, the trend of larger wet seasons is still evident. This trend is also observed in both Katherine and Douglas Daly, though to a lesser extent in Dalbeg.

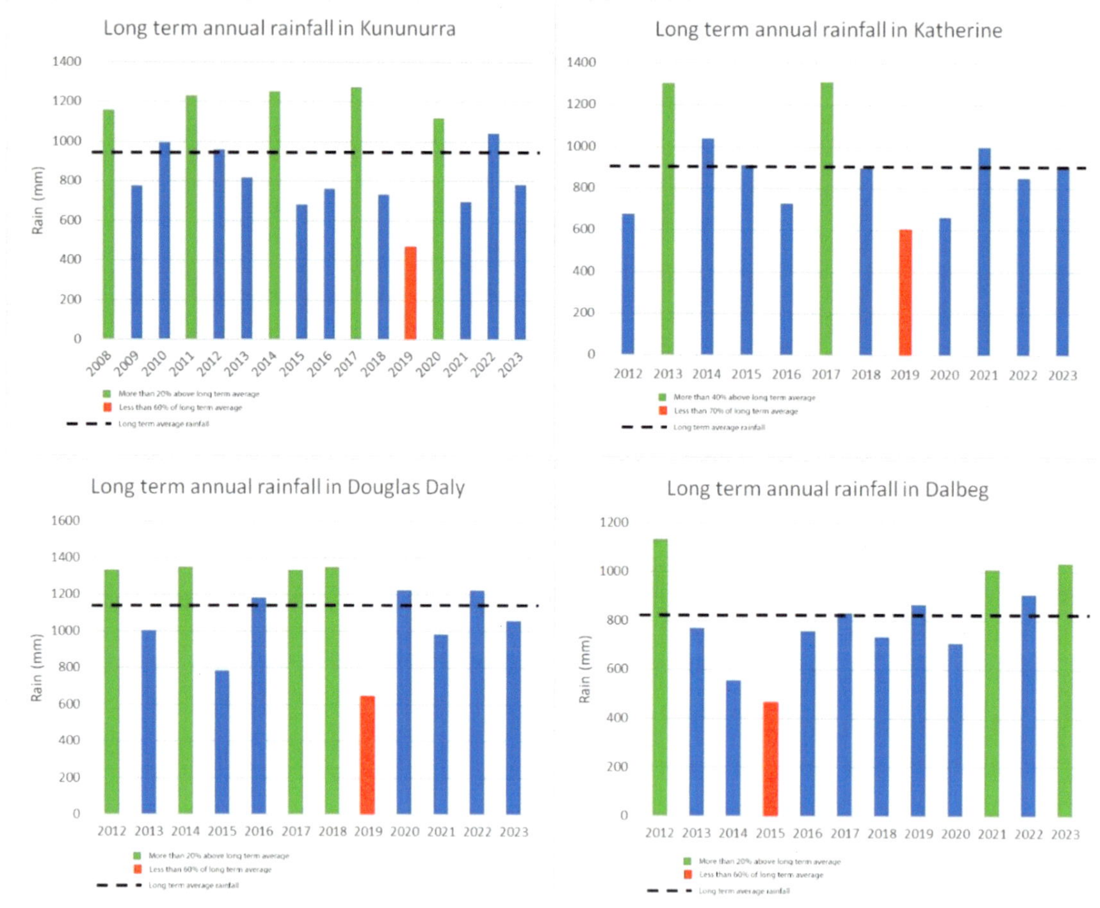

Figure 57: The mean annual rainfall for the four Quintis plantation site. Each graph shows annual rainfall, the long-term mean rainfall is indicated with the dotted line, those years with rainfall that exceed the average by more than 20% in green and those years with 70% to 60% less than the long-term average is shown in red, (Meteorology, 2023).

Rainfall in Australia is influenced by the occurrence of El Niño and La Niña events. These climate patterns across the Pacific, specifically ocean temperatures, occur when the Pacific Ocean and the atmosphere above it deviate from their neutral state for several seasons. El Niño events are associated with a warming of the central and eastern tropical Pacific, while La Niña events involve sustained cooling of these same areas (Meteorology, 2023). Recorded La Niña years in 2007/2008, 2011/2012 correspond with higher-than-average wet season precipitation. However, a La Niña phase has persisted since 2020. While wet seasons have continued to occur, rainfall has generally been only around average. In contrast, the El Niño periods of 2015–2016 and 2018–2019 were marked by significantly below-average wet season rainfall.

c) Humidity

Relative humidity is a measure of the natural levels of water vapour in the air. The maximum amount of water vapor held by the air is dependent on the air temperature. Warmer air holds more water than cooler air. To a lesser extent the air barometric pressure can also affect humidity. Relative humidity (RH) is expressed as the percentage of water vapor in the air in comparison to the total amount of water that could be held by the air if it were saturated. The lower the RH, the drier the air.

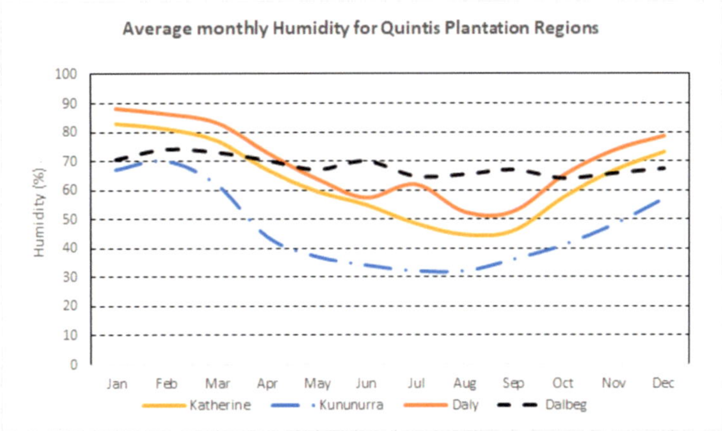

Figure 58: Mean monthly Rh% of each plantation region, (Meteorology, 2023).

The changes in the relative humidity and temperature have a direct impact on the photosynthesis process, therefore, it influences the growth and development of plants. Low relative humidity and high wind velocity can cause water stress and stomatal closure (Shibata , Iwao, & Takano, 1995). It is well understood that the optimal relative humidity for maximum plant growth lies within 60 to 70% relative humidity (Chia & Lim, 2022). If conditions are maintained within this desired range, the photosynthesis capability of the plant is not compromised.

The effect of increasing humidity can be observed in the plantations. There is a major flush of new growth in all the species between September to November as humidity increases.

In both Northen Territory regions, the relative humidity increases by 20% in a six week period between September to November. The *Dalbergia lanceolaria* which has been leafless over winter, starts to flush with new growth. Although not deciduous, the Cathormion umbellatum also responds with new growth in this period. The relative humidity increase in Kununurra is smaller in this period, an average of 12%, and while not as dramatic as the Northen Territory, the flush of new growth can still be observed.

The Dalbeg site has little to no change in humidity over the year. Even though this site has the coolest winters it has the highest relative humidity and is only surpassed by the Northen Territory regions in

December. Due to this lack of humidity increase, there is no obvious annual flush of new growth. Additionally the Dalbeg site maintains relative humidity in that optimal range throughout the year, whereas the other sites drop below that 60% optimal line.

d) Evapotranspiration

Evapotranspiration (ETo) is an estimation of water loss from both plants and soil, influenced primarily by sunlight, wind, humidity, and temperature. It offers an objective and dependable gauge of water needs for actively growing plants in plantation settings. This information aids irrigators in scheduling irrigation more accurately, thereby enhancing yields and water efficiency.

Understanding ETo is crucial in developing irrigation schedules, as it reduces guesswork in watering practices. Regular ETo monitoring provides valuable insights into plant water requirements. Moreover, the calculation of ETo offers the advantage of providing locally relevant values applicable across entire compartments, plantations, or districts.

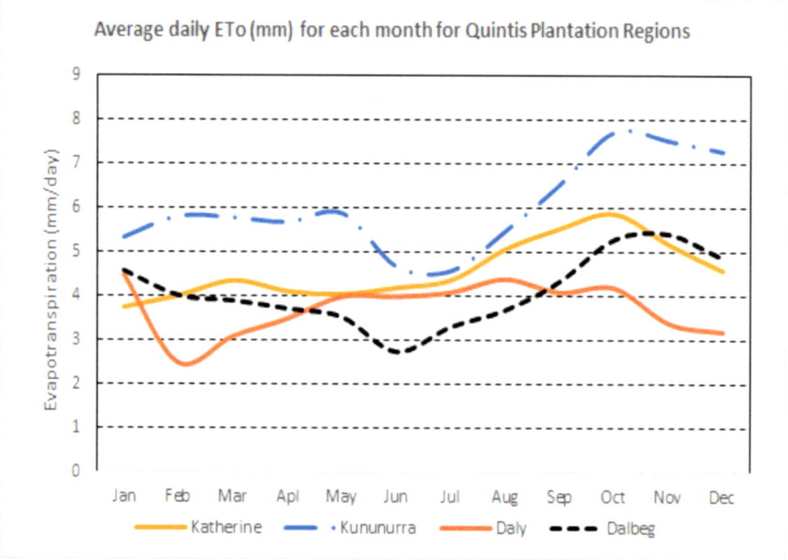

Figure 59: Mean daily ETo for the plantation regions.

The Penman-Monteith equation, endorsed by the United Nations Food and Agriculture Organisation in Irrigation and Drainage paper 56 (Allen, Pereira, & Smith, 1998) is used to calculate evapotranspiration. This method compares ETo relative to a standard "reference" crop, typically a well-watered, actively growing green grass covering the ground. Adjustments for different crops are made using crop factors. The Australian Bureau of Meteorology automatic weather stations (AWS) use standard wind speed measurements ten metres above ground level, air temperature, humidity and daily solar radiation values from satellite data to calculate the ETo.

ETo values vary significantly across plantation districts. Kununurra plantations experience notably higher daily ETo compared to Daly and Dalberg regions, particularly in summer (see Figure 59). Consequently, higher ETo necessitates greater irrigation water application. In Dalberg, higher humidity during cooler winter months reduces ETo, while in Daly, higher humidity in summer months decreases daily ETo.

4. SITE CLASSIFICATION INDEX

To predict growth and yield of plant biomass under different management regimes, a reliable and efficient estimation of site quality is required. Site quality may be defined as the physical and biological factors that characterise a site's ability to support tree growth (Skovsgaard & Vanclay, 2008).

A site quality index should possess certain characteristics: it must be reproducible and consistent over long periods of time, indicative of the site, independent of management and stand conditions, correlated with the productive potential of the site as a cause-effect relationship, and in line with the current knowledge of tree physiology.

The problem with using traditional site index methodology as developed in forestry, is that it is based on tree height, and height is actively managed in Sandalwood plantations. Traditional methods cannot be used to calculate a value for site index in a Sandalwood plantation and thus cannot be linked to productivity. All it will reveal is that a particular site has taller trees than another, without indicating whether the site with the taller trees is productive.

Consequently, there was a need to develop a suitable index that can predict site productivity in Sandalwood plantations. This was approached in two ways, the first was to quantify the site on which Sandalwood has been established to gauge how well they will perform, and the second was to review tree growth data against an expected maximum growth potential.

a) Site capability Index

The site capability Index is a methodology that ranks the capability of a site to grow a plantation. Plantation managers can compare various stands to determine potential management inputs for optimal growth. The ranking index uses three critical components of the site, readily available water (RAW) in the soil profile, Phosphorous Buffering Index (PBI) in the top 30cm of the soil and pre management of the site. Within each component there is variation, and that variation is grouped and given a weighted score and the levels of each are added to give a site score. These are then ranked from best to poorest. When 2 scores are the same but with different attributes, they are ranked with RAW given the highest priority, then PBI and then previous management.

RAW

RAW is the calculated amount of water in the soil profile (to 140 cm depth) that plants can easily extract from the soil. RAW is calculated by determining the soil texture and percentage stone/gravel of each layer down the profile. Quintis methodology was developed by Mackenzie et al. (2008) and used tables developed by Moore (2004).

When determining how to calculate RAW in the profile, two soil depths can be used. For most agricultural crops, only the rooting depth is used. This is the top 40 cm to 60 cm of the profile. For deeper rooting tree crops, the profile depth is used (140cm). Although the tree roots within the Quintis plantations were mostly within the top 60 cm of the profile, the host trees do send roots much deeper and have access to the available water at depth. For this, the Index score did use the profile RAW, as this is the more accurate indication of the site water availability for trees.

Weighting

To calculate the index, the amount of RAW across all the sites was grouped into three groups, high, medium and low, and these were given a weight. As the more available soil water is an important component to continued growth and health, the high RAW soils with more than 90 mm available were given a maximum weight of 10.

The medium RAW soils hold about 18% less water than the high RAW soils and have been given a weight of 8. The low RAW soils hold 36% less than the high RAW soils and have been given a weight of 4.

Table 12: The RAW SCI weighting.

Level	SCI Weighted Score	RAW
High	10	More than 90mm
Medium	8	70 to 90 mm
Low	4	Less than 70 mm

Example

Assuming that only half of the RAW in the profile between irrigation events is to be utilised. A plantation with a crop factor of 1, and a daily Et of 6.5 mm, with a soil RAW of 110mm will hold eight days of water. A site with the same crop factor and daily Et with a RAW of 88 will hold six days of water. A site with the same crop factor and daily Et with a RAW of 68 will only hold three days of water.

Note that crop factors in Quintis plantations range from 0.3 to 0.7. The Crop factor of 1 in this example was only used for ease of calculation and maximum usage.

PBI

The Phosphorus Buffering Index (PBI) approximates the soil's ability to 'fix' phosphorus (P). Low PBI values indicate that the soil P has limited ability to tie up applied P and therefore should indicate higher fertiliser efficiencies in the short-term and more P is available to the plant. However, soils with very low PBI have the potential to leach and lose phosphorus in irrigated crops and high rainfall areas. Conversely high PBI values (>280) indicate soil types that have the ability to quickly bind up P and make it unavailable to the plant (Moody, 2007).

In the Quintis plantation sites, soils with low PBI typically have less available phosphorus, which means plants can deplete phosphorus more quickly due to limited replenishment capacity (low solution P buffering potential). In contrast, moderate PBI soils can replenish phosphorus when the soil solution is depleted, thanks to their higher buffering potential. In the Northern Territory, none of the soils in Quintis plantations have a PBI above 280, so those with PBI values greater than 140 are considered high locally, even though the literature would classify these as moderate PBI soils.

Table 13: The PBI SCI weighting.

PBI value	Described Category (Moody 2007)	Quintis Category	SCI Weight
> 140	moderate	High	8
70 - 140	low	Medium	6
36 - 70	Very low	Low	4
<15	Extremely low	Very low	2

For the SCI, the PBI is an indicator of the amount of response that can be expected with additions of P and can be used with a Colwell P analysis. Soils with a BPI of over 140 will buffer applied P but not to the extent it will be unavailable for plants. They can resupply buffered P to plants and will also hold P from leaching under irrigated conditions. Soils with a low SCI PBI indicate that they will not hold applied P and potentially will require reapplication at some point as P is depleted and or leached.

Weighting

To calculate the index, soil with a PBI of over 140 are weighted as the best (8), due their ability to buffer and resupply over time. The weighting decreases as the PBI decreases, due to the explanation above. As the PBI is primarily determined by the soil texture, increasing from sand to clays, only minimal changes in PBI values are seen when adjusted for current or increasing P fertility.

Previous management

Pre plantation land use is taken into consideration because of the potential effect on future plantation trees. Most of Quintis plantation sites had some type of commercial pre use, most being ex pasture and grazing. Some having intensive agriculture such as melons, peanut, sugar cane, banana and mango cropping. A few, prior to plantation had native vegetation regrowth that was removed.

The SCI divides these pre management types into three groups, Intensive management, Managed and Unmanaged, and ranks them in that order.

Intensive management

The intensive management sites all have had some sort of agriculture prior to Quintis acquisition for plantation establishment. Agriculture usually involves annual cropping, and this requires yearly site cultivation and additions of fertiliser. Most of these sites had a nutrient bank that was an advantage for tree establishment. However, due to the constant use of heavy machinery, on some sites soil compaction and poor soil structure are potential issues.

Managed

The land that was managed for cattle grazing is classed in the SCI as Managed. These sites generally have had some improvement, albeit not as intensive, with either improved pasture seed sown, potentially some additions of fertiliser and years of added cattle manure and other organic matter. There is the potential compaction of soils, erosion due to over grazing and high levels of weed species.

Unmanaged

Some sites did have recent native vegetation cleared prior to plantation establishment. Many sites were not cleared or had regrown back with regrowth vegetation because they were failed or unsuited for cropping. The main reason for this unsuitability was soil, it is either too shallow (rock layers) or had poor RAW.

Table 14: The previous management SCI weighting.

Quintis Category	SCI Weight
Intensive Management	3
Managed	2
Unmanaged	1

Index and ranking

The index consists of the added score of the components described above, with the highest weighting on RAW, PBI and then previous land management. The scores range from 21 down to a minimum of 7.

The ranking is a numerical ranking of the best Index score to poorest score. A score of 21 receives a ranking of 1 and a score of 7 receives a ranking of 36.

Where several sites count the same score, the ranking is determined by RAW primary, PBI secondary, then Pre management. For example, the four sites below all have a SC Score of 15, however they are ranked according to attributes of the site by weight.

Table 15: Sites with a SC score converted to SCI using priority weightings.

RAW	RAW score	PBI	PBI score	Previous land management	PLM score	SC Score	SC Index	SC Classification
High	10	Low	4	UnMgt	1	15	15	Good SCI
High	10	V.low	2	Intens Mgt	3	15	16	Good SCI
Med	8	Low	4	Intens Mgt	3	15	19	Good SCI
Low	4	High	8	Intens Mgt	3	15	25	Average SCI

Grouping

The classification breaks the SCI into 4 general classifications, High SCI (12), Good Sci (9), Average SCI (9) and Sub optimal SCI (6).

High SCI sites

The High SCI sites have good water hold capacity soils with high to medium RAW. They have good Phosphorous Buffering with high to medium PBI soils. Most have had some intense management or pastural management. These sites require the least amount of input to produce optimal plantation growth.

Table 16: The high SCI classifications.

RAW	Raw score	PBI	PBI score	Previous land management	PLM score	SC Score	SC Index	SC classification
High	10	High	8	Intense Mgt	3	21	1	High SCI
High	10	High	8	Mgt	2	20	2	High SCI
High	10	High	8	UnMgt	1	19	3	High SCI
High	10	Medium	6	Intense Mgt	3	19	4	High SCI
High	10	Medium	6	Mgt	2	18	5	High SCI
High	10	Medium	6	UnMgt	1	17	6	High SCI
Med	8	High	8	Intense Mgt	3	19	7	High SCI
Med	8	High	8	Mgt	2	18	8	High SCI
Med	8	High	8	UnMgt	1	17	9	High SCI
Med	8	Medium	6	Intense Mgt	3	17	10	High SCI
Med	8	Medium	6	Mgt	2	16	11	High SCI
Med	8	Medium	6	UnMgt	1	15	12	High SCI

Good SCI sites

The Good SCI sites have good water holding capacity soils with high to medium RAW but tend to have poor Phosphorous Buffering ability, low to very low. Previous management is a mix of all previous management. The management of these sites requires more attention to nutritional inputs rather than water.

Table 17: The Good SCI classifications.

RAW	Raw score	PBI	PBI score	Previous land management	PLM score	SC Score	SC Index	SC classification
High	10	Low	4	Intense Mgt	3	17	13	Good SCI
High	10	Low	4	Mgt	2	16	14	Good SCI
High	10	Low	4	UnMgt	1	15	15	Good SCI
High	10	V. low	2	Intense Mgt	3	15	16	Good SCI
High	10	V. low	2	Mgt	2	14	17	Good SCI
High	10	V. low	2	UnMgt	1	13	18	Good SCI
Med	8	Low	4	Intense Mgt	3	15	19	Good SCI
Med	8	Low	4	Mgt	2	14	20	Good SCI
Med	8	Low	4	UnMgt	1	13	21	Good SCI

Average SCI sites

The Average SCI sites have medium to low water holding soils but mostly good Phosphorous Buffering ability. The management of these sites requires more attention to irrigation and water supplements than nutritional inputs.

Table 18: The average SCI classifications.

RAW	Raw score	PBI	PBI score	Previous land management	PLM score	SC Score	SC Index	SC classification
Med	8	V. low	2	Intense Mgt	3	13	22	Average SCI
Med	8	V. low	2	Mgt	2	12	23	Average SCI
Med	8	V. low	2	UnMgt	1	11	24	Average SCI
Low	4	High	8	Intense Mgt	3	15	25	Average SCI
Low	4	High	8	Mgt	2	14	26	Average SCI
Low	4	High	8	UnMgt	1	13	27	Average SCI
Low	4	Medium	6	Intense Mgt	3	13	28	Average SCI
Low	4	Medium	6	Mgt	2	12	29	Average SCI
Low	4	Medium	6	UnMgt	1	11	30	Average SCI

Sub optimal SCI sites

The Sub optimal SCI sites have soils with both low water holding capacity and low to very low Phosphorous Buffering ability. These sites will require the most inputs. Management inputs to achieve optimal growth outputs will be additional water supply and fertiliser.

Table 19: The suboptimal Sci classifications.

RAW	Raw score	PBI	PBI score	Previous land management	PLM score	SC Score	SC Index	SC classification
Low	4	Low	4	Intense Mgt	3	11	31	Suboptimal SCI
Low	4	Low	4	Mgt	2	10	32	Suboptimal SCI
Low	4	Low	4	UnMgt	1	9	33	Suboptimal SCI
Low	4	V. low	2	Intense Mgt	3	9	34	Suboptimal SCI
Low	4	V. low	2	Mgt	2	8	35	Suboptimal SCI
Low	4	V. low	2	UnMgt	1	7	36	Suboptimal SCI

b) Site Density Index

Site Density Index (SDI), developed by Andrew Brown, uses four Sandalwood productivity indicators: age, ground-level heartwood area (estimated by diameter at 20 cm), bole height, and photosynthetic activity (currently new and not fully understood).

The SDI uses a similar mathematical formula to Site index except it utilises diameter and number of trees rather than age and height.

SDI allows calculations to work out crop potential or SDImax, the maximum expected capacity of a site, to allow comparisons between sites to assess the relative potential performance (using relative density (RD) which is SDI/SDImax). This will enable managers to plan and implement management activities and monitor performance.

The following figures will help to explain. Figure 60 is a graph of Log (Diameter over bark (DOB) at 20cm) vs Log stems per hectare (SPH) for the SDIs for the plantation Pioneer. It shows to have increased over the last four years with the SDI for compartments increasing each year as the compartments grow out to the max crop potential which is defined by the line joining the three points SDImax. This is to be expected as the compartments are young and filling up the space available.

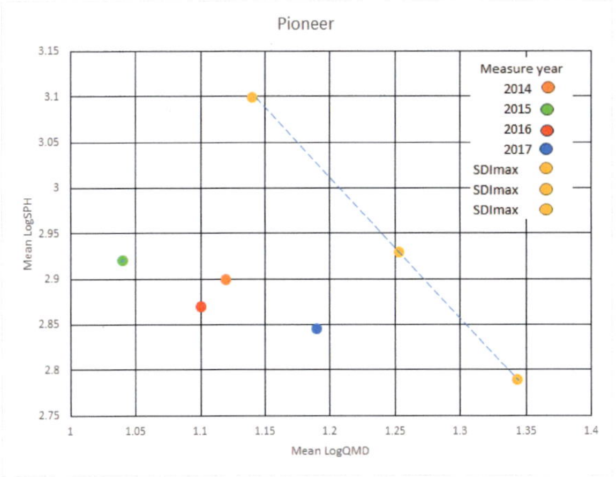

Figure 61: Performance of the Pioneer plantation over the 4 years. The plantation is growing out towards SDImax, show good growth for now.

Figure 60: Performance of Smiths plantations. Compartment following the SDI max line down and growth stalled in the last three years.

Figure 61 shows the same but for the plantation Smith. The situation is a bit different here. These compartments have run into SDImax and can be observed that they are following the line defined by SDImax down. This is showing that the only way that these compartments will keep growing in size, is through adding the required resources, such as free up spaces for growth. Here control of host and possibly Sandalwood density are important. Not surprisingly the last three years have observed these compartments essentially stall in growth of QMD.

The final figure shows a different scenario again, is for the Chapmans plantation. It shows that the last three years the compartments have slowed down in growth. Essentially it is financing growth in QMD through the mortality of individual trees. This is indicating that the site is not enabling the crop to reach its potential. The site is lacking nutrition or water, or disease is present, or the plantation is grossly overstocked.

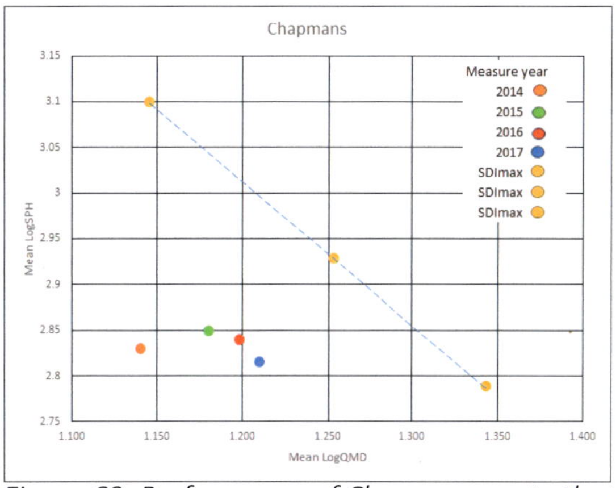

Figure 62: Performance of Chapmans, note that the compartments appear to be pulling up short of SDImax.

These three compartments tells two different stories.

How does SDI relate to Sandalwood.

SDI is an index that relates stocking and size of trees now when compared to those of trees with a mean diameter of 22 cm. 22 cm is relevant because it the mean diameter that we are looking for at harvest.

Using the Chapman plantation as an example, the relative density (RD) in 2017 was running at 67% or 403, with the SDImax at 610. To calculate where the site will be if nothing is done from now until harvest with the SDI of 403 the following calculations are used.

Total basal area for crop = 403 x 3.14 x (22/2)^2 = 15.3 m

Amount of that BA that is Sandalwood is 67% equating to 10.2 m

With stocking likely to be to be about 395sph at harvest mean QMD will be 18.1 cm at harvest, equivalent to a yield per tree of about 10.6 kg. This yield is below what is expected.

Table 20: The growth data for Chapmans and out comes at harvest is no silviculture intervention is undertaken.

YOE	Area	Age	Plantation	SDI2017	RD	SPH 2017
2009	140	8	Chapmans	420	67%	417

SA BA 2017	Tota BA 2017	QMD 2017	SA BA @ harvest	Total BA @ Harvest	QMD @ harvest	Yield kg/tree t/ha
7.4	15.3	14.2	10.2	15.3	18.1	10.6 4.2

How to Use SDIs for Planning and Monitoring

In this Chapman example, the SDI is clearly indicating that intervention is required. This could be adding extra water and nutrients, or silviculture management such as thinning. By the addition of these inputs and increasing the RD to 95% and SDI to 580. A recalculation will indicate an increase in DOB at 20 cm to 21.7cm which translates to 15.7kg per tree.

Table 21: This table showing sample plan for Chapmans plantation to reach yield expectation. This plan is between 2017 and harvest, it could be easily broken into yearly buckets.

SPH 2017	SPH at Harvest	Bole Ht (m)	Change in host BA	SA CAI (cm)	Proportion SA @ harvest
417	395	2	-1	1	67%

HW/ tree at Harvest 2017	Target RD @ harvest	SA BA at harvest	Total BA @ harvest	QMD SA @ harvest	Yield kg/tree t/ha
15.7	95%	14.8	22	21.7	15.7 6.2

Data Acquisition

Data is currently acquired from inventory plots but in the future satellites offers a promising tool. There is a good relationship between % crown cover, SPH and the mean QMD of a plantation compartment.

With the addition of Permanent Sample Plots to calibrate the satellite images, and by running the inventory program concurrently, the satellite program can gradually scale this back as confidence in the data from satellite and PSPs.

5. NEW PLANATION SITE IDENTIFICATION

Establishing a commercial-size plantation of Indian Sandalwood outside of the well-tried region of the Ord River Irrigation Area (ORIA) has been a long process of research and due diligence on many areas throughout northern Australia. To locate suitable areas within Australia, the following systematic six-Step Due Diligence process is suggested. This involves a series of filters, the application of which allows the progressive exclusion of unsuitable areas, and by default ensures the acquisition, as far as possible, of only suitable areas (Underwood, 2024) (Robson & Barnes, 2021).

Step 1: Eliminate areas with climatic risks.

Indian Sandalwood and most of the host species used in the Sandalwood plantations are tropical species and therefore intolerant of heavy, frequent frosts, especially when the plantation is less than 2-year-old. The BoM produces a "frost line" for Australia which can be consulted to map out areas which are generally likely to be frosty and to identify those which generally will be frost-free.

There is a history in tropical Australia of plantation projects wiped out by cyclones (Melville Island, coastal North Queensland). Cyclone damage can be minimised by locating plantations more than 70 kms from the coast and by strategic separation of plantation sites.

Step 2: Water availability

Areas selected must have a developed rural water supply system, or a potential water supply which can be legally accessed in sufficient quantities, under license, for plantation irrigation. There are relatively few areas in northern Australia where water is available for irrigation and where it is inexpensive. The Ord and Burdekin are arguably the most secure water supplies in northern Australia and almost never have allocation restrictions. There are very few others. The security over water is not only to overcome risk associated with growing the plantations but also to satisfy investors who have their own criteria to mitigate risk.

Step 3: Scale

Sufficient land must be available in discrete geographic areas to make investment in staff, infrastructure, equipment and nurseries viable for a self-contained plantation program and to provide economies of scale at time of harvest. A plantation estate of at least 1500 hectares in the one general region is desirable (other than for experimental purposes).

Step 4: Soils and topography

The soils in the target area must be generally suitable for Indian Sandalwood in terms of nutrition and physical structure, or capable of being ameliorated using fertiliser or gypsum and appropriate cultivation. Soil suitability is determined firstly by consulting soil maps and then by actual field survey by a soil scientist. Although Sandalwood has been shown to grow on a range of soil types, a major constraint for successful establishment and survival is drainage. The sites must drain well both from the overland flow and through the soil profile.

Step 5: Potential properties and detailed property survey.

For areas that meet all the previous criteria, real estate experts should be hired to find properties currently for sale or likely to become available after negotiations with landowners, as well as to maintain an up-to-date database of potential plantation sites and their estimated costs.

The land to be acquired should be freehold and zoned for plantation development, or if crown land it must be available for long-term secure lease. Plantation projects should not be commenced in areas where State or Local government policies are antagonistic to commercial forestry. Converting farmland into plantations is subject to different levels of regulation depending on location and council areas. In the ORIA, Sandalwood is classified the same as horticulture and requires no formal applications to convert from field crops or tree crops to Sandalwood plantations. In the Burdekin Shire, an application for a material change of land use is required and must be approved by the local council. Some councils may not allow this conversion, and the code assessment process may be such that other stakeholders can have input or protest the changes.

Once identified properties are "on the market" they must be subjected to a detailed and comprehensive survey looking at soils, topography, native vegetation, infrastructure, water supply and fences, problem spots (e.g. frost hollows or areas subjected to inundation) to meet established company protocols for acceptable sites for Indian Sandalwood; at this stage liaison commences with landowners and Local Government.

Step 6: Cost/benefit analysis:

The cost of purchase is examined in relation to the cost of plantation development and likely production and returns, taking into account predicted yields, distance to ports and markets, processing mill, distillery or other endpoint for plantation produce.

Chapter 4. PLANTATION DEVELOPMENT

Every forest grower should be clear on what products they aim to produce. Unlike most commercial forestry, Sandalwood plantations are specifically grown to produce heartwood for the valuable oil constituents. The design of the plantation should consider how best to efficiently achieve the desired products. The forestry methodology to produce the required stem form and size should be considered as part of the design. These systems will be discussed in detail in this Chapter. As heartwood is a critical product, systems should be included to ensure maximum its development by the end of rotation.

1. PRODUCTS

The plantation yields two primary products: whole logs and wood designated for oil production or chipping. The latter category includes below-ground root balls, curved sections between straight portions of the bole, and other logs that do not meet the standards required for sale as whole logs.

Straight lengths of the tree bole can be used as a log product. Dependent on the end market, in most cases there would be minimum lengths and diameters. Logs that meet these requirements can be used in Traditional Medicine, such as Ayurveda and Traditional Chinese Medicine, carving and furniture. Depending on the customer, the bark and sapwood can either be removed or retained.

For wood that does not meet the log measurement requirements but has good heartwood, this can be chipped and utilised for chip, oil and powder sales.

To extract Sandalwood oil, the wood must first be chipped to separate the heartwood from the sapwood. The heartwood is then subjected to steam distillation, which involves a four-step process: boiling, steaming, condensation, and separation. Water is heated to high temperatures (above 100°C) to produce steam, which is passed through the wood. The steam's high temperature releases the tightly bound oil within the cellular structure of the wood. The resulting mixture of steam and oil is subsequently cooled and separated, allowing the essential oil to be collected.

Figure 63: Sandalwood oil and powder produce, (Quintis website, 2024).

Sandalwood oil is a high price raw material for the perfume industry. It has a characteristic sweet, woody odour which is widely employed in the fragrance industry, but more particularly in higher priced perfumes. It has excellent blending properties and the presence of a large proportion of high boiling point constituents in the oil also makes it valuable as a fixative for other fragrances. The oil is used extensively in the cosmetics industry in the manufacture of soaps, face creams, toilet powders and air fresheners. The residual wood powder, after oil extraction, retains traces of oil and is used in the manufacture of incense sticks, which are commonly employed in rituals and religious practices, as well as in various cosmetic applications (Quintis, 2024).

Chipping the heartwood also presents numerous applications ranging from skincare to fragrance and food. Heartwood chips can be blended with other aromatic ingredients and powdered to create incense or distilled to produce essential oils or hydrosols for use in various fragrance and personal care products.

With the commencement of the Sandalwood plantation industry in the 2000s, the primary product in demand was Sandalwood oil due to a global shortage (Ananthapadmanabha, 2022). However, due to changes in Indian legislation in 2022 which legalised Sandalwood harvesting (The Times of India, 2022), the supply of oil has significantly increased. Despite this increase in supply, which has consequently led to a decrease in value, the demand for Sandalwood wood remains steady. Consequently, the product supply has shifted from predominantly oil derived from chipped wood to logs containing heartwood.

2. FORESTRY METHODOLOGY

When planning a new plantation, host species, configuration, spacing, and silviculture need to be considered for the end product that is to be produced.

Tree form

Most of the Sandalwood plantations are from an unimproved, small genetically diverse population. This means that there is a large variation in individual tree size and form. To improve the tree's form, trees are planted at a high stocking where they compete for light. This is standard practice for many plantation species. The methodology is to plant at a high stocking to push the trees up with good form and then thin those trees that have lost the race. This leaves only the best, straightest trees as the final crop.

As explained in Chapter 2, Species characteristics, Sandalwood, when young prefer to grow in a shaded environment and when open grown will produce many apical leaders.

For this reason, Sandalwood plantations have always been established at a high stocking of approx. 1500 stems per hectare, of which 33% are long-term hosts, 33% are short-term hosts and 33% are Sandalwood. The spacing is 1.3 metres within the Sandalwood row and 3.6 metres between rows.

Figure 64: Ten year old Sandalwood plantation in Dalbeg, QLD (Robson, 2024).

This strategy does improve Sandalwood height and reduces multi stemming and for the first 18 months the Sandalwood is reasonable uniform in height and size.

However, the short-term host start to die off from age three years and the Sandalwood must go through a transition phase, switching to the long-term hosts. With the short-term host mortality, the tree stocking is reduced to approx. 1000 stems per hectare.

The configuration, spacing and species design will be discussed later in this chapter.

Silviculture

Although the stocking along the Sandalwood row is dense, without stem pruning the Sandalwood will still produce multi stems. Observation at harvest has shown that a single stemmed tree will produce more heartwood and a higher quality log. If the aim is to produce predominately for the log market, then stem pruning is essential.

The stocking density for Sandalwood plantations post-transition is typically 1000 stems per hectare. However, many of the sites planted in Australia are on soils with low nutritional value, which limits the site's capacity to support such a high number of trees while optimising their growth potential. Consequently, prescribed thinning of both Sandalwood and host plants will be necessary to enhance their growth conditions.

The effect of competition and the required silviculture of thinning and pruning is discussed in Chapter 5.

Rotation length

Managed investment schemes were integral in the development of the Sandalwood plantation industry (Barbour, Plummer, & Norris, 2011). This scheme had a maximum investment period of 15 years, and this decided the rotation length of Sandalwood plantations. Although there is heartwood present in most trees after 15 years, research has shown that heartwood volume and oil quantity will significantly increase if plantations are left to grow on longer.

A Quintis trial which ran a small part of a plantation until age 20 showed a statistically significant increase in yield. The 20 year trees contained on average 68% more heartwood than they did at 15 years, even though there was only a small increase in diameter at 20 cm above ground level. This was a 13.8% increase in heartwood per year (Brown, 2020).

Plantation Site Selection

Indian Sandalwood, a tropical species well-suited to northern Australia (Done , Kimber, & Underwood, 2005), thrives on a variety of soil types but prefers well-drained soils ranging from light-medium clays to lighter . Due to its parasitic nature, Sandalwood requires compatible host tree species that also match the soil conditions.

Establishing plantations can involve up to five different species, making it challenging to find sites that suit all.

Choosing suitable areas for Sandalwood cultivation across northern Australia requires thorough diligence. This is crucial due to significant pressures on land resources from competing crops, limited availability of land and water, and the need to mitigate risks such as cyclones, flooding, fire, pests, and climate change impacts—including temperature extremes and rainfall variations. Moreover, the strong demand and projected increase in Sandalwood oil prices add to the importance of selecting optimal sites (Clark, 2006).

Several factors must be considered when selecting a site for Indian Sandalwood plantations, including climate variables like temperature and humidity, rainfall patterns, soil types, water availability for irrigation, risks from cyclones and termites, and the scale of suitable planting areas. Meeting all these criteria restricts the suitability of much of northern Australia, posing a genuine challenge in finding sufficiently large and appropriate sites. Climate risks are discussed in detail in Chapter 5: Environmental Stress. For the existing plantation estate, extensive due diligence was undertaken to identify suitable plantation sites. This process involved initial region selection based on geographic and water access considerations, followed by detailed investigations into climate, land characteristics, and soil profiles. Over a two-year period, site visits and comprehensive soil testing were conducted across various locations in the Northern Territory and North Queensland to ensure the selection of the most suitable areas met all specified criteria (Robson & Barnes, 2021).

Location

Indian Sandalwood thrives in Australia's coastal climates, though these regions are susceptible to cyclones during the wet season. Balancing proximity to the coast for protection against cyclones with adequate rainfall and warmth without risking cold temperatures significantly limits available land. Millaroo, one of the closest coastal plantations, is approximately 60 km inland.

The oldest Sandalwood plantations in Queensland and the Northern Territory were established in 2012. Since then, cyclones have not directly affected Northern Territory plantations, unlike those in Queensland.

Water Supply

Water availability for irrigation is limited in northern Australia, with few areas providing reliable and cost-effective access. The Ord and Burdekin regions are among the most secure water sources, rarely subject to allocation restrictions. Ensuring water security not only mitigates plantation risks but also meets investor criteria for risk management.

Climate

Summers in Queensland and Northern Territory sites are typically hot and humid, with temperatures often reaching the low 40's degrees Celsius. Winters can be cool, occasionally experiencing light frosts. Some damage to host trees like *Senna siamea* and *Cathormion umbellatum* have been observed due to frost, though the impact has been minor and short-lived. Sandalwood trees have shown resilience to cold weather, albeit with slower growth compared to those in warmer regions. It is anticipated they will catch up or surpass their counterparts in other regions by the end of their rotation.

Soils and Drainage

Successful establishment and survival of Sandalwood plantations depend significantly on well-drained soils. Quintis meticulously selects soil types and dedicates considerable effort to drainage planning in plantation designs. The Burdekin region, with its heavy cracking clays and sodic soils, poses challenges due to dispersion during high rainfall, compromising soil drainage. Flat terrain exacerbates flooding risks during floods, making much of the area unsuitable for Sandalwood cultivation. In contrast, Northern Territory soils, ranging from sandy clay loams to well-drained light clays, generally support plantation growth with minimal flooding issues, despite competition from other industries like mango and cotton.

Scale

Establishing viable operations requires substantial land parcels meeting all specified criteria. Currently, plantations cover 1600 ha in the Burdekin, 2430 ha in Katherine, and 2660 ha in Douglas Daly, with potential for expansion at each site.

Figure 65: Aerial photo of Sandalwood plantation in the Ord River Irrigation Area and potential for expansion, (Robson, 2018).

Local Regulations

Converting farmland to plantations is subject to varying regulations across locations and council authorities. In the Ord River Irrigation Area (ORIA), Sandalwood is classified akin to horticulture, requiring no formal change-of-use applications from field or tree crops. In contrast, the Burdekin Shire Council mandates a material change of land use application, subject to council approval, potentially involving stakeholder input or protests. Quintis had successfully navigated this process, converting purchased land from sugar production to forestry in the Burdekin Shire Council through the code assessment process.

3. LAND ASSESSMENT

Assessing the land involves a detailed examination of the site intended for plantation, focusing primarily on soils and drainage conditions. This process requires comprehensive land resource information gathered through a land resource survey. Land resource surveys categorise various landscape elements such as soils, land units, and land systems, depending on the survey's scale.

Land evaluation typically encompasses an integrated approach that includes both land resource assessment and land suitability/capability analysis. This integrated process involves iterative feedback between these components as they are conducted concurrently (DSITI & DNRM, 2015).

Legacy data on land resources is frequently available for many sites. However, it has become increasingly common for resource assessment and land evaluation phases to be performed at separate times and by different personnel. Consequently, the effectiveness of any land evaluation process is dependent on both the quality and quantity of existing land resource data. For a thorough examination of land resource surveys and evaluation methodologies, McKenzie, Grundy, Webster, and Ringrose-Voase (2008) provide a comprehensive overview.

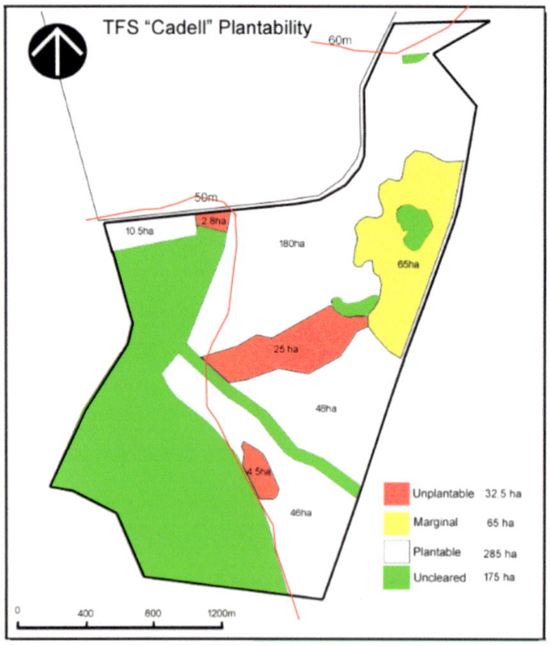

Figure 66: Plantability map showing area suitable for plantation establishment.

a) Description of soil and land attributes

A detailed soil survey will determine the soil attributes and the plantability of the site. This process is described in Chapter 3: Site characteristics.

A contour survey should also be undertaken across the site to determine drainage points and risk of erosion. This survey will assist in development of the plantation design, specifically the position of drainage points, row direction and the need for contour banks.

Limitations

The assessment may find a number of limitations to the site. These can be grouped into soil limitations or topography.

Soils

The soil physical factors such as water infiltration, permeability, water holding capacity (texture and structure) can all potentially have limitations. These soil characteristics have been discussed in Chapter 3: Site characteristics. Soil chemical factors such as nutrition, sodicity and salinity have also been covered in Chapter 3: Site characteristics.

Topography

The assessment will determine any potential flooding/waterlogging and erosion (slope) issues. These have also been covered in Chapter 3: Site characteristics.

4. PLANTATION DESIGN

The plantation design considers the planting configuration, the row direction and the irrigation design.

a) Planting layout and spacing (configurations)

Host species are strategically planted in proximity to Sandalwood seedlings to promote rapid root unions. This practice involves a sequence of hosts: initially, pot hosts are followed by short-term hosts planted within one to 1.5 metres from the Sandalwood seedling, which typically live for up to three years. Subsequently, long-term hosts are planted to provide host services throughout the plantation's lifespan. Dependant on the configuration type, these are generally planted in a separate row or in the same row as the Sandalwood trees. To mitigate risks, at least two species of short-term hosts and a minimum of two (preferably three) species of long-term hosts are established.

The spacing between rows is determined by the planting layout. Traditionally, Sandalwood plantations have row spacings of 3.6 metres, although configurations with 4.5 metres between rows are also observed. Selecting the appropriate configuration is crucial as it dictates tree growth throughout the rotation. Row and tree spacings influence tree stocking, which affects intra-row (within the same species) and inter-row (between varied species) competition. The design of the configuration not only determines overall tree stocking but dictates the species selection and initial planting numbers.

Configurations are classified based on various characteristics and are categorised into two types: Standard (STD) and Mixed (MIX). Configuration types are determined irrespective of species included in the design. Within each type, variations exist due to differences in the cycle of long-term host used across the stand, short-term (STH) and long-term host (LTH) placements, as well as intra-row and inter-row spacings.

The STANDARD configuration type places Sandalwood with short-term hosts in separate rows, typically adjacent to rows of long-term hosts. The STH row configuration remains consistent within the row and alternates with two to three rows of different LTH species across the block in cycles. This layout, widely used in initial ORIA plantations, facilitates effective management of varied species within the plantation. Managing more vigorous LTH species separately allows for targeted crown management, which enhances Sandalwood form and height through light competition, as discussed in competition section of chapter 5.

The MIXED configuration type integrates all species within the same row. Initially known as "Chapmans," this system maintains uniformity across every row. Originally designed to reduce the transition period by having all hosts in the same row, it was implemented in challenging, drier sites where Sandalwood might struggle during the transition to LTH. The wider spacing between trees and higher ratio of Sandalwood to host at establishment characterises this configuration. Sandalwood grown in MIX configurations typically exhibit shorter heights compared to those in STD configurations. However, the reduced transition period discussed in the Stress section of Chapter 5 appears to correlate with lower amounts of oil-bearing heartwood in trees of the same age in MIX configurations compared to STD configurations.

Configuration diagrams

Most STD configurations are 3.6 metres between the tree rows. Sandalwood spacing historically has been at 2.75 metres. This is to achieve a Sandalwood stocking of 505 stems per hectare (SPH). However, evidence has shown that at this spacing there is early within row competition. An increased spacing of three metres between Sandalwood and 463 SPH, reduces or at least delays early age competition.

Figure 67 shows a typical STD configuration at establishment, using two LTH species and two STH species. The total stocking used in this configuration gives 1389 SPH. Figure 68 shows a typical MIX configuration at establishment, using two LTH species and two STH species. The total stocking using this configuration gives 1111 SPH.

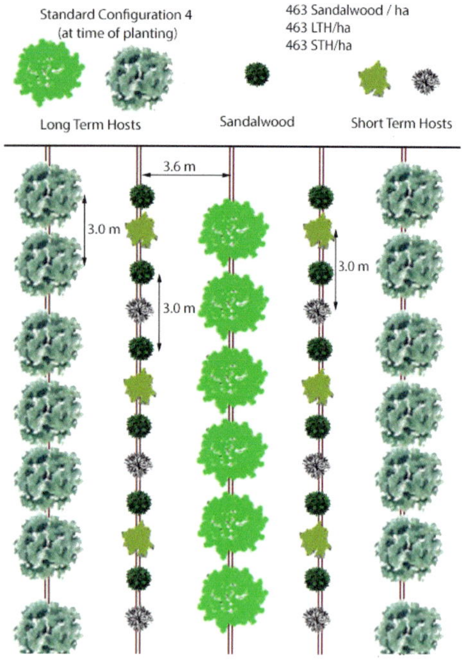

Figure 67: STD configuration at establishment. Showing Sandalwood, STH and LTH.

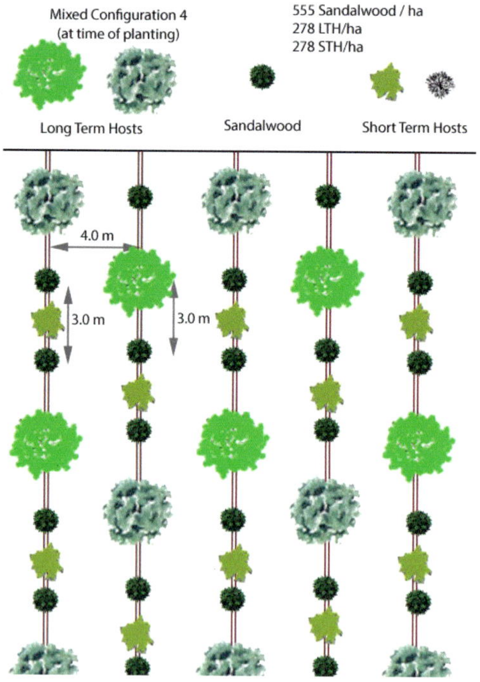

Figure 68: MIX configuration at establishment. Showing Sandalwood, STH and LTH.

Note that at establishment each Sandalwood, in both configurations, has STH near each pair of Sandalwood. In the MIX example in Figure 68 there is only one STH for two Sandalwoods. Some variations do have STH planted between the Sandalwood and the LTH. On wetter sites, some variations in STD configuration do not have the Acacia and only Sesbania between every second Sandalwood. The variations in the use of STH in both configurations have all worked well and is really dependent on the site.

From age three, the STH will start to die off and in the STD configurations for the next two to three years the Sandalwood will be transitioning across to the LTH. This transition time is much shorter in the MIX configurations.

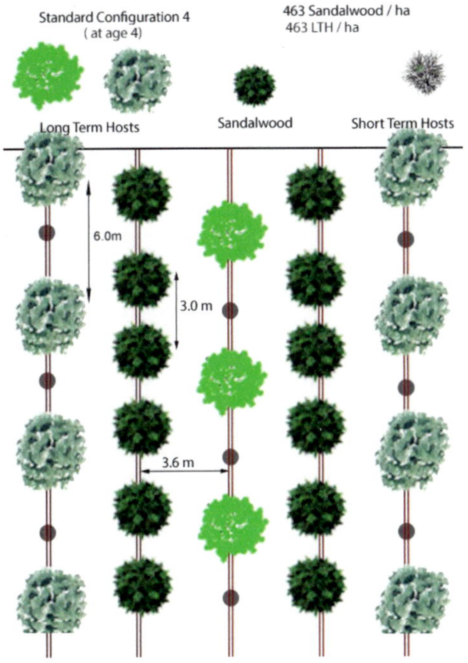

Figure 70: STD configuration at age 4. The STH have died off and the LTH have been thinned by 50%.

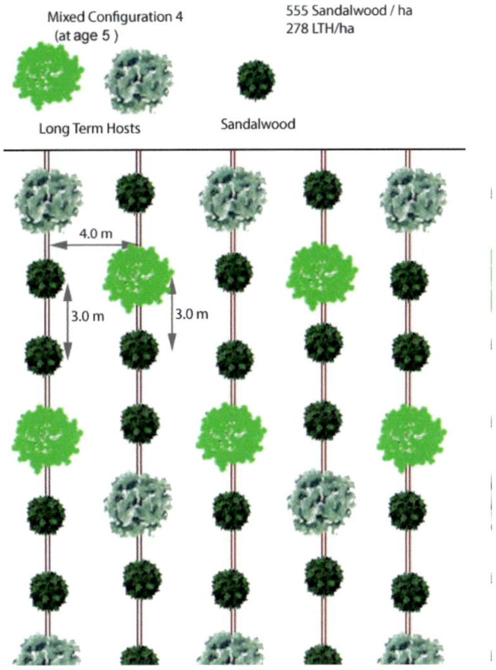

Figure 69: Mix configuration at age five, after all STH have died.

In the example shown in Figure 70, the STH have all died and the LTH stocking has been reduced by 50% in each row. A noncommercial thin of the LTH at age three to four is the optimal time to do this. The higher within row stocking has caused competition and increased growth and improved form. Thinning to reduce the number is optimally completed during the transition period, before Sandalwood attaches.

b) Research configuration trials

Trial 1 (0702). In 2008, several configuration experiments were established in ORIA plantations. The results for the trial in Voyager 14 will be discussed here. Four varying STD configurations using species previously used elsewhere in the estate were designed in consultation from various staff members and planted at Voyager. The areas of each configuration differed to achieve the same Sandalwood stocking rate for the stand as required under the Managed Investment Scheme Product Disclosure Statement. The treatments were grown for full rotation, 15 years, and harvested. Treatment yields were then compared. The experiment the treatments replicated three times across the stand.

Table 22: Configuration treatments stocking of Sandalwood and LTH.

	Between tree spacing (m)		Stems/ha Stocking				Ratio
	SA	LTH	Sa	CU	DLN	Tot LTH	SA:LTH
Trt 1	3 x2	3	617	154	154	308	2.0
Trt 2	3 x 2	3 x2	617	231	231	462	1.3
Trt 3	2	2	694	252	252	504	1.4
Trt 4	3	6	617		154	154	4.0

Configuration 1

This alternating tree and access row configuration is the most like the conventional commercial configuration with similar species. Conventional spacing of 3.6m between rows and 3.0m between trees, with a double row of Sandalwood (SA). Alternating *Dalbergia lanceolaria* (DLN) and Cathormion umbellatum (CU). With twice as many Sandalwood this has a Sandalwood to LTH ratio of 2 to 1.

Configuration 1

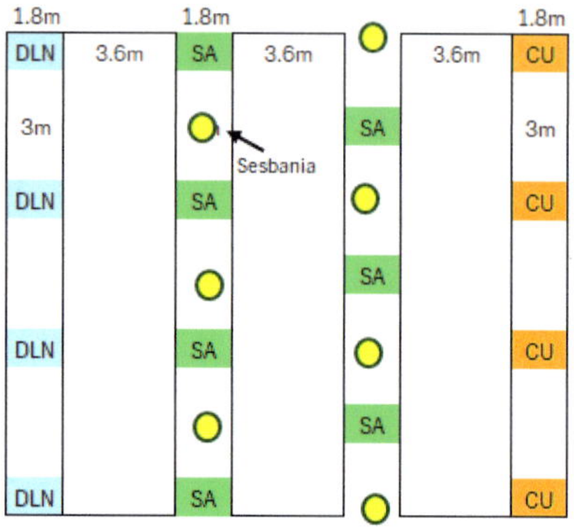

Configuration 2

This double row configuration was previously established in an unreplicated trial at Packsaddle 15. Early feedback from the plantation manager was positive, with their observations being that shading on three sides of the Sandalwood reduces the risk of sunburn. The form of the Sandalwood is improved, which in turn reduces the amount of pruning required. The flood irrigation can also be limited to the middle of each double row. It is possible to separately irrigate hosts, many of which require more water than the Sandalwood. The double rows also reduce weed growth due to suppression by shading. This configuration starts with a Sandalwood to LTH ratio of 1.3 to 1.

Configuration 2

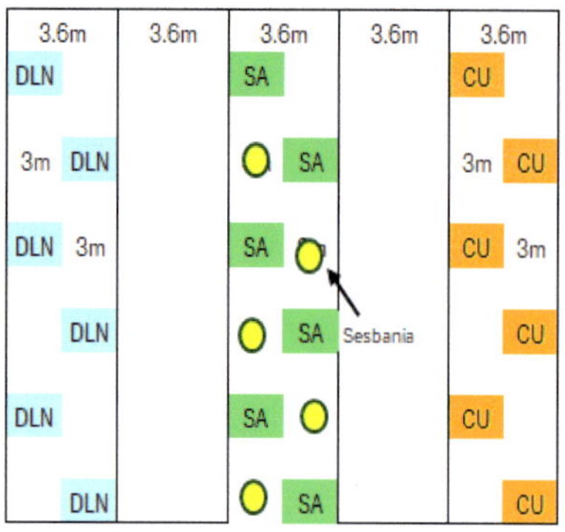

Configuration 3

This configuration was designed to provide shade to the Sandalwood, particularly in the first few years, whilst planting hosts further from the Sandalwood. The thought was that this may have multiple advantages as shading the Sandalwood should encourage the trees to grow up rather than out, which will improve their form and reducing the need for pruning. Sesbania should be at a sufficient distance from the Sandalwood so that they will not need pruning. This starts with a Sandalwood to LTH ratio of 1.4 to 1.

Configuration 3

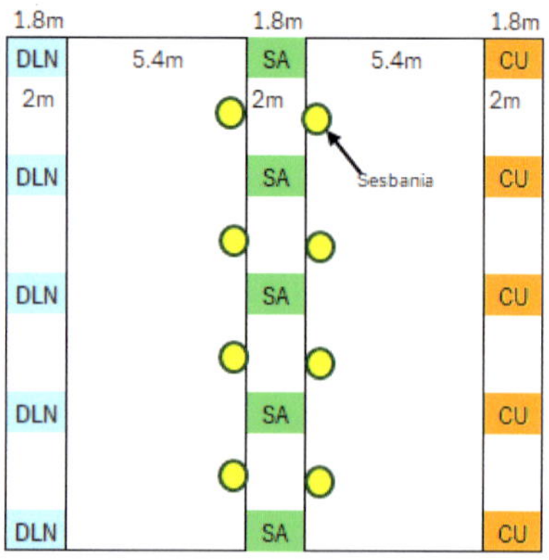

Configuration 4

Configuration 4 investigated the growth potential of Sandalwood at low stocking densities. With only an *Acacia trachycarpa* STH. Planting at a lower density would reduce establishment and management costs, particularly as the Sandalwood to LTH ratio is 4 to 1. Furthermore, harvesting and processing costs could be reduced if a similar volume of wood can be produced, by growing fewer, but larger trees.

Configuration 4

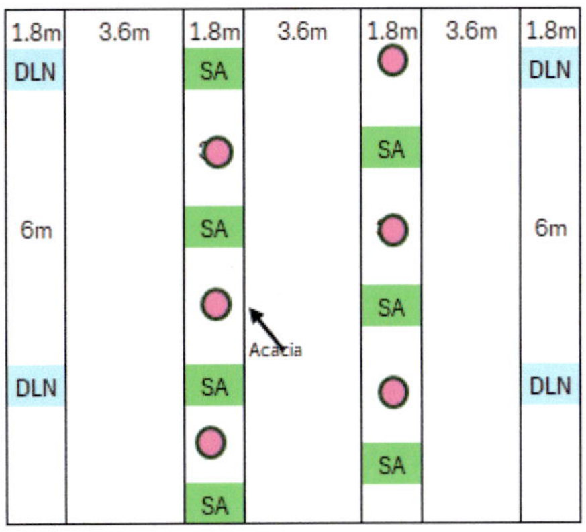

Harvest.

The commercial harvesting regime of machine felling and data recording of stumps was followed. Every tenth stump was photographed and measured for over bark and heartwood area of Stump (cm^2), area of Heartwood at Ground Level (cm^2) and area of Rot (cm^2).

Calculations for area of heartwood at ground level / area of stump, count of trees / ha, trees with Heartwood (%), total area of stump per treatment unit / area of treatment unit, total area of heartwood at ground level / area of treatment unit and total area of rot / area of treatment unit. The measure from each tree went to form the average, total, or count per treatment unit, and formed the dataset that was analysed via Analysis of Variance tests.

Results and discussion

Though some measures differed significantly between blocks, none were found to do so between configurations at 95% confidence ($P < 0.05$). Substantial variation in site quality across the experiment area was increasingly evident from satellite imagery as the stand matured. Though the effect of dividing the area into blocks and replicating each configuration within have some effect on cancelling subsequent responses in growth and yield, it was not an ideal site for such experimentation, and this likely was not apparent prior to design. This site effect makes it more difficult to detect any significant differences in the responses due to configuration. Any variation in the yield of each configuration may have been masked by that due to the site. There may be an order of profitability of the configurations if yields could be considered alongside management costs, though this was not available.

The results for this trial were typical for most of the early large scale configuration trials established. There tended to be too many variables to determine if one type was better than another.

Trial 2 (R003). This trial was established in Taylors Park 6 in 2015. It consisted of two large blocks with two configuration types, STD and a MIX. The STD configuration is Quintis STD 8. This is two rows of *Dalbergia lanceolaria* and then a single row of Cathormion. The STH hosts were Sesbania with alternating *Acacia trachycarpa* and *A. simsii*.

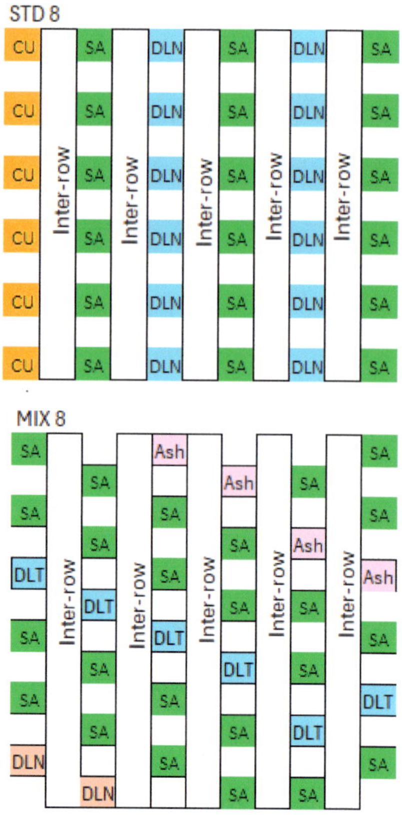

The MIX configuration block had the same layout as Quintis MIX 8 except for the STH alternated between Sesbania and A. *simsii* (rather than *A. trachycarpa*). The LTH were *Acacia shirleyi* (Ash), *Dalbergia latifolia* (Dlt), and *D. lanceolaria* (Dln).

The initial Sandalwood stocking in the STD config was 505 and in the MIX was 520 stems/ ha. The initial LTH stocking in the STD config was 278 and in the MIX was 212 stems/ ha.

An assessment for diameter at breast height and a five point health score for all Sandalwood was undertaken at age four. This was during the transition period when most of the STH had died, and the Sandalwood was generally between hosts.

The results displayed that the mean diameter in the MIX plots was 5.3 cm, with the largest being 9.3 cm. The mean DBH in the STD plot was 4.7 cm with the largest being 7.4 cm.

The health score showed that in the MIX plots, 70% of the trees were considered to have maximum health, with less than 7% considered to be poor to fair health. In the STD plot only 5% were of maximum health, with just over 50% deemed to be poor to fair health. It was noted that in the STD treatment, where there was still *A. simsii* alive, the health score was a full point better for the Sandalwood on each side.

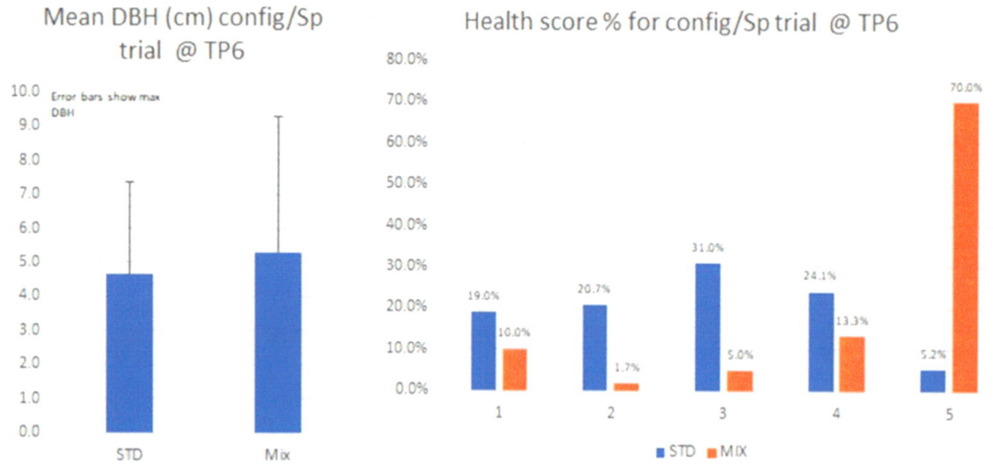

Figure 71: The results for the year four assessment of Sandalwood in the configuration trial. Note that with the health score, 1 is dead and 5 is a healthy deep green crown.

There was some mortality with both Sandalwood and LTH in the plots. In the MIX treatment there was only 16 host trees to support the 54 alive Sandalwood, a ratio of 3.4 alive Sandalwood for every host. In the STD treatment, there was 26 host trees to support the 46 alive Sandalwood a ratio of 1.8 alive Sandalwood for every host.

Discussion

During the year of the assessment, the site experienced drought conditions due to a poor wet season and technical issues with pumps unable to provide irrigation. Across much of Taylors Park 6 there was recorded poor health and mortality due to lack of water. The consequences of these issues can be found in the health assessment in the STD treatment plot in this trial but not in the MIX treatment.

This trial clearly demonstrates the health and survival differences between the two configuration types and the benefits in the MIX configuration for reducing host transition and therefore reducing that risk period. Although there has been 10% mortality in the MIX treatment, the droughting issues during the year have doubled that figure in the STD treatment. It was also noted that in the STD treatment, where there was still STH (*A. simsii*), the Sandalwood health was significantly better.

The number of hosts required to sustain a specific number Sandalwood is still a relatively unspecific figure. However, this trial shows that with a ratio of 3.4 Sandalwood for each healthy host, the Sandalwood is predominantly healthy and is growing significantly.

The two new species trialled at this site, *Dalbergia latifolia* and *Acacia shirleyi,* indicated potential in large pot trials but were previously untested in the field. *A. shirleyi* is a dryland hardy acacia with a 'pyramidal' shaped crown. The *D. latifolia*, unlike *D. lanceolaria* is also a predominantly single stemmed tree with a moderately compact crown. When the mean DBH data for Sandalwood is compared for trees planted next to these species, the *A. shirleyi* hosted trees average 5.5 cm DBH and the *D. latifolia* hosted average 6.0 cm DBH.

Trial results summary

Research indicates that variations within a configuration type do not significantly impact tree performance, whereas site variation may have a greater effect. There are notable differences in tree performance between the two configuration types, Standard (STD) and Mixed (MIX), particularly under environmental stress during transition periods. A recent study by Duc Bui (2026) highlighted soil type and configuration as the primary factors influencing tree performance, though this may be more attributable to species effects than configuration alone. This conclusion was drawn from Quintis PSP data on plantations aged six to ten years, which included various configuration types, species mixes, and their variations.

Although the variations within the STD type may not provide much performance variation, anecdotal evidence suggests that 4.5 metres between rows is too wide and takes Sandalwood too long to make reconnections. Sandalwood usually starts to recover physical crown improvement from transition after 12 months in a 3.6 metre inter-row. When planted with 4.5 metre inter-rows this period is extended for a percentage of the population and the recovery from transition can be uneven across the stand.

Trees in MIX configurations, with wider spacing, tend to be shorter with larger diameters at 20 cm. In contrast, the closer spacing in STD configurations generally results in taller trees with smaller diameters. This suggests that STD configurations require silvicultural manipulation, such as thinning, to optimise tree growth.

Regarding heartwood development, the STD configuration significantly outperforms the MIX configuration, producing more oil-bearing heartwood, as detailed in the Environmental Stress section of Chapter 5.

Each configuration type has its pros and cons. STD configurations are generally optimal for most sites, promoting early heartwood initiation and long stem boles, making them suitable for long-term (20+ years) Sandalwood production. However, for sites with lower water availability and harsher environmental conditions, the MIX configuration is preferable as it reduces early stress on the trees.

c) Row direction

The orientation of the rows can be crucial to maintain a healthy stand over the period of the rotation. Ideally, rows should be aligned approximately northeast to southwest to minimise the number of Sandalwood stems exposed to the afternoon sun. As discussed in Chapter 5, exposure to the afternoon sun can lead to sunscald, particularly if neighbouring long-term hosts die off. Field evidence shows that Sandalwood stems exposed only to the afternoon sun are prone to burning. The figure below illustrates the sun's movement in northern Australia and its impact on Sandalwood stem exposure.

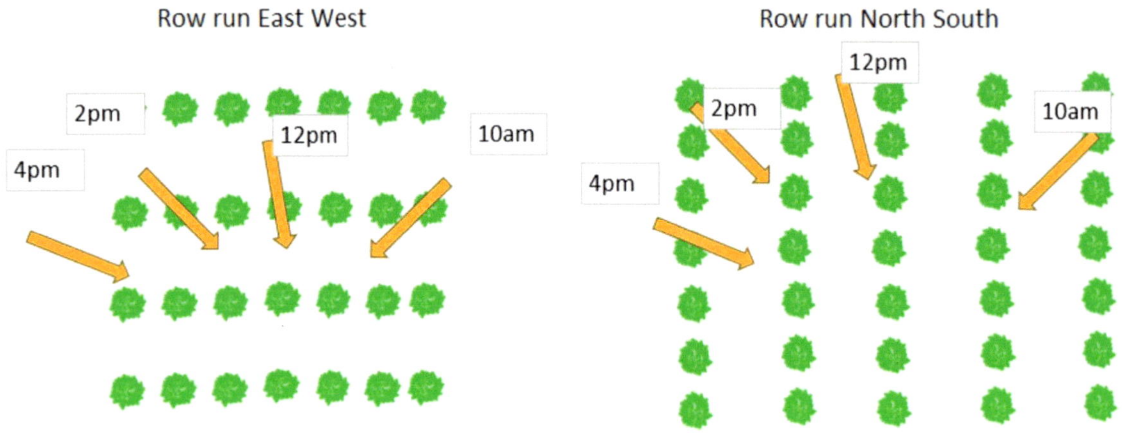

Figure 72: Two plantation layouts, east to west and north to south. The arrows show the sun direction at various time of the day.

Rows oriented north to south received both morning and afternoon sun, with self-shading occurring around midday. Rows running east to west receive sunlight throughout the day, but they self-shade to some extent and are primarily exposed around midday. Angled rows, northeast to southwest (180° – 100°), receive morning sun but are self-shaded during the more damaging afternoon sun.

Historically, the orientation of rows in most Sandalwood plantations were not considered and were primarily determined by irrigation design. However, due to the mortality of hosts from Sandalwood hosting and disease, the damaging effect of sunburn has become significant. Therefore, new plantations must consider row direction as a risk mitigation factor in case of host mortality.

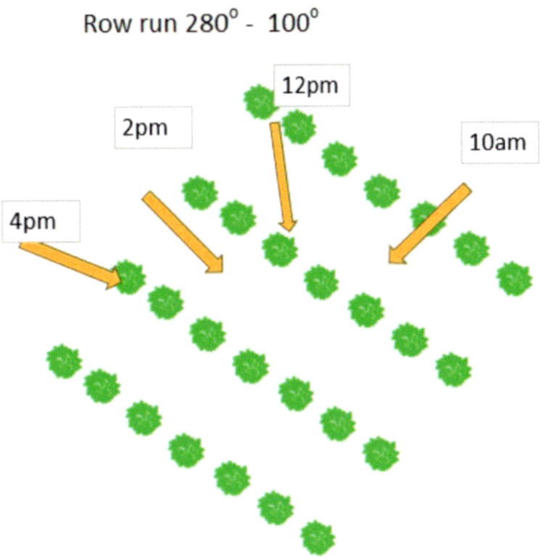

Figure 73: Row angled at 180° – 100° self-shade from the afternoon sun.

d) Irrigation design

Irrigation design and management significantly impacts the performance, yield, and quality of the trees. The efficiency and effectiveness of water use are largely determined by the irrigation design. It is recommended to engage a certified irrigation designer for system planning.

The initial design choice depends on the type of irrigation system, whether flood or drip, as each requires a different approach. Two key aspects of irrigation affecting system performance and efficiency are design and operation.

Regardless of the irrigation type, the design must consider the following:

Seasonal Water Requirements: These are based on crop stage demand and environmental conditions specific to the location.

Soil Structure: This influences the infiltration rate of water into the soil, determining the maximum irrigation application rate. Applying water faster than it infiltrates can cause runoff.

Water Delivery: The system must supply the correct amount of water within the required period. For instance, if 12 mm of water is needed over 12 hours, the application rate is 1 mm per hour. In practice, water may need to be applied in several shifts to match the soil's water holding capacity in the crop's root zone. This, combined with meeting the demand for various irrigation shifts throughout the plantation, may necessitate different application rates.

Topography: The area's topography affects the hydraulic design of the irrigation system. Elevation changes impact pressure: a one-metre change equates to approximately a 10kPa pressure difference.

Operational Costs: Over the irrigation system's life, running costs typically exceed capital expenses. Proper hydraulic design minimises power waste from unnecessary pressure creation and improperly selected pumps.

Future Expansion: Plans for future expansion may influence the size and layout of pipes and the system's capacity to deliver a certain volume of water over a specified time.

5. GROUND PREPARATION

The sites designated for Sandalwood plantations in tropical Australia have a rich history of varied land use. These lands were previously utilised for agriculture, growing crops such as sugar, bananas, melons, and peanuts. Additionally they have been used for improved pasture, hay production, and cattle farming. Typically, these uses spanned over a decade or more. The benefit of such extensive agricultural history is the enhanced fertility of the topsoil, enriched by years of fertiliser application. However, a significant downside is the varying degrees of soil compaction found across these sites. This compaction results from natural hard-setting common in sandy Kandosol soils and anthropogenic factors, such as the constant use of machinery and the presence of hooved animals like cattle (Fuller, 1958).

For many forestry species, compacted soils present a unique challenge, where soil type and cultivation interact to affect growth responses. Increased cultivation generally leads to better growth (Costantini, Podberscek, & Nester, 1995). For instance, an experiment with *Eucalyptus pellita* in North Queensland demonstrated significant growth increases with each level of enhanced cultivation, ranging from no cultivation to ripping and small mounding (Robson K. , 1995).

Figure 74: Mounding at a 2nd rotation site at 1.8m spacing in ORIA, (Robson, 2022).

The levels of cultivation across the Sandalwood estates vary, influenced primarily by the irrigation system in place and the degree of drainage required or potential for waterlogging. Sites intended for flood irrigation undergo extensive manipulation, including laser leveling and high mounding, to effectively channel water down the rows. Many of these sites, with their long histories of agricultural use, have been extensively cultivated over time.

Figure 75: Deep ripping with dozer in Taylors Park, (Robson, 2012).

Due to the amount of soil moved when leveling, this should be done at least two years prior to planned establishment. This will allow the wet season rains to 'settle' the soil over two years. Without this time, there is the risk of slumping. This is where low areas occur due to sections of soil settling, which will reduce the flow through of the irrigation water. These low areas will cause water logging and poor plantation performance and increased mortality.

After mounding in the flood irrigated stands, a temporary drip line is laid out on the planting rows. This feature significantly increases seedling survival.

In the Northern Territory, plantation sites exhibited a range of previous uses, from annual cultivation to irregular or no recent cultivation. Most of these sites, particularly those on clay loam Kandosol soils, required deep ripping to a depth of 60 cm to break up topsoil compaction. The necessity for deep ripping is reliant on the compaction score derived from soil profile analysis.

Soil compaction analyses dictated the level of cultivation required for each site. Highly compacted sites underwent deep ripping to at least 60 cm, usually in two directions. Post deep ripping, the sites received overall cultivation with off-set disks or rolled with a heavy roller to break up large clods and fill rip trenches. Ripping should be carried out when the soil is dry to avoid smearing.

Row directions are surveyed and then cultivated with a speed tiller, an implement featuring a small tyne ripper to 30 cm and either small cultivating disks or a rubber tire to flatten the ripped line.

Figure 76: This implement (Gessner single row crop tiller) was used for most of the site preparation cultivation at the Douglas Daly, (Robson, 2015).

For sites with less compaction, only the planting line is cultivated. This process involved a deep rip to 60 cm, followed by a one metre wide cultivation over the rip line to fill the trench. Sites identified as having flooding or waterlogging potential were mounded to prevent water accumulation and ensure optimal growing conditions for the Sandalwood trees.

In the tropics, when newly cultivated soil is exposed to monsoonal rains there is a considerable risk of overland erosion. Although most of the sites planted with Sandalwood plantations are relatively flat, newly cultivated soil will erode in heavy rainfall events. To mitigate this risk, it is advisable to minimise soil disturbance whenever possible and to establish a cover crop before the onset of the wet season.

a) Timing

The site preparation occurs two to three years before planting. There are many operations that need to be timed.

Table 23: This table shows approximately when various operations need to occur.

	Dripper		Flood		
	year 1	year 2	year 1	year 2	year 3
Soil survey	March - June		March - June		
Topography survey	March - June		March - June		
Plantation design	July to Oct		July to Oct		
Irrigation design	July to Oct		July to Oct		
Laser levelling			July to Oct		
ripping	July to Oct				
cultivation	July to Oct				
irrigation infrastructure	Oct to Nov			Oct to Nov	
cover crop sowing	Oct to Nov				
mounding	Oct to Nov			Oct to Nov	
tape laying		Feb to March			Feb to March
Pre weed control spray		March to Apl			March to Apl
Planting		May to July			May to July

b) Sandalwood root depth research trial

There have been no experiments that have looked specifically at site preparation for Sandalwood plantations. However, this experiment examines the architecture of the Sandalwood below ground root structure. This is critical information to know for development of the planting row and any cultivation within the inter-row.

Sandalwood needs to develop extensive root systems to achieve sufficient haustorial connections to host roots. This study was undertaken to investigate the location of most of the Sandalwood roots and understand the Sandalwood root architecture. Conducted in 2019 in a four year-old plantation in a STD configuration on four trees. The soil surrounding the trees was removed with a high-pressure hose to expose the root system of both Sandalwood and host.

Soil was removed to a depth of 40cm between the Sandalwood and host. The distance between the trees was marked in one-metre square sample sections. Using water, the soil depth in 10cm increments in each quadrate was removed. All roots in each 10cm increment were divided into species and collected for weighing. At harvest, a count of haustoria in each soil profile and quadrat was recorded. Soil was removed at depth until no more roots were found.

Figure 77: Sandalwood feeder roots were only found to a depth of 20cm, (TP6. Robson, 2019).

All the roots collected were partitioned into groups from the quadrat where they were collected. Transects 1 and 2 (furthest from tree >1m), transects 3 and 4 (closest to tree < 1m), Root butt (section of stem below ground), and roots below 20 cm (structural roots all found in the watered zone). Sandalwood and host roots were identified and collected separately. The process was repeated for the 10 to 20 cm depth zone. No roots, apart from those under the tree were found deeper than 20 cm.

Host roots were collected during the harvest of Sandalwood roots because it was too difficult to determine the difference between each species in the field. After drying it was easier to cull out the host roots so that Sandalwood roots could be weighed separately. All the Sandalwood roots were in the top 20cm of the soil profile. It was noted that there was significant volume of host roots below this depth, down to at least one metre across the centre of the row.

Results

The Root to Shoot ratio (R) was 1 for the three trees in the STD configuration. The fourth tree was the largest tree and had the largest root to shoot ratio with two times above ground to root mass (R=0.5). There is not a lot known about what the R ratio for Sandalwood should be. Barbour et al (2011)) found the mean R ratio in a 12 year old Sandalwood plantation trees to be about 0.3. Numerous studies show a range in R values from 0.05 to 0.63 (Sanquetta, Corte, & da Silva, 2011), albeit this is for tropical pines. Harris (2011) suggests that most tree species have an R of 0.2 (1r:5s) to 0.16 (1r: 6s). These reported values are much smaller roots systems compared to shoots than what was found in this study. Sandalwood being hemi parasitic, has a greater reliance on nutritional additions from being able to connect to a host root. Therefore, it is conceivable and would require a greater root to shoot ratio than other non-parasitic trees. These are also young trees and ratios may change as they age.

it is considered that the numbers of haustoria are important for survival of the Sandalwood. The data shows that, as expected, the larger the tree there is more haustoria, with almost twice as much haustoria as the other trees but also twice the volume. If the Sandalwood can get their feeder roots over to the centre of the row, where the chance of more host roots are, there is a higher probability that they will have more haustoria.

Figure 78: The majority of Sandalwood roots tend to be in the top 20cm of the profile, (TP6. Robson, 2019).

Root architecture

The general root shape for Sandalwood in both configurations is a truncated ellipsoid (like a cut off AFL football). Roots predominantly follow the row direction or area of most water. This is where the roots for short-term hosts are likewise located.

Conclusions

This excavation work has shown there were no Sandalwood feeder roots deeper than 20cm in the soil profile. In STD configurations, Sandalwood needs to quickly get their roots across to the middle of the row to interact enough with host roots to increase haustoria number. As their above ground biomass increases, the area of highest haustoria (67%) was more than one metre away from the Sandalwood tree. The ratio of haustoria to above ground biomass is more critical than total root mass ratio to above ground biomass.

6. NURSERY

The nursery phase for Sandalwood involves specialised treatments to ensure optimal growth. Sandalwood and host species require different potting media and seed pretreatments. Sandalwood seeds have natural dormancy and need to be treated with Pro Gibb (Gibberellic acid) to ensure consistent and timely germination. The seeds are soaked in a solution of gibberellic acid at a concentration of 0.1 to 0.25 g/L active ingredient (100 to 250 ppm) for 24 to 48 hours, with agitation from a bubbler to enhance the treatment's effectiveness.

The potting media differs for Sandalwood and host species. For Sandalwood, a free-draining mix of 75% 3-6mm and 25% 0-6 nursery blend is used. For the hosts, a 0-6mm nursery blend is preferred. Ammonia levels should be below 30 mg/L, and the electrical conductivity (EC) below 0.7 dS/m. The Sandalwood mix includes NPK Max (Osmocote Exact 5-6 month) at 3 kg/m^3 and trace elements (Flowtrace) at 1 kg/m^3. For the host mix, NPK Max is added at 1.3 kg/m^3 along with trace elements at 1 kg/m^3.

Sandalwood seedlings are susceptible to bacterial wilt and fungal pathogens, particularly root rots. Regular inspections and Integrated Pest Management (IPM) techniques can help mitigate these issues. This includes maintaining good nursery hygiene by sterilising tools, equipment, and surfaces, taking measures to reduce plant stress, and adjusting the pH to 5.5 to minimise pathogen impact.

Figure 79: Sandalwood seedling after being dibbled into tubes, (Robson, 2009).

Traditionally, Sandalwood seedlings are grown in 90mm square bottomless pots, which are 15 cm deep. Operational trials using 50 mm forestry tubes (50x50x120 mm) have shown promise with Sandalwood survival and growth like the seedlings planted from the larger pots. There is no evidence that switching to small pots affects establishment and can provide significant saving in nursery costs and time.

a) Post germination

Sandalwood seedlings take three to four weeks to germinate. Prior to geminates developing their first set of true leaves, they should be carefully pricked out. Seedlings with "J" roots, twisted or snapped roots, double heads, or mutant characteristics should be discarded. Good seedlings should have their roots trimmed to five to six cm and placed in containers with water and Trichoderma solution before being dibbled into pots with Sandalwood potting mix.

To reduce transplant shock, seedlings benefit from an addition of Seasol and Super Stand Phos applied three times a week for the first two weeks after dibbling.

Dalbergia species should receive an application of Siratro Group inoculant two weeks after germination.

Pot host

The pot-host, Alternanthera, is introduced to the potted Sandalwood seedling at 150 to 170 mm. Stock plants of Alternanthera are harvested, washed and cut into segments each with at least two nodes, and planted with care to maintain the polarity of the cutting. The apical end remains upright, and the distal end being pushed into the pot and firmed with the index finger ensuring the cutting is firm in the potting mix. Being mindful not to compact the potting mix around the Sandalwood seedling.

Figure 80: Sandalwood seedlings with recently planted pot host, (Robson, 2016).

The cuttings are planted two cm from the bottom right corner of the pot with at least one node beneath the soil surface, ensuring a secure root structure and faster hosting. These cuttings need to be well-watered until roots develop, and applications of Humaboost (Tropical Trichoderma) and Seasol applied to promote root development.

Consolidation

As the seed source is generally unimproved, it is important to undertake several consolidations. Consolidation sorts the trees by height and health, with the first consolidation occurring before pot hosting. When 30% of the seedlings reach 200 mm, a second consolidation should be conducted. All host seedlings should be consolidated at least three times during the nursery growing period to ensure smaller seedlings get adequate light and develop strong stems and roots.

Pot host pruning

Alternanthera is fast growing and will require on average three prunes. The timing of pruning is determined by the height of the Alternanthera in relation to the Sandalwood. Pruning is conducted when the alternanthera is nearly the same height as the Sandalwood.

Hardening off

A few weeks before planting, seedlings need to be hardened off by removing the shade. Seedlings over 200 mm can be consolidated and pruned before being placed in the sun. They must be hardened for at least a month prior to dispatch.

Specifications

Sandalwood seedlings must meet specific criteria for transport. They must be 250 to 350 mm tall, with a basal diameter of at least 3.5mm, upright without support, and have dark green leaves. The lower 35% of the stem should show brown bark. Alternanthera ought to be one-third the height of the Sandalwood or at least 70mm tall with multiple leaders. The potting mix must be consolidated so that when the plant is removed it does not fall apart.

Figure 81: Four month old Sandalwood seedlings after first consolidation, (Robson, 2022).

Watering

It is important to keep seedlings moist but not let them dry out before the next watering. Moisture levels can be affected by weather conditions. Effective watering restores what was lost since the last session, while avoiding excess that spills from the pot, as this can promote fungal root rot.

Dispatch to field

A week before dispatch, consolidate all seedlings, Sandalwood and hosts, to ensure they meet planting specifications. Apply liquid fertiliser two days before dispatch to promote vigour and protect against Phytophthora spp. Host seedlings should receive a liquid fertiliser application of Agrimax or similar. All seedlings will receive at least two applications of Seasol. Immediately prior to the dispatch of the plants to the field, all seedlings should receive thorough watering. This should be heavy enough to cause water to run through the bottom of the pots.

b) Research nursery trials.

Trial 1 (R030): Most native N-fixing trees have a symbiotic relationship with native Nitrogen fixing bacteria available in the natural environment. However, this is not necessarily the case with exotic N-fixing trees such as Dalbergia. Prior to this experiment the *Dalbergia lanceolaria* seedlings being planted were small and survival was sporadic. This experiment investigated two commercially available inoculants on *Dalbergia lanceolaria,* a *Leucaena leucocephala* group inoculant and a Siratro Group 'M' CB756 FDV. These inoculants were applied with water and assessed 14 weeks after application. To assess for the effectiveness of the inoculation, measurements for indirect nitrogen levels were taken by using a chlorophyll content meter (CCM-2000), stem basal diameter and height of the seedlings were also measured. Total wet and dry weight of each seedling was also recorded to compare growth.

At 14 weeks after inoculation the absolute amount of chlorophyll in the leaves of the seedlings inoculated with 'Siratro' was significant (P<0.005). with 25 CCI when compared to 7 CCI in the control and 7 CCI in the 'Leucaena' inoculated treatment The visual differences between the Siratro inoculated seedlings and control were obvious. The 'Leucaena' inoculant was clearly ineffective in all measurements and had no significant differences to the control group.

At 14 weeks, stem diameters for the 'Siratro' inoculated seedlings were significantly (P<0.05) bigger with an average 2.3 mm compared to an average of 1.79 mm for the control. The average height of seedlings from the 'Siratro' group was 16.4cm, on average 5.6 cm greater than those of the control group (P<0.005). The average wet weight of the 'Siratro' treated seedling was 3.5 g compared to 2.4 g for the control group (P<0.05) whilst the average total dry weight per seedling was 1.15g for the 'Siratro' inoculated and 0.8g per seedling for the Control group (P<0.05). Root to shoot ratios were constant across treatments (Manson, 2014) .

A small number of nodules observed on the Control and 'Leuceana' treated group were of different morphology to the 'Siratro' nodules and they were assumed to be a background Rhizobium species (N2-fixing bacteria) native to Kununurra soils and water supplies. Whilst this native Rhizobium was effective in producing nodules in the root systems of Dalbergia seedlings in small numbers, the effectiveness of this Rhizobium in fixing Nitrogen for Dalbergia was obviously limited and the 'Siratro' inoculant produces a far superior symbiosis.

Figure 82: Comparison of the Siratro-inoculated dark green Dalbergia with the yellow control, (Manson, 2014).

The effective nodulation of Dalbergia using the Siratro Group M inoculant has produced larger, healthier, more vigorous seedlings that will be easier to establish and manage.

<u>Trial 2</u> (R051): This trial tested treatments with Sandalwood seedlings that were hosted later than the standard 140 mm average height in the nursery, to see if later hosting increased Sandalwood root mass. The key objectives were to compare biomass of all treatments and determine if alternanthera placement affected Sandalwood root development.

The seedlings used for this trial were kept under 50% shade for the first 12 weeks then moved to open sunlight. The hosting treatments began when the tray of 12 Sandalwood reached an average size of 140 mm (as per standard operations), 170 mm and 200 mm. Each treatment consisted of 36 seedlings; twelve seedlings (one tray) were un-hosted with alternanthera as a control. All treatments were kept within the operational benches and treated the same as all operational seedlings. The 140 mm seedlings were hosted and all seedlings were measured for total height weekly for 11 weeks while under shade and then fortnightly while in full sunlight. After six weeks in full sun, three average size seedlings from each treatment were selected for destructive biomass assessment.

For the biomass procedure, all potting media was carefully removed from the roots by lightly shaking and washing with water. Sandalwood and alternanthera roots were carefully separated. Measurements for Sandalwood stem collar diameter and total height were taken. Below ground biomass (roots) was removed from the above ground biomass. Counts of haustoria were taken for each seedling and any hosting noted. All stems were separated into roots, leaves and stem then weighed. Each biomass group was bulked for each treatment and placed into paper bags for oven drying. Samples were oven dried at 75^{0}C for 48 hours and then weighed again.

The results demonstrated that there was no difference in height growth over the 18 week trial period between the seedling hosted when the seedlings were 140 mm tall compared to 170 mm tall. However, when hosting was left until the seedlings were 200 mm there was a significant drop-in growth rate. Moving seedlings into full sun also slowed height growth for all treatments.

When the difference between the 140 mm and 170 mm treatments is considered, the following can be observed.

- 140 mm has significantly more leaf,
- They had similar root biomass, with 170 mm slightly smaller,
- They had similar numbers of haustoria, with 170 mm having slightly more,
- The 140 mm treatment had the smallest alternanthera, this was possibly due to heavy shading by Sandalwood leaves in this treatment.

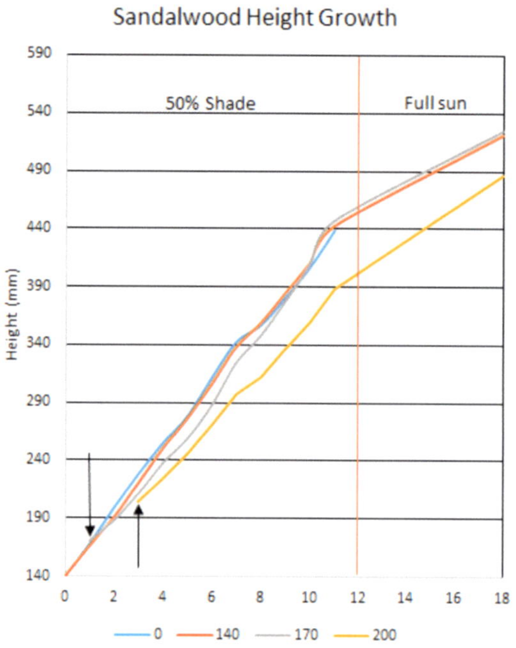

Figure 83: Chart showing growth of seedlings in each treatment. Arrows indicated when treatment were hosted.

The growth data indicated that there may be some timing leeway when hosting Sandalwood seedlings, particularly between 140 mm and 170 mm. However, it should be noted that the largest alternanthera root biomass, in the 170 mm treatment, also had the smallest Sandalwood root biomass. There has been some suggestion that there was the possibility that early hosting with alternanthera could lead to alternanthera over colonising the potting mix, leaving no room for the Sandalwood to develop a substantial root system. This data seems to suggest in some cases this may be so. The most interesting discovery during the destructive biomass sampling was the haustoria count, with a considerable number of haustoria attached to the slow-release fertiliser granules. This may be the reason that the zero hosted treatment grew at a similar rate to the hosted treatments.

The conclusions from the trial are that the timing of the hosting of small Sandalwood seedlings with alternanthera in the nursery is not critical for good above ground growth and development. In fact, the alternanthera did not increase the size or health of the seedlings. However, if hosting is to be done, then all seedlings should be hosted with alternanthera by the time the average seedling height is 170 mm (Robson & Mende, 2016).

Trial 3: This experiment was about investigating ways to increase germination specifically for mechanised sowing. The was two areas of research, seed size and seed colour. The seed size work looked to confirm the trial work undertaken by Robson (2012) looking at the variation in Sandalwood seed diameter and its effect on nursery growth. Seed from a single batch was sorted into four treatment groups of 0.5cm sizes (6 to 6.5, 6.5 to 7, 7 to 7.5 and 7.5 to 8mm) in diameter. All seed was pretreated by bubbling with 0.05% gibberellic acid for 24 hours.

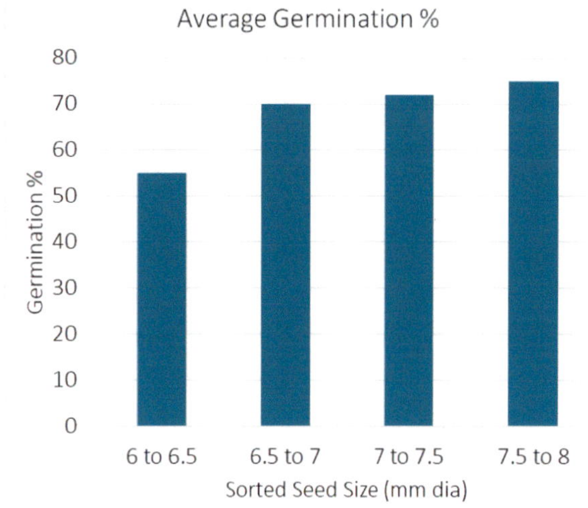

Figure 84: The larger seed germinated at a higher rate.

The results clearly show there is a trend for the larger seed to germinate better. Although the average germination for seed when sown operationally is around 50%, this trial realised an average germination of 68%. This may be due to the seed quality, coming from a new clonal seed orchard. The best treatment was the largest size 7.5mm to 8mm with an average germination rate of 76%.

Figure 85: Seed was grouped into 6 colour variations from lightest to darkest, (Nortier, 2015).

In each kilogram of Sandalwood seed there is significant colour variation. This is mostly due to the age variation of the dupe when collected. When the fruit reaches mature size, it rapidly changes from green to red to purple. As the dupe skin colour changes so too does the seed colour. The earlier the seed is collected and cleaned the paler the seed. The fully mature purple fruit tends to have darker coloured seed.

The germination results clearly show a trend for the lighter colour seed to germinate at a higher rate. The very pale seed germinated at 85%, with the darkest coloured seed at only 30%. Once again, the overall average germination in this seed batch was much higher than the operational norm, 62% versus 50%. Although the lighter the colour of the seed the better the germination, it was only the real dark colours that produced less than satisfactory germination (Robson, McCallister, & Nortier, 2016).

7. ESTABLISHMENT

Plantation establishment takes place from late April to the end of July, taking advantage of cooler daytime temperatures to minimise transplant shock for young seedlings. With irrigation available, reliance on rainfall is unnecessary.

a) Planting seedlings

When seedlings reach the required planting specifications, they are transported from the nursery to a field storage nursery until they can be planted. Typically, the planting site is a considerable distance from the nursery, necessitating transportation by truck.

Field storage

Field storage areas, though temporary, must be well-planned and durable to handle the required turnaround and volumes of seedlings. Adequate space must be designated to allow large trucks (sometimes road trains) and farm machinery to operate safely. Efficient irrigation, preferably automated, is essential to maintain optimal media moisture levels in all seedling pots, including those on the edges of the storage area, without under or overwatering. The storage area must be close to the planting site, fully exposed to sunlight, have access to clean, pressurised water for irrigation, be slightly sloped for drainage, and have room for delivery trucks.

Figure 86: Planting trailer, (PSR. Robson, 2022).

Prior to loading onto the planting trailer, seedlings should be drenched with a mixture containing Super Stand Phos. Any seedlings delivered from the field storage area to the field with dry potting mix should not be planted and must be returned to the field storage area. Only seedlings with moist potting mix should be planted, as transplant shock is greatly increased if seedlings are planted dry.

Marking out

Proper spacing along the row is crucial. Planting too close results in over-stocking, increased costs, and heightened competition between plants, reducing growth and vigour. Under-stocking leads to inadequate coverage and excessive gaps. Small discrepancies in tree spacing can affect overall stocking, potentially resulting in a shortage or excess of seedlings. Therefore, Sandalwood tree rows are pre-marked at the correct distance between trees. This is usually done using a pole marked to the required distance, with paint marking the spot where each tree needs to go. Initially, host distances are marked, but experienced planters can accurately step out the required distance without marking after a period of practice.

Figure 87: planting Sandalwood into a pre-marked hole, (PSR. Robson, 2022),

Planting

Given the large pot size of Sandalwood seedlings, holes are dug with a shovel. Sandalwood seedlings must be planted to a depth of approximately one cm above container level. It is crucial that the alternanthera (pot host) is not completely buried during planting, as it will die. If the alternanthera is not tall enough to allow planting at the correct depth, the seedling should be returned to the field nursery. All Sandalwood planted must have a healthy pot host.

Host plants, usually in smaller pots (hyco tubes), can be planted using a pottipuki tree planter (Figure 88). Host plants are to be planted to a depth of no more but at least ten mm above the container soil level. Seedlings with exposed potting mix after planting are likely to die, as potting mix dries out quickly when exposed to air. It must be adequately covered with soil to prevent water loss.

Figure 88: Host planting with puttiputki tree planter, (PSR. Robson, 2022).

All seedlings must be firmed in and planted vertically. The soil must be well-firmed around the planted seedling to eliminate air pockets and ensure direct contact between the root ball and the soil. The planter must ensure the seedling is planted as vertically as possible to prevent leaning and epicormic growth.

Trees must be planted in a straight line and as close to the drip line as possible. Trees planted out of line risk being damaged, drying out and dying or killed during future plantation operations.

Edge tree rows must be planted with a long-term host. Edge trees are more exposed to elements such as sun exposure and high winds, requiring extra protection.

b) Infilling

The initial survival rate after establishment is typically high, with over 95% of trees surviving past the first 12 months. However, after the transition of the Sandalwood to the long-term host, some sites have experienced significant host mortality due to hosting pressure, making them susceptible to pathogens, as discussed in the Environmental Stress section.

When high host losses occur, it becomes necessary to replace the dead hosts with new ones. Initial attempts at infilling small seedlings into transitioned plantations proved ineffective. Sandalwood quickly began hosting on the infilled seedlings, causing nearly all to die within six weeks.

Figure 89: Sandalwood haustoria on small infill host root, (Robson, 2023).

A research project aimed at increasing survival rates explored the inclusion of aggressively growing species such as Sesbania, native acacias, and *Albizia lebbeck*, along with increasing the size of the infilled seedlings. Both treatments showed some success, improving survival rates by up to 50%. Planting large, aggressively growing species as infill was successful enough for intermittent infilling across a stand but not for more intense infilling.

For sites where whole rows of hosts have died or are at risk of dying, an integrated approach is necessary. Planting large numbers of large seedlings is uneconomical, so an approach allowing the establishment of smaller infill was required. This involves sowing a nitrogen-fixing cover crop, such as Wyn cassia (*Chamaecrista rotundifolia*), into the inter-row in November the year prior to infill establishment.

Aim for a May establishment, as this period has the mildest weather, reducing planting shock. To prevent Sandalwood from hosting on the small seedlings, a rip line is needed to cut all Sandalwood roots, giving the small seedlings adequate time to establish before Sandalwood hosting begins. Aggressive species should still be used, and all seedlings will require a tree guard to facilitate chemical weed control.

Figure 90: Infilled row in eight year old plantation. Note the ripped line in the inter-row, (Midway B. Robson, 2023).

Chapter 5. PLANTATION MANAGEMENT

Once the plantations are established, effective silvicultural and operational management becomes crucial to maximise productivity throughout the rotation period. Regrettably, a lack of knowledge has led to suboptimal management practices across much of the plantation estate.

This chapter aims to explore current knowledge regarding the management of established plantations. It will address various aspects, including managing competition within the stand to achieve a balanced canopy and optimise productivity. Historically, stand competition management was often neglected, with managers relying on natural mortality to alleviate competition. Moreover, there was reluctance to thin Sandalwood due to their perceived individual value, resulting in many stands being overstocked near rotation. Consequently, Sandalwood basal area was frequently lower than it might have been had competition been reduced through timely thinning.

Additionally, the history of stem and host crown pruning has evolved over time. Minimal stem pruning was undertaken on Sandalwood between 2005 and 2012, leading to many stands near rotation having double leaders or large pruning wounds prone to rot.

Fertiliser application has been a consistent management practice since the establishment of the first plantations, with ongoing research trials dating back to 2005. This section will examine research findings and current operational methodologies regarding fertiliser application.

Weed control methodologies have evolved considerably since the inception of plantations, especially since 2012 when forestry systems of weed control became widely utilised. This chapter will explore the various plantation weeds and the research aimed at controlling them.

Irrigation scheduling methodologies have also evolved, particularly with the widespread adoption of dripper systems across a considerable proportion of the estate. This section will compare different irrigation types and delve into how scheduling works.

Lastly, the chapter will discuss the main stress factors that pose risks to tree health and plantation productivity, providing insights into mitigating these risks for optimal plantation management.

STAND COMPETITION

The aim for every forester is to grow the tree under their care to the maximum possible size in the specified rotation length. Tree growth is defined as, the increase in its dimensions over a specific period, while diameter growth is the increase of diameter of a tree in centimetres per year (Vanclay J. K., 1994).

Stand competition can be defined as the tendency for neighbouring plants to compete for the same resources such as light, water, nutrients, and space (Fround-Williams, 2008). This competition can occur between varied species or within the same species, both above and below ground. While stand density refers to the number of trees per unit area at the stand level, competition generally occurs between individual trees (Avery & Burkhart, 2015).

During stand establishment, seedlings can be planted closely together due to their limited requirements for water, light, nutrients, and space. However, as they grow, their needs increase, leading to competition among trees. This competition often results in the domination of larger or healthier trees over others, with suppressed trees dying due to intense competition (self-thinning). Self-thinning is a natural process by which a forest reduces its resource demand to match site capacity, resulting in a temporary reduction in stand canopy size and improved resource availability.

Self-thinning has been observed across various plant forms and populations and is represented by a self-thinning line. This defines the maximum possible combination of mean tree size and density for a given species composition, stand age, and site quality (Soloman & Zhang, 1998). The self-thinning line represents an upper boundary to all possible combinations of mean tree size and density and becomes evident as tree growth slows down as plantations approach it (Vanclay & Sands, 2009).

The site capacity for a plantation corresponds to a point on the self-thinning line that maintains full site occupancy at the minimum possible stand density (Ge, Zeng, Ma, & Meng, 2017). Site capacity is discussed in Chapter 2. When approaching site capacity, tree growth slows down, leading to unproductive and unhealthy plantations. To maximise Sandalwood growth, plantations should be artificially thinned before reaching critical proximity to the self-thinning line, ensuring healthy and productive growth.

While this concept is typically applied to even-aged monocultures, it holds true for mixed-species plantations as well, with each species having its self-thinning line based on its resilience to competition (Hernandez, Comeau, & Bokalo, 2013).

Without proactive management, stand density will naturally reach site capacity, leading to self-thinning, where the largest trees, often host trees in Sandalwood plantations, are more likely to persist. However, adopting proactive silvicultural interventions such as thinning ensures the retention of high-value crop trees while culling those that draw more resources than they add value. This not only protects healthy trees from pests and diseases but also promotes their growth, particularly in terms of diameter growth (Salih, Saied, & Wali, 2019).

1. PLANTATION

There is a requirement to have Sandalwood and several host species, in a planting layout that is a mixed species planting due to the parasitic nature of Sandalwood, (see configurations in Chapter 4). There are two configuration types, Standard (STD), which has Sandalwood and long-term hosts in separate rows and Mixed (MIX) which has Sandalwood and long-term hosts in the same rows. With the STD configurations almost all of the classes of configurations start with 505 stems Sandalwood per hectare (SPH) and the LTH stocking range from 278 to 447 SPH. The initial planting also has a short-term host for each Sandalwood, so a further 505 SPH. This equates to between 1345 and 1470 initial SPH for STD configurations and 853 to 1235 SPH in MIX configurations.

Sandalwood

The competition that occurs with stands is between Sandalwood and hosts and between Sandalwood along the row. The spacing for Sandalwood is 2.75 x 7.2 metres which is 505 stem per hectare (SPH). This stocking of Sandalwood was determined by economists and not by foresters and has generally continued to be the initial stocking for all Sandalwood plantations. In the Ord River Irrigation Area (ORIA), over the length of the rotation on most sites there has been significant natural mortality, especially in Sandalwood. The final stocking of Sandalwood being between 25% and 45% less than at establishment (Haberkorn, 2023).

The start of self-thinning of Sandalwood can be visually observed in many sites across the entire estate where there has not been much establishment mortality, and the initial high stocking remains into mid rotation (six to eight years). There is significant along the row competition between Sandalwood and a proportion of the Sandalwood at this age in the unimproved population could be considered runts. These are trees that have 'lost the race' and will eventually self-thin. A visual observation along any six year old Sandalwood row shows the competition. An example would be the obvious large trees with small trees on either side which is common to see.

There has been a reluctance to thin Sandalwood due to the potential high value of each stem and the lack of understanding of the interaction between the stand competition and the parasitic relationship between Sandalwood and hosts and how that will affect productivity.

Short-term hosts

On most sites, the short term hosts (STH) have been established at the same stocking as the Sandalwood and start to naturally die off from year three, which reduces the overall stems per hectare, in most cases, by a third. However, this is still a stocking of between 850 and 950 SPH for STD and 640 to 740 for most MIX configurations.

There have been several occasions where the Sesbania short-term host has not died off when anticipated. Research data has shown that if sesbania does not die off at two to three years after establishment, it will significantly affect Sandalwood growth (Manson, 2013).

Figure 91: By age five, along the row the Sandalwood sort themselves into dominate, co-dominate and sub dominate trees. In this photo from Florina Road plantation at age six, it clearly shows large trees and the much smaller tree between the larger one, (Florina Rd, Robson, 2017).

Long-term hosts

Although it has been clear to the foresters that proactive management of stand density is required to improve productivity, there has been reluctance for this to happen. Partially the reluctance is due to the uncertainty of the benefits of a thinning program and the lack of data to show these benefits. The other reluctance is because of the under hosting risk. The removal of hosts could tip the ecology and balance of the plantation and cause under hosting which in turn could cause a domino type effect. As host numbers decline, the remaining trees are required to support an increasingly heavy parasitic load. This can lead to elevated host mortality, further intensifying the burden on those that remain. If this cycle continues unchecked, it may ultimately result in the complete loss of hosts within the plantation.

a) Research trials.

Trial 1: (Experiment R028a) To assess the impact of sesbania mortality on Sandalwood growth in plantations, four measurement plots were established in Pioneer plantation areas affected by sesbania mortality and four plots where sesbania persisted.

A comparison of mean diameters revealed a highly significant difference between plots with and without sesbania mortality. Plots without sesbania exhibited a 23% larger mean basal area compared to plots with persistent sesbania, strongly supporting the theory that high competition levels in unthinned plantations significantly hinder Sandalwood growth. Persistent sesbania beyond two years likely indicates undesirably high competition levels affecting Sandalwood growth in most plantations.

To measure Sandalwood's morphological response to high competition from persistent sesbania, data on bent trees, bends per bent tree, and leaning trees was recorded. Plots with sesbania displayed significantly more bent trees (average of 49%) compared to plots without sesbania (average of 32%), more bends per bent tree (1.4 bends per tree) than plots without sesbania (1.1 bends per tree), and more leaning trees. These differences can be attributed to lower light levels created by sesbania, causing trees to lose apical dominance and search laterally for higher light, resulting in bent stems. Sandalwood trees under sesbania invest more energy in leaf area production, leading to larger, heavier crowns and higher percentages of leaning trees.

Using Pioneer plot data, a model was developed to explore the timing of sesbania thinning and its impact on Sandalwood growth in different scenarios. The model revealed that delaying sesbania mortality results in greater growth reduction from competition and a departure from maximum possible growth. Longer delays in sesbania removal exacerbate the growth lag as Sandalwood re-hosts, especially due to the proportional sub-model used for this phase, Figure 92.

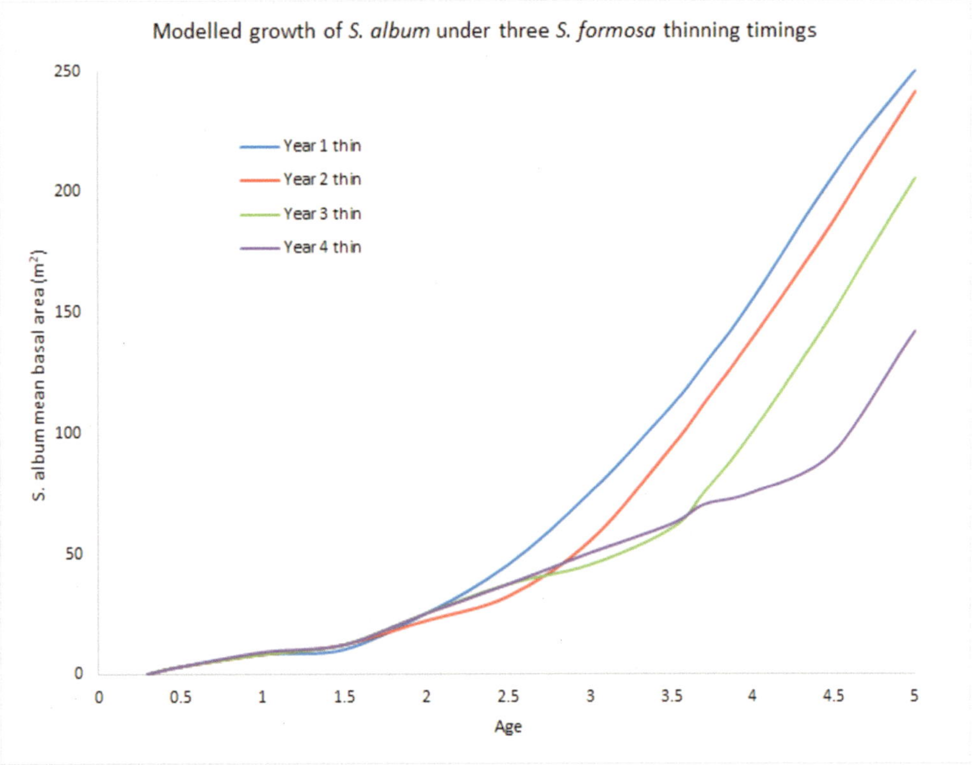

Figure 92: Modelled trajectories of mean basal areas under three different thinning timings based on Pioneer plantation data. The 'lag' due to re-hosting following sesbania mortality is modelled here by a reduction in growth by 90% in the first month following mortality, 80% the next month continuously until returning to normal growth ten months after sesbania mortality. The reduced growth of Sandalwood in un-thinned plantations due to competition from sesbania is accounted for by reducing growth by 2.5% per month until the plantation is thinned. This is the magnitude of the growth reduction observed in the unthinned Pioneer plantation, and it is assumed that this would continue at this magnitude if sesbania mortality does not take place, (Manson, 2013).

<u>Trial 2 & 3</u>: The short term host *Sesbania formosa* is essential for the successful establishment of Sandalwood. However, the sesbania on some sites in NT with superior quality soils and high-water tables did not appear to be showing any deleterious effects of being parasitised by the Sandalwood. Unlike stands in the ORIA region where most sesbania tend to self-thin and be all gone by age three, the sesbania on some NT sites continued to dominate the site.

This experiment examined various approaches to remove the sesbania from the trial site at the end of the second year. A preliminary trial conducted in mid-2013 (R026) observed several methods to kill Sesbania without any detrimental effect on the Sandalwood. These were, ring barking, stem injection with 100% Glyphosate for every ten cm DBH, basal bark application with diesel & Starane (Fluroxypyr) @1%, girdle with wire, and a severe top hedge.

Five weeks after treatment application, a health assessment score on both Sandalwood and sesbania was undertaken. The results showed that as expected, the ringbarking, wire girdling, and top pruning of the Sesbania did not affect the Sandalwood. The two chemical treatments did influence the health of the Sandalwood under the Sesbania treated. However, with the stem Injection of Glyphosate treatment only 3.1% of the Sandalwood showed slight to mild effect. The basal bark treatment with Starane had the greatest effect on the Sandalwood under the treated Sesbania. However, only 25% of the Sandalwood trees where severely affected.

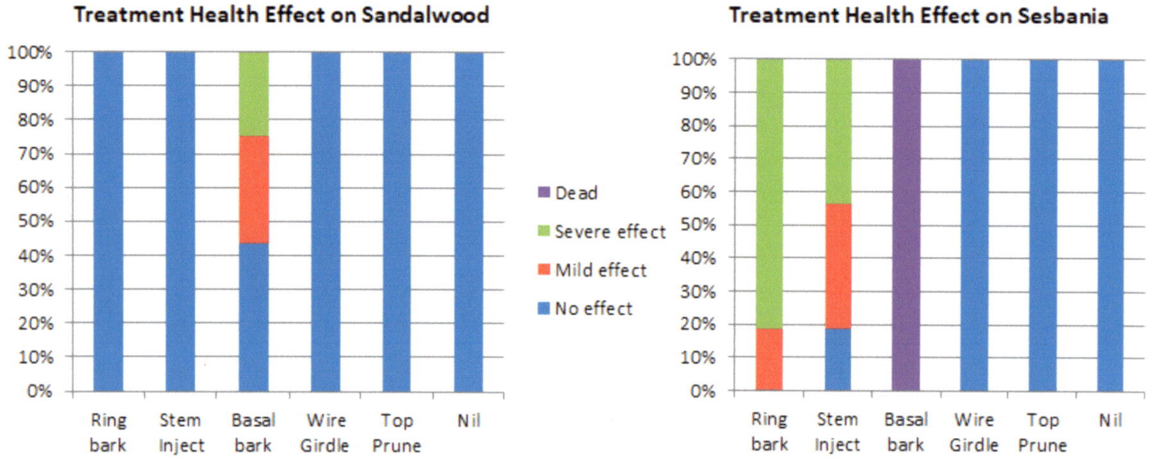

Figure 93: The health results for Sandalwood near the treated sesbania five weeks after treatment application.

For the sesbania, the results demonstrated that the wire girdling and top pruning had not affected the health after five weeks. The Ring barked and Stem Injected (majority of) trees all exhibited symptoms of declining health. However, 18% of the stem Injected sesbania did not show any signs of the treatment and none of the stem injected trees were dead. The Starane diesel basal bark treatment was lethal with 100% of the trees dead.

The chemical thinning treatments were the most efficient and this trial indicated that more investigation was required. Although the stem injection of Glysophate did not kill any trees (sesbania or Sandalwood), it did influence the health of the sesbania. The basal bark application system of applying a poison is easy and relatively cheap.

A common risk in the use of chemical thinning is flashback. This is the unintended negative impact of chemical application on trees adjacent to those treated during the thinning process. This occurs when a chemical translocates from the stem into the root system and via root grafting, moves into an adjacent tree (Walter, Garret, & Godsey, 2004). As Sandalwood is connected to nearby host trees via haustoria the risk of flashback is extremely high. However, Trial 2 indicated that possibly flashback does not always occur and the use of chemical to thin may be an option.

To test this, a second trial (R042) was established in Taylors Park 4 stand 6 in 2015. This was a large-scale randomised complete block trial comprising three treatments: stem injection of 100% glyphosate into 25% of the sesbania stems, stem injection of 100% glyphosate into 50% of the stems, and an untreated control. The trial area was 4.8 hectares which was evenly divided into 6 units of 0.8 ha. Each tree received two millilitres of 100% Glyphosate for approximately every 3cm in DBH.

The sesbania started to show signs of poisoning two weeks after poison application. The crowns turned yellow, gradually losing leaves and after six weeks, 90% of all treated sesbania were dead. The first three months after treatment application there were no visible effects of the treatments on the Sandalwood. This technique initially appeared to be a practical and effective operational method for removing unwanted sesbania from plantation blocks. However, a health assessment conducted four months after treatment revealed a sudden "flash" phytotoxic effect in more than 70% of Sandalwood trees adjacent to injected sesbania..

Figure 94: Photo 1 shows Sandalwood severely affected by flashback from Glyphosate. Photo 2 shows mild symptoms in affected Sandalwood, leaf loss only on the side of the tree near the treated sesbania, (TP4, Robson, 2015).

Over 35% of the Sandalwood beside a treated Sesbania exhibited severe flash effects, this showed as loss of more than 80% of the leaf area (figure 94-1). A further 40% of the Sandalwood beside a treated sesbania were somewhat less affected with up to 40% leaf loss (figure 94-2).

There were several sites where the sesbania had naturally died and therefore not chemically treated, but there was a Sandalwood nearby. The Sandalwood beside these missing host also exhibited some detrimental effects with 34% exhibiting some crown loss. However, the expression of physical symptoms on the Sandalwood was quite different between natural Sesbania mortality and the poisoned Sesbania mortality.

The injection of Glyphosate into the cambium layer of the Sesbania initially transported the poison via xylem into the crown of the tree. There was moderately slow die off of the foliage and crown, up to Six weeks. It was anticipated that the Glyphosate would be bound up in the affected crown and confined in this part of the tree. This was obviously not the case although it is unclear why it took four months for symptoms to show in the attached Sandalwood. It may be related to the moisture status as symptoms did not show until the wet season had finished, and soil profiles had reached a certain dry point.

Although no Sandalwood died, as all those that showed some affect eventually recovered, it was clear from this trial that the chemical thinning of hosts is not suitable in Sandalwood plantations.

Trial 4: Despite the results from Trial 2 and 3, a further trial was established in 2018 to see if the use of a basal bark treatment could be used to remove sesbania wildlings from within the plantation. The experiment looked at the use of three MOA Group 4 herbicides (Starane, Garlon and Tordon) with two levels of diesel sprayed on the bark at the base of the tree. The assessment was a health score on sesbania and nearby Sandalwood.

Figure 95: Sesbania trees 3 days after application of Chemical 3 at a rate of 4ml/L diesel, (TP15. Mende, 2018).

The sesbania started to show signs of poisoning within three days of application. The rates did not make any difference to the effectiveness of the kill. The Sesbania die-off at low rates were similar to the higher rates. Although there was no flashback effect on the nearby Sandalwood in this trial, this is most likely due to the small application of chemical within the trial area or the small unhosted Sesbania.

Trial 5: As chemical thinning has shown to be unsuitable and manual thinning is an expensive operation due to their size, the ringbarking as a way to remove long-term hosts has been tested several times. Ringbarking, or girdling, involves removing a strip of bark and severing the cambium layer, thereby interrupting the flow of water and nutrients to the leaves and upper parts of the tree. As a result, the tree gradually declines and eventually dies. Because this process occurs over an extended period, it is thought that the gradual loss of the host allows Sandalwood sufficient time to establish connections with alternative hosts, thereby minimising the risk of shock associated with the sudden loss of its existing host.

Although not within a designed experiment, ring barking trees to remove them was first tested in 2017 on large *Dalbergia lanceolaria* branches at Smith in ORIA. Several large branches were ringbarked by removing a section of bark and cambium. An assessment two months after showed no visible signs that the large branches were affected by the girdling, except that the wound had developed callus and was starting to heal (Figure 96A). At a further assessment 12 months after application, the wound had completely healed, and it was difficult to determine where the ringbarking occurred.

A second trial, this time ringbarking six year old Senna and Cathormion at Florina Road in the Northern Territory by removing a large section of bark (at least 20 cm) was completed in 2018 and had more success (Figure 96B). Within two months it was obvious which trees were ringbarked and after four months all ringbarked trees were dead.

Figure 96: A - D. lanceolaria two months after large branches ringbarked with callus starting to repair damage. B – six year old ringbarked Senna (arrow) next to control tree two months after treatment application. It is clear the ringbarking is detrimentally affecting the treated tree, (Robson, 2018).

A second ringbark trial with large seven year old *Dalbergia lanceolaria* was undertaken in Midway A in 2020. Using the same technique as the successful Florina Road trial removing at least 20cm of bark from the base of several trees (Figure 97A). Once again, this damage did not seem detrimental to any of the trees, other than the growth of several coppice shoot below the gridle, However, none of the treated trees died.

Figure 97: A - D. lanceolaria in Midway A, three months after girdling (2020). B – leafless D. lanceolaria in Midway D with two lines cut around the base on the base of the tree, (Robson, 2023).

A third attempt to remove large nine year old *Dalbergia lanceolaria* was implemented in 2023 in Midway D. This time the trees did not have bark removed but the stem was girdled by a chainsaw cut 5cm deep. At the time of treatment, the trees were leafless. After two months there was visually less leaves returning to the girdle trees.

<u>Trial 6:</u> This trial investigated the response to Sandalwood and hosts by thinning long-term hosts at variable thinning intensities. This thinning design is based on a variation of the Nelder Fan Design (Nelder, 1962) and is a wave that increases in intensity until it reaches the peak, based on stocking. There are six long-term host (LTH) thinning treatments, 0 (hedged hosts), 17, 33, 50, 67 and 83% (based on original stocking). This therefore means that a row of Sandalwood has the following LTH remaining, 100, 91, 75, 58, 41, 25 and 17%.

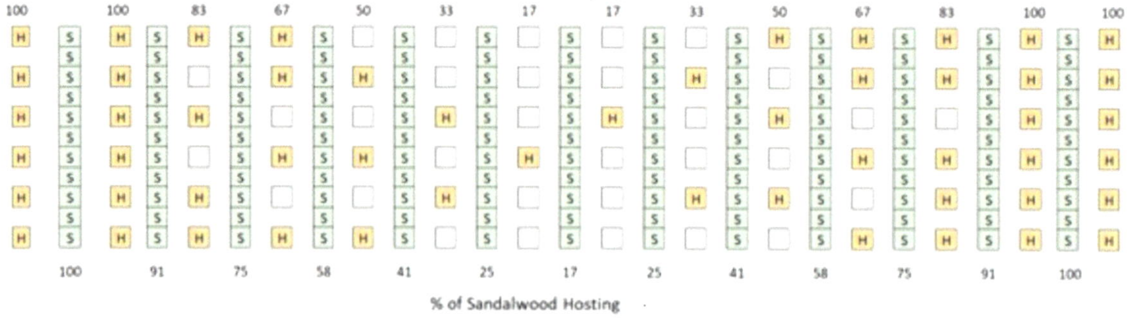

Figure 98: A diagrammatic representation of the variable spacing design.

This experiment was established at Taylors Park 4, Stand 6. The thinning was completed at the end of July 2018 at four years old. Assessments were undertaken at treatment establishment, 10, 20 and 32 months after treatment application.

After 2.5 years since thinning, there was no significant difference between any of the thinned treatments and the control (no thinning). This was very surprising considering up to 83% of hosts had been removed in the highest treatment.

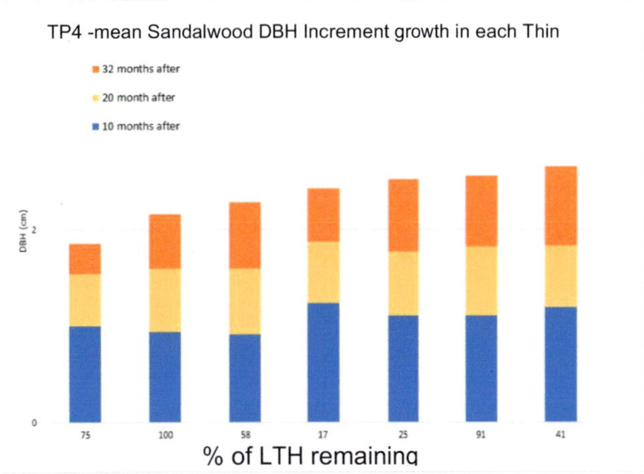

Figure 99: The total initial mean DBH and subsequent increase at 10, 20 and 32 months after thinning.

However, over the 2.5 years there was some LTH mortality which has changed the treatments. For example, in the 100% hosting treatments, the mortality has reduced this to 83% in one row and 67% in another.

Due to the change in LTH stocking from the original treatments, it is difficult to make a definitive conclusion with these results. However, after 2.5 years, the data shows that the thinning of hosts has not significantly increased diameter growth. For that matter, the results have also shown there is no negative effect on Sandalwood tree growth due to host removal, either by enforced thinning or natural mortality.

This trial only reduced competition from the LTH across the row. The Sandalwood stocking is still approximately 90% of original, so the competition release was only across the row and not along the row. The lack of response could indicate that the along row competition is greater than the across row competition.

Trial 6: This experiment was the first to investigate the benefits to the overall stand productivity by thinning both Sandalwood and LTH. It was established on two sites in the Douglas Daly, Maneroo, a four year old plantation with an unimproved seed source, and Midway B, a seven year old plantation with an improved seed source.

Anecdotal evidence suggests that a 1 LTH to 2 Sandalwood ratio can be sufficient for optimal Sandalwood growth. However, this is dependent on the health of the hosts and the mean size of the Sandalwood in comparison to the size of the hosts. The establishment ratio is generally 1 to 1, with over time the basal area of larger faster growing hosts filling the space and restricting Sandalwood diameter development.

For initial selection for thinning there are two main methods, 'from below' or 'systematic'. With thinning from below, trees are removed primarily from the lower canopy, these are the suppressed sub-dominant trees with smaller diameter and/or height. The aim of below thinning is to concentrate potential for growth on the larger diameter trees by removing competing smaller trees. This method of thinning tends to result in stands of evenly distributed trees, although clumps of dominant trees can develop. All of the trees thinned in this trial were selected using the from below method.

Table 24: Treatments for thinning trial. Treatment 1 is the control with no thinning and treatments 4 and 7 have no Sandalwood thinning but host thinning.

Treatment Number	Sandalwood / HA	Cathormion / HA	Dalbergia/ Ha	Total LTH/Ha	Total Stems/Ha	Sandal to Host Ratio	% Sa remove	% LTH remove
1	505	174	174	348	853	1.5 to 1	0	0
2	420	145	131	276	696	1.5 to 1	17%	21%
3	378	131	131	262	640	1.4 to 1	25%	25%
4	505	131	131	262	767	1.9 to 1	0%	25%
5	420	87	131	218	638	1.9 to 1	17%	37%
6	378	87	104	191	569	2.0 to 1	25%	45%
7	505	131	87	218	723	2.3 to 1	0%	37%
8	420	87	104	191	611	2.2 to 1	17%	45%
9	378	87	87	174	552	2.2 to 1	25%	50%

The results three years after thinning in the Maneroo site show a 25% thinning of Sandalwood has a significant increase in mean diameter at 20 cm above ground. The larger Sandalwood tended to be in the 50% host thinned treatments. However, after three years, the unthinned Sandalwood treatments still have more basal area than the thinned plots. The higher thinning treatments had the most diameter response with Cathormion in the 50% treatment 16% larger than the unthinned treatment. The Dalbergia were larger but only 6% bigger than the unthinned. There was no difference for growth when the host to Sandalwood ratio was compared.

The results at the Midway site, only 12 months after thinning showed a quick response to thinning for Sandalwood with all thinned treatments significantly larger than the unthinned treatments regardless of host thinning. However, the total basal area of Sandalwood per hectare reduced the most in the 25% thinned plots.

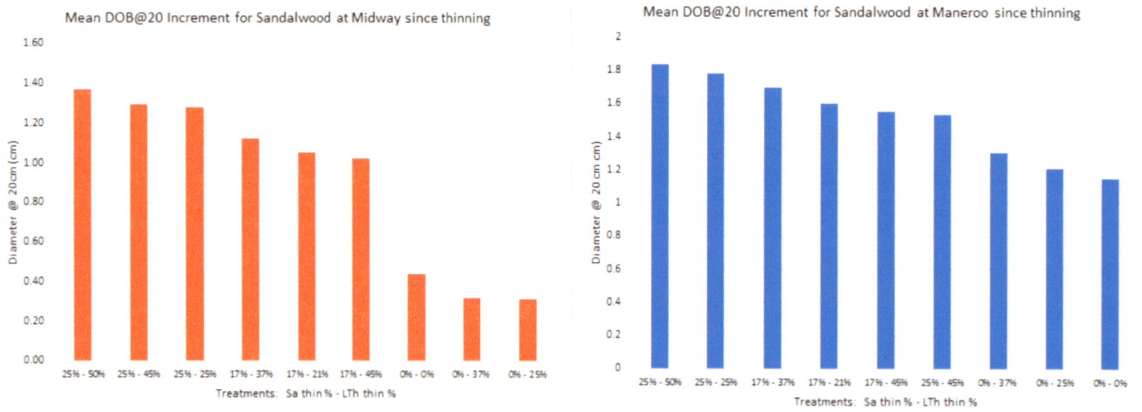

Figure 100: Charts show the diameter at 20cm increment of Midway and Maneroo. The increment for the Midway trial is for 12 months after thinning, whilst the Maneroo increment data is combined three years after thinning.

Although not definitive, these trials have given a particularly good indication that the thinning of both Sandalwood and hosts to reduce competition can significantly increase productivity.

Trial summary and conclusions

The host species Sesbania has shown that it is the best short-term host for Sandalwood at establishment. When grown with sesbania the Sandalwood establishes quicker and develops better stem form. However, if sesbania is still growing close to the Sandalwood after three years the competition between the larger sesbania and the smaller Sandalwood becomes detrimental to the Sandalwood. If sesbania does not start to die off after three years, the manager must manage the species out.

Thinning by herbicide injection when Sandalwood is attached is not suitable due to potential severe flash back. However, basal bark application of herbicide could be used for young wildlings before any hosting occurs.

The most effective method for removing mature sesbania is severe hedging, which involves cutting back the upper portion of the tree. This is best carried out during the build-up period, typically from August to September, followed by withholding irrigation until the sesbania shows obvious signs of desiccation, after which mortality generally occurs. Care is required, however, as this treatment can also affect adjacent Sandalwood. Because Sandalwood may still be physiologically connected to the declining host, it can experience transitional shock, often expressed as temporary leaf loss. If irrigation is reinstated once the sesbania begins to decline, the Sandalwood will recover.

The thinning of only hosts will partially relieve the stand competition. The larger dominate Sandalwood trees will respond but the co dominates and smaller Sandalwood will not. Only thinning hosts and not reducing the numbers of Sandalwood also increases the hosting pressure on the remaining hosts. This can cause undue stress and make remaining hosts susceptible to pathogens and pests.

To reduce stand competition, both Sandalwood and hosts must be thinned. This is partially due to site capacity as well as individual tree competition. Although all the trials that have thinned both Sandalwood and host when they are relatively young, early results indicate that a ratio between 1.5 to 2 Sandalwood per host is satisfactory. On both thin trial sites, the most productive treatments reduced original Sandalwood stocking by 25% and host stocking by 25% to 50%. This is a total reduction in tree stocking by 25% to 35%.

b) Operational competition reduction

The competition within the Sandalwood plantation stand is between the various trees within the stand. As it is a mixed species plantation with three to four varied species that grow at different rates it is difficult to develop a 'one size fits all' approach to competition.

The parasitic nature of Sandalwood necessitates the requirement for a high stocking. The need to supply a host that the Sandalwood will quickly attach to and a host that can survive for the period of the rotation. Additionally it must have a commercial stocking of Sandalwood. This can equate to initially a lot of stems per hectare. As the stand ages it develops through a range of stages as explained in Chapter 2 and 4, and with the selected species the stand will self-thin to a certain point.

Establishment phase

This stage is from establishment to four to five years, when all the tree species are developing quickly. The species groups consist of the Sandalwood, the short-term host and long-term hosts. The initial stocking at establishment varies dependent on the configuration type and group. Stems per hectare vary between 1536 and 853, reducing to 922 to 629 when the short-term host die off. Although there is a lot of stems per hectare, the competition between individuals in this stage tends to be beneficial.

It is a standard forestry practice to establish at a high stocking as this has a range of benefits to the establishment of the stand. High stocking rates are used to encourage early canopy closure specifically along the row. This closure will reduce the level of weed control needed. An elevated level of competition has the advantage of minimising each tree's canopy area thereby reducing the size of the branches and ensuring that most of the wood is produced in the stems (Farm Forest Line, 2009).

Stems may become more cylindrical as stocking increases (Muhairwe, 1994) resulting in a reduction in stem taper (Neimisto, 1995), high stocking also influences stem form and crown characteristics. Studies have found that branch size and crown length decrease with increasing stocking (Pinkard & Beadle, 1998).

Predominately the seed source for all the species planted is unimproved, and due to this, there is a high variation in stem form and growth. Starting with a high number allows for a phenotype selection to happen, which will remove inferior quality and underperforming individuals for the stand.

Although competition is most beneficial in this phase of the plantation, there is a requirement for some silvicultural intervention.

This is mostly crown size management, to control growth of some long-term hosts. It would be preferential to be in the form of stem pruning. The removal of multistems during the first and second year of growth will reduce both crown size and overall basal area. This crown management has been discussed in the pruning section of this chapter.

The configuration design is such that the short-term host die off at the end of this phase of the rotation. However, occasionally this die off does not happen, specifically sesbania and the die off for this species needs to be assisted.

Mid Rotation phase

All Sandalwood plantations start off at approx. 500 Sandalwood stems per hectare with a further 350 to 400 long-term hosts per hectare. The within row competition between Sandalwood can be shown late in the establishment phase, from age two to four and is obvious by age six. Across row competition between Sandalwood and long-term hosts can also start from early mid rotation (age six to seven) on good sites. The high basal area of long-term hosts is a contributing factor to reduced annual increment in Sandalwood. Dependant on site, the transition of Sandalwood to long-term hosts is mostly or partially completed from about age six and fully completed by age eight. Thinning to reduce stand competition should also be completed by this age.

Any long-term hosts that are planted closer than four metres along the row will require reduction in stocking as their basal area will dominate the site and inhibit Sandalwood development. With different host species alternating across the stand, there is usually one species that grows faster and taller than the other. A high stocking along the row and lack of crown management will cause complete shading that will force the smaller Sandalwood to lean out away from the host toward the light (Figure 101).

Figure 101: Sandalwood leaning away from the Senna row at Smiths. Stem overcrowding is not only obvious with leaning stems but also lack of light between the rows, (Robson, 2018).

Trial results indicate thinning from below, at least for Sandalwood is the best option to increase Sandalwood stand volume. Although this thinning methodology is more than likely optimal for hosts, it is not yet clear. Removal rather than retaining the largest dominate long-term hosts may be beneficial to the remaining hosts by letting them compete with comparable size trees. In turn this will see the stand become more balanced in terms of crown sizes and spatial proximity within the canopy for the various species.

The early removal of very poor-quality Sandalwood stems from the plantation will reduce early within row competition and increase the time before yearly increment slows. In mid rotation stands with high Sandalwood stocking a 16% to 25% Sandalwood thinning may be required. This operation will not be difficult as the plantation is naturally sorting outgrowth patterns and on most sites with an unimproved seed source at least 16% to 25% of trees are small and not contributing economically to the basal area of the stand.

On the better-quality sites with unmanaged hosts crowns, the long-term hosts may have overgrown the Sandalwood and are the dominate species. This will be having a strong effect on Sandalwood health and growth. In many cases it may be necessary to remove up to 50% of these trees. However, due to the parasitic nature of the Sandalwood, removal of this many long-term hosts could have negative impact on the Sandalwood. Long-term host removal may be a regime of several steps.

The use of crown hedging is essential during this phase, at least until Sandalwood become fully transitioned. The benefits to Sandalwood in host crown management are discussed in the Pruning section of this Chapter.

On some sites, this period of the rotation can also see one sided release of competition through mortality of weak hosts. The long-term hosts on one side of the Sandalwood die off and the Sandalwood is reliant on the host on the other side. This die off on one side has several consequences for the Sandalwood and remaining host. In many instances, the ratio of Sandalwood to host plants is reduced to one host for every two or more Sandalwood plants. When the Sandalwood is particularly vigorous and large, this imposes significant stress on the remaining host plants, which can sometimes result in the die-off of these remaining host plants.

Figure 102: Long term host Senna row die off in Kingston Rest, (Robson, 2019).

The loss of a row of hosts exposes the Sandalwood stem to the tropical sun and this can cause sun scald. Sun scald is detrimental to the health and wood quality of the tree and will be discussed in more depth later in this chapter, under Stress. To prevent this, regular monitoring should identify the die off risk and an infill of fast-growing resilient hosts can be established. Infilling methods have been discussed in Chapter 4.

Later Rotation

This is the period when most Sandalwood have their crown tops alongside the hosts. Transition is complete and the stand, if mid rotation competition has been managed, will have sufficient resources, light, water, space, nutrition, to grow well until harvest.

With a rotation length of 15 to 17 years there may not be a need for any further competition management. However, there may be a requirement for some crown side management of the more vigorous long-term hosts during this period, particularly if the Sandalwood have been slow to get their crowns into a shared space in the canopy.

PRUNING

This section will discuss branch and stem form pruning in Sandalwood and hosts.

Silvicultural pruning of a plantation tree is a way to remove defects, this may be large branches and secondary stems that can affect straightness or bole length. For forestry, the main aim of silvicultural pruning is to produce knot free clear wood. Branches are removed from the lower crown to a pre-determined height either in a single or multiple lifts. This is usually only necessary if the species does not naturally shed lower dead branches (Pinkard & Beadle, 2000).

Usually pruning can be an expensive operation and foresters have several options to improve stem form to reduce pruning. Initial planting at high stocking to develop a system where the trees compete for light, improving both straightness and stem form, with this high density stands close canopy faster and reduces a trees requirement for lower shaded branches. This methodology works well for some tropical species like Teak, and it also improves stem form for Sandalwood as well. Sandalwood will shed small branches at canopy closure, and the pruning of smaller branches is not required.

Therefore, the main aim for pruning Sandalwood is not to remove the branches but to remove any co dominant leaders to force the tree into a single stem form. This makes the timing of the pruning operation critical. The seedling grows rapidly for the first two years after establishment, and this is the period to manipulate the stem to a single apical form. Any pruning creates a wound and in the moist humid tropics fungal infection is a real danger if the tree does not include over the wound quickly.

1. SANDALWOOD PRUNING

The pruning prescription for Sandalwood has evolved as the final product has changed. Initially aiming to produce an oil product, to now primarily producing a log product. Between 2001 and 2008 the thoughts on stem and branch pruning varied from, not required to a bare minimum.

Figure 103: Photo 1 - Natural growing S. album in an emergent forest environment. The tree is single stem with bole length of over four metres. This tree is about six years old (Cardwell, Robson 2007). Photo 2- Plantation grown Sandalwood with no pruning. This tree is about four years old. Note many codominant leaders and large branches, no distinct bole is obvious, (Eagle Park, Robson, 2016).

After 2008, a series of research trials were established to try and determine the optimal pruning prescription for Sandalwood to ensure a single bole of up to 2.4 metres.

Sandalwood naturally occurs as an understory/emergent species. Growing in these environments, it develops predominantly a single stem as the main leader chases light. The crown shape is dominantly rectangular (Figure 103-1).

However, when grown in open environments, such as plantations, Sandalwood very rarely maintains a single leader. The abundance of light encourages many leaders and larger branches. The crown shape tends to be more oval with multi forking (Figure 103-2). This oval crown shape can clearly be found when driving around young plantations and in some older unpruned or poorly pruned plantations.

There are several detrimental consequences to not pruning Sandalwood in plantations. Heartwood production data from plantation harvests have shown that a single bole will produce significantly more heartwood than those trees that have multi-boles (forking or large branches). As the unimproved population of Sandalwood, without pruning, will more than likely multi stem, the production of heartwood would be reduced.

Figure 104: Heartwood from harvested stems show reduced productivity in forked and poorly prune stem, (Nortier, 2014).

The multi stemming effect also produces larger crown sizes, which can lead to unbalanced trees that are susceptible to wind throw and stem breakage. During the short-term host phase of plantation rotation, the Sandalwood root system is not spread out but confined due to availability of nearby host roots. The confined root mass is not large enough to hold up a large crown in a wind event. Large crown sizes can cause management issues with inter-row machinery. When crowns invade the inter-row space, tractors can catch and cause damage to stems or pull them over.

Figure 105: Large crown Sandalwood falling over in slight wind due to unbalanced below and above biomass, (Ruby Downs, Robson, 2014).

a) Research Sandalwood pruning trials.

Between 2008 and 2013, several pruning trials were conducted to determine an effective methodology for ensuring a single stem in Sandalwood trees and to assess the amount of crown that can be removed without negatively impacting growth.

Trial 1; This trial was undertaken in 2009 in Voyager 3 in ORIA. Comparing three treatments, treatment one was not prune control, treatment two was a form prune that thinned out upper branches to promote one main leader and cut back to one main leader at the bottom where possible, and treatment three was to only remove branches from the bottom of the tree, leaving all upper branches. Pruning was performed once in each of the two years of the trial and assessment was via height comparisons between treatments and counts of damaged Sandalwood through breakage or leaning. Any trees resurrected via tying to others were considered damaged.

Differences were found to be significant (P<0.0001) the control and treatment three in a greater damage group to treatment two, however the treatments made no difference in tree heights. The trial found that by ensuring Sandalwood was pruned back to a single stem significantly reduced damage and stem breakage.

Trial 2: Substantial crown removal is well known to reduce tree growth by limiting photosynthetic capacity. However, most research has focused on tropical pines and subtropical hardwoods. These studies consistently show that removing 50% to 70% of the live crown can significantly reduce tree productivity (Pinkard & Beadle, 1998) (Pinkard E. A., 2003).

However, there is some indication that trees may respond to pruning by changing their patterns of biomass partitioning to favour stem growth at the expense of branch growth (Morris, 2013) and it may be possible that Sandalwood is one of the species that do that.

This experiment looked at Sandalwood trees with increasing amounts of their crown removed (0, 40, and 80%) and assessed the recovery by measuring diameter and height growth. The experiment was conducted on two sites, Florina Road in Northern Territory and at Dalbeg 32 in Queensland in 2013.

The results showed that at both sites, seven months after pruning occurred that there were no significant differences (P<0.05) in height or diameter between any of the treatments (figure 106). A further analysis was completed at the Florina Road site 30 months after pruning and found no significant differences (P<0.05) in height or diameter between any of the treatments.

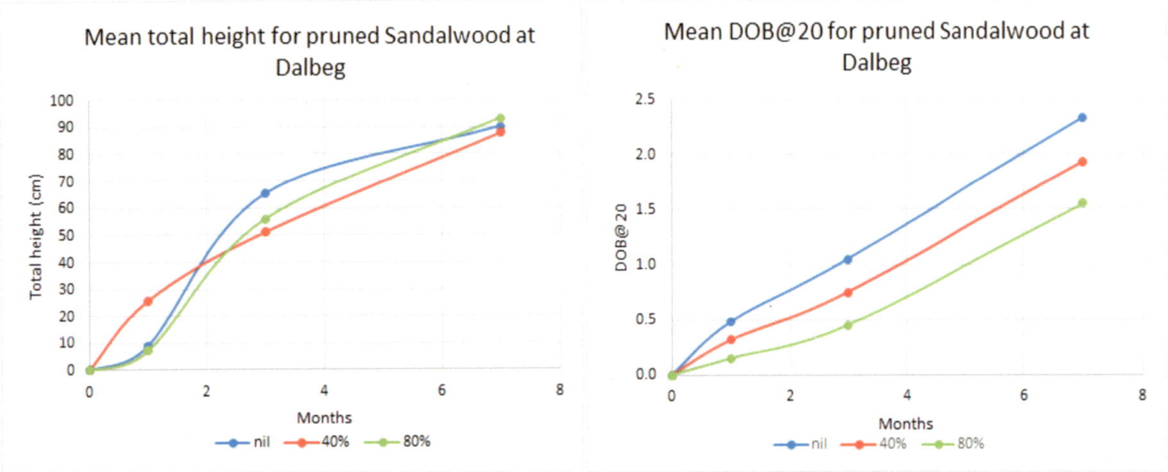

Figure 106: Growth parameters for pruned Sandalwood at Dalbeg site up to seven months after trees were pruned show that crown removed did not influence either height or diameter growth.

Trial 3: This trial was established in an eight year old previously under pruned stand in Kingston Rest 3. The trees on this site are not performing as well as would be expected. The site has soil nutritional and drainage issues. The treatments were, treatment 1 was a no prune (control), treatment 2 was remove up to 20% of crown, treatment 3 was remove up to 40% of crown and treatment 4 was remove up to 60% of the crown.

The results indicate that on this hard site, large, pruned trees will still grow as fast as the unpruned trees.

All trees ranged between seven and eight cm DBH at the start of the trial. There was no growth trend due to the level of pruning, all treatments grew at a similar rate (figure 107).

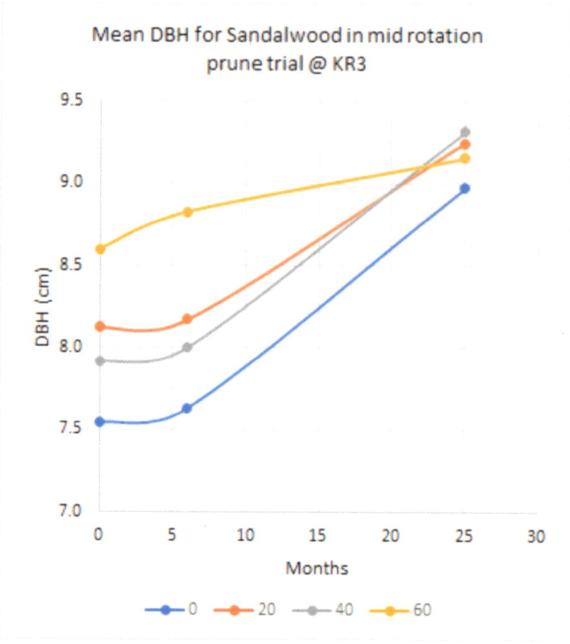

Figure 107: DBH increment over a 25 month period with increasing crown removal in the trial at KR3.

Trial 4: A key risk associated with pruning Sandalwood is the introduction of wood rot via the pruning wound. This is not a concern if pruning is timed so that only small branches are removed, however a high proportion of stands were never pruned on time and required a remedial prune to remove larger branches and leaders. A study looking at wood rot in ORIA Sandalwood indicated that some of the heartwood rot found was from prune wounds (Barbour, Norris, & Burgess, 2010). This concern about wood rot influenced the pruning prescriptions and this trial aims to assess the effectiveness of different cut sealants to determine if they protect the wound sufficiently to reduce wood rot.

Trial 4: This experiment was established in early March 2016 in five year old trees at Ruby Downs. It investigated the potential of four commercially available prune wound sealants. Two products, Steri-prune and Ryset Tree Sealant were bitumen rubber based, one product, Omnia Greenseal was an acrylic resin and was the only product being tested that includes a fungicide (10g/l tebuconazole), and there was a non-bituminous sealant, ATCS Tree wound dressing trialled.

The assessment looked at the level of wound healing after a 12 month period. All the sealant treatments had occluded more than the control. The other assessment was to see if there was evidence of wood rot after the stub was removed. With all the sealants there was no evidence of wood rot, unlike the untreated prunes.

Figure 108: Untreated large prune, showing the wood rot under the cut, (D.Firth, 2016).

The results from the trial demonstrated that the application of a sealant does not inhibit the prune wound healing, in fact in all cases, on average the treated wounds were healing faster. Due to the consistence of the various products (some were more viscous than others) the application varied from a spray application to a painting application. The spray application was much faster and more efficient. The Omnia product was the easiest and quickest to apply. The only issue that was raised was the inclusion of the fungicide. It was unclear whether this may influence heartwood in older remedial prunes. The Omnia sealant without the fungicide additive, was recommended to be used operationally for all remedial pruning. A further outcome was that flat and or rough cuts that retained moisture had a higher chance of wood rot than steeply angled and smooth cuts.

Figure 109: All the sealants significantly reduced rot in the prune wound, (D.Firth, 2016).

Summary of trials

The experiments demonstrated the importance of early stem formation in developing a single-stemmed Sandalwood tree. Trees with multiple-leader stems were prone to stem breakage and produced less heartwood compared to single-stem trees. Additionally, the trials highlighted the significance of early and timely pruning to avoid the need for remedial pruning of large branches and stems. In cases where remedial pruning was necessary, the application of a sealant was essential to prevent early infection by wood rot in the wound. To minimise the risk of rot following pruning, cuts should be smooth and sharply angled to prevent moisture accumulation. Furthermore, the trials indicated that Sandalwood growth is not heavily dependent on crown size, as up to 60% of the crown could be removed without significantly slowing growth.

b) Operational Sandalwood Pruning methodology

The operational pruning methodology is based on shaping the Sandalwood crown early and frequently, with timing being critical to remove small branches when the tree is growing at its fastest rate.

First year

The management in first year of growth and development of Sandalwood seedlings is critical for setting it up for the rest of the rotation. It is also the time when rapid height growth determines stem form. The pruning in the first year uses a tip prune. This is not removing the entire branch but only the growing tip. This removes any apical growth from the pruned branches and shapes the crown into the desired shape.

The 1st prune is when seedlings are 600 to 1000 mm tall with a round crown shape (figure 110-1). This is usually about three to five months after establishment depending on site vigour. This is a manual labour operation with staff using secateurs or grape picking snips. The aim is to define a clear leader by removing tips off co-dominants. Any wide lower branches should also have their tips removed.

Benefits of this prune include increased height growth, which is important to get trees out of the weed spray zone and to reduce crown size and weight. Unpruned seedlings with large crowns are susceptible to windthrow.

Figure 110: Photo 1 shows an unpruned seedling, the drawn shapes indicate the desirable shape after pruning. Photo 2 show the seedling after prune. Any co-dominant leaders have had their tips removed and lower branches tipped to ensure the chosen leader remains.

A 2nd prune should occur when trees are 80 to 160 cm tall and show a rounded crown shape. This is also a manual labour operation with staff using secateurs or grape picking snips. The aim is the same as the 1st prune, to define a clear leader and remove tips off co-dominants. Lower whorl of lateral branches can also be removed in this prune. This will ensure Sandalwood is not sprayed during weed control operations.

Second year

In the second year of growth, a 3rd prune is necessary when trees are 1.5 to 2.5 metres tall to eliminate persistent co-dominant leaders and larger lower branches. This is also a manual labour operation with staff using secateurs and loppers.

The aim is to still tip the co-dominant leaders but completely remove branches and not just tips like the previous prune events. Several rules are followed to ensure the health and continued growth of the tree. Thes include to avoiding cutting branches larger than 30% of the main leader or over 40 mm in diameter with secateurs. Removing only one cut branch per whorl to minimise the risk of stem breakage and pruning off any torn or damaged branches at the union to the stem, regardless of size.

Figure 111: Photo 1 shows a ten month old, pruned tree. Photo 2 shows a closer look at the pruned branches. The red lines show where the second tip prune has reduced the vigour of most lateral branches, and the main stem has become clearly dominant as a result, (Dalbeg, Robson, 2015).

From age three

By age three, with correct pruning the tree has developed a single bole. However, there may still be some branches that will become potential multi leaders above 2.4 metres. These may still need a branch prune, which would follow the same rule as the year two prune.

Remedial Pruning

Remedial pruning is for trees that missed a scheduled pruning in the first and second year. This should never happen as it is detrimental to the health of the trees and is an expensive operation.

Nonetheless, it does happen, and the objective of this prune is to try bringing each stem back in line with the standard prune prescription. However, depending on the development of the stand this may take two to three prunes to do this as it may require too much crown removal in one hit.

As they will potentially have several stems, these trees have large round crowns, and this can make it difficult to see into the stem to decide what must be removed. As there is multi leaders or steep branches of a similar diameter to the main stem the bole tends to be wavy or bendy.

The prescription is to remove all steeply angled branches and double leaders (co-dominant stems) until approximately 50% of the crown is removed. All cuts must be clean and smooth. Branches with a flat angle are retained. Any double leaders are to be removed.

Until this point, the pruning has been with secateurs or loppers. Most of the branches and stems to be removed during remedial pruning tend to be too large for these implements and electric chainsaws are usually used.

With these larger branches, a high percentage of heart rot is caused by torn or broken branches rather than by open prune wounds. Whether or not there is wood rot from an open prune wound seems to be determined by cut angle rather than cut size. Vertical cuts have no to little wood rot, were as more horizontal cuts have significantly more rot. Another factor is how clean the prune cut is, if it is smooth there is less chance of wood rot, rather than if the wound is uneven.

All cuts made in a remedial prune must be coated with a sealant as soon as possible after the cut has been complete.

c) Host stem pruning

The removal of branches or stems from hosts has never been an operational commitment. The main reason for this is because the host has no commercial use, so there is little enthusiasm for the extra expenditure to remove branches. It has always been presumed there is no commercial use for the host species however, there has never been any research into the marketability of short rotation timber from these tropical trees. Regardless, there are advantages in stem pruning hosts when young.

The rapid growth and crown structure of Cathormion, Senna and *Dalbergia lanceolaria* prior to the Sandalwood parasitism is vigorous and wide. The current operational method to control the crown width and growth in the first three years after establishment for these species is to have a proportion of their crown removed by hedging. This will be discussed in the next section. However, stem branch and multi leader removal and a thinning at age five could potentially do the same thing.

Research has implemented a few small operational trials looking at stem pruning of Cathormion, Senna and *Dalbergia lanceolaria.* The outcomes from this work have shown that, like Sandalwood, if any pruning is to be done, it must start in the first 12 months, while stems and branches are small. Trial work with Sandalwood indicated that up to 60% of the foliage could be removed without affecting growth, this does not seem to be the case for Cathormion and Senna. A trial comparing 15 stems pruned Cathormion against unpruned, showed that the pruned stems were up to 40% shorter after six months. For the Senna, although the difference was not as large, the results were the same, with the pruned trees shorter after six months.

Although it appears that the removal of lower branches and singling of extra leaders does slightly slow growth, the modified form provides better access for machinery and people into the plantation and Sandalwood responds to the extra light.

Dalbergia latifolia is a high value cabinet timber species that was introduced into the host mix in 2016. While this species tends to be a single stemmed tree, nonetheless it can occasionally have a double leader form. Even though there is no current commercial use in Australia for this species, it is grown for timber in 20 year rotations in Bangladesh (Bhandari, 2023). With little effort and expenditure, this species should be pruned to a single stem by the first ten months after establishment.

Although it is not operational procedure for the long-term hosts to be stem pruned, branch removal from the short-term host *Sesbania formosa* is important. Sesbania is only planted 1.2 metres from the Sandalwood and can develop large lower branches that can affect the growth of Sandalwood.

Figure 112: Photo 1 shows an unpruned sesbania, the arrows indicate lower branches to be removed. Photo 2: Shows the same tree after pruning. These branches would have become large and impeded the Sandalwood stem development, (Melon Farm WA. Robson, 2009).

2. HOST CROWN MANAGEMENT

Most host species are rapid growers and can develop large crowns, which is crucial for their quick establishment before Sandalwood begins to host on them. During this phase when they are unhosted, failure to manage the size of the crown can result in them outgrowing the slower-growing Sandalwood and dominating the site.

Unlike pruning and thinning, hedging causes only a temporary reduction in canopy, briefly increasing the availability of nutrients, water, and light before the recovering host once again draws on site resources to support regrowth. Alcorn et al (2013) found that the effects of hedge pruning on transpiration in two tropical and subtropical eucalypt species were negligible after just 36 days. Under such conditions, any improvement in site resource availability is likely to be limited and short-lived.

This short recovery time in the management of Sandalwood hosts also is true, particularly during the unhosted phase for host species. Generally, in the first two to three years, with up to 50% of the foliage removed in a hedge event, the recovery is very quick. On some sites, the Senna is required to be hedged twice in a single year. This recovery time does slow down once transition to being hosted occurs. On most sites, ongoing crown management is necessary every two to three years.

Host crown management does not necessarily reduce long-term competition, but it can improve canopy structure by allowing light penetration that benefits Sandalwood form. In addition, hedging the crown presents a method to shape the host crown in a way that benefits the Sandalwood. The goal, by reducing the host crown size is to improve Sandalwood stem form and size to achieve a balanced canopy shared by both species.

Initially to achieve this, a side-mounted hedge trimmer was employed in the plantations between 2012 and 2014 (figure 113-1). Although effective at removing branches, it was found to be slow and time-consuming. As a result, it was replaced with the Mango topper hedger (figure 113-2), as it can cut quite large branches from varying heights.

Figure 113: Photo 1 is a side mounted hedge trimmer. This was used to take off the small side branches from hosts (Taylors Park. Robson 2013). Photo 2 is the Mango topper hedger that is used for crown management in larger trees in Sandalwood plantations, (Dalbeg. Robson, 2014).

a) Crown Management Theory

When growing a mixed species plantation, it is beneficial to have species that have complementary structural and functional traits as it can be useful in improving their resource efficiency and yield.

Research suggests that there are growth benefits in a mixed species planting. It can result from combining light demanding, deep rooting, nitrogen-fixing, wide-crowned and more laterally expanding species like the nominated host species with a shade tolerant species, shallow rooted, non-nitrogen-fixing, slim-crowned and height oriented species like Sandalwood (Pretzch, 2014).

Figure 114: Light evenly reaching trees in a monoculture.

When considering Pretzch's idea, Sandalwood and the host do complement each other. However, it can take a while before the plantation is balanced. During the development phase of the plantation, the varied species are growing at varying rates and with diverse crown shapes. Without silvicultural intervention, the more vigorous hosts can prevent light from reaching the slower growing Sandalwood. The evidence of this is where Sandalwood can be observed leaning away from the heavy shade looking for light (Figure 101).

As mentioned previously, Sandalwood prefers a dapple shaded environment when it is a seedling. Like an understory species, and with most understory species they wait for an opening in the canopy and then grow towards that light. When large host crowns are not controlled, the Sandalwood will lean away from them towards the light.

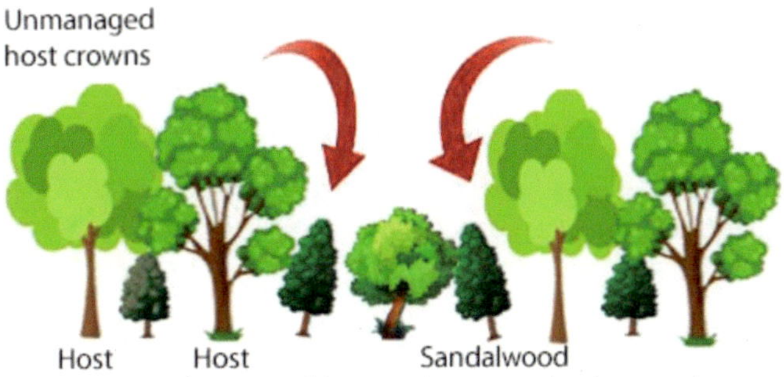

Figure 115: As host get bigger, they can shade out the Sandalwood restricting its growth or force them to lean to chase the light.

This can be remedied by manipulating the crown of the larger trees to create 'light wells' within the plantation, that can mimic fallen trees in a canopy to 'release' Sandalwood and 'pull' the Sandalwood up towards the light instead of chasing the light sideways.

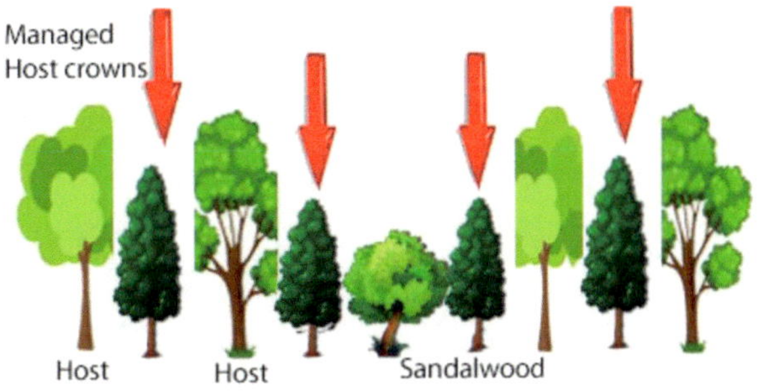

Figure 116; By removing a portion of the host crown lets light in for Sandalwood.

This only works if the host are taller than the Sandalwood. If the host are topped and sided, making the Sandalwood the tallest tree, the Sandalwood will reduce height growth and develop several apical leaders.

b) Hedging Methodology

Box out a section around the Sandalwood to create the light well. In essence, the number of cuts is not the critical part of the specification, as the cuts may change depending on the structure of the plantation. The main aim is to set up the light corridor in the diagram (dimensions may vary a little).

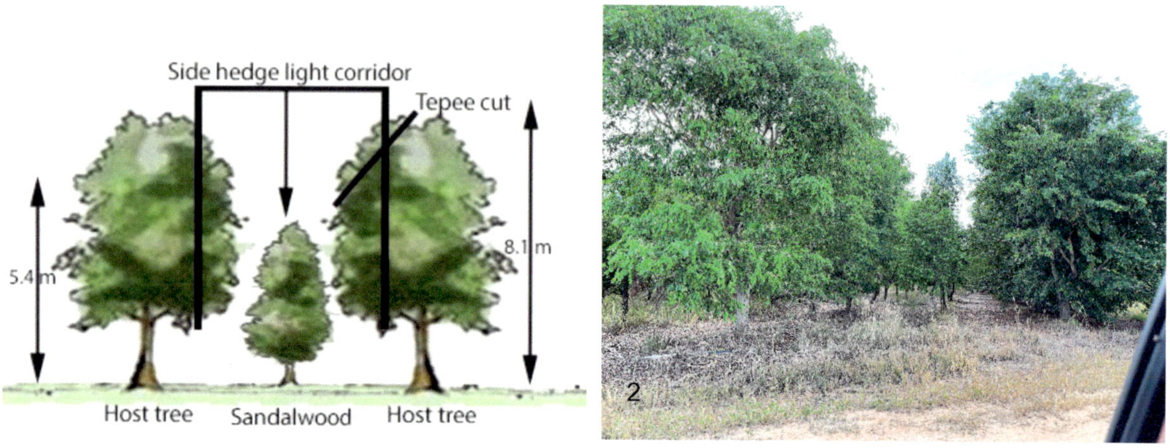

Figure 117: By removing the side branches encroaching over the Sandalwood, it creates a well of light. Photo 2 demonstrates this in the Dalbeg plantation, (Robson, 2024).

Sites with small Sandalwood and large hosts

On the better quality sites where there is tendency for the host to grow much faster than Sandalwood, they can quickly overtop Sandalwood. As a result the Sandalwood will slow growth and/or start leaning away from the host.

To rectify this, the branches reaching across the row can be removed by hedging them back to approximately one-third of their reach. This will remove the shading but retain enough foliage on the host without causing it too much stress.

Dependent on the height of the host this may require two passes with the machine, resulting in two cuts on each side of the tree. The first cut removes the bottom branches and the second could be angled to the leading top of the host tree (sometimes called a teepee cut). The teepee cut should reduce the overall height of the host tree.

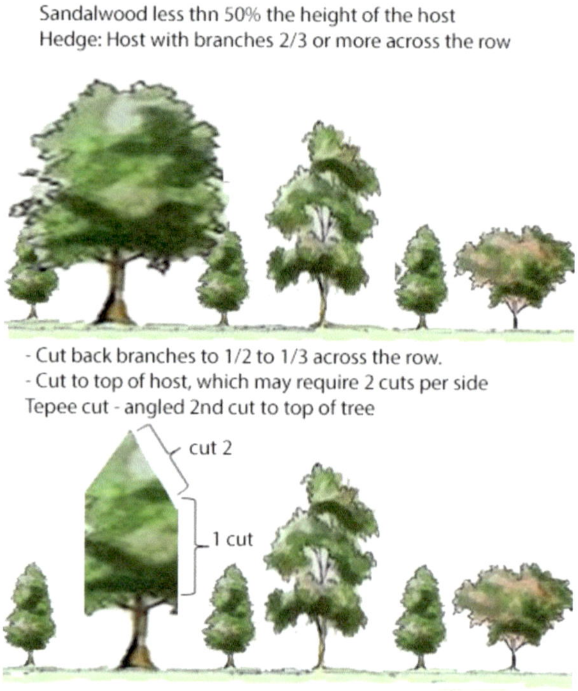

Figure 118: Hedge off the branches on the overtopping host and teepee the top.

Sites with Sandalwood 50% to 75% height of hosts

For most areas in mid rotation (five to eight years), it is common to see hosts taller than Sandalwood and they can have branches spreading across inhibiting the Sandalwood growth.

To address this, branches extending across the row can be hedged back to roughly one-third of their length. This reduces shading while maintaining sufficient foliage on the host.

Dependent on the height of the host this may require two passes with the machine with resulting two cuts on each side of the tree. The first cut removes the bottom branches, with the second cut straight to the top of the host tree.

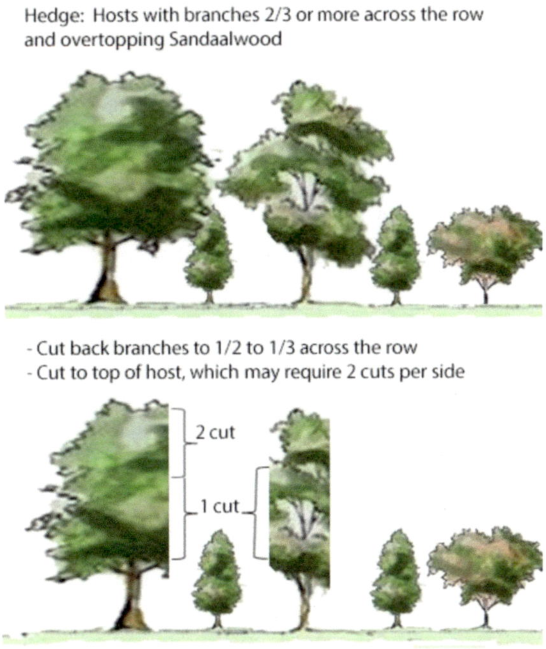

Figure 119: Hedge off the branches on the overtopping host with two passes.

Sites with Sandalwood more than 75% height of hosts

In older plantations, typically six to ten years of age, Sandalwood and host crowns often intermingle within the canopy. However, smaller Sandalwood trees may still be affected by excessive shading.

To address this, branches extending across the row can be hedged back to approximately one-third of their length, reducing shading while retaining sufficient foliage on the host.

Depending on host height, this usually requires only a single cut on each side. As the lower branches should already have been removed, a straight cut across the upper crown is generally all that is needed.

Figure 120: Hedge off the branches on the overtopping with one high pass.

There are some rules when not to hedge. In areas with Lasiodiplodia dieback in the Dalbergia or Senna as the blades can spread the disease from one tree to the next. The tropical regions in Northern Territory and Western Australia have a period before the wet season, colloquially known as the buildup, this is from September to December. It is recommended to never hedge in this period, particularly October as this is when the buildup is at its peak. The hedging program should aim to be completed by the end of August. In addition, do not hedge if the stands irrigation system is not working, as drying out after hedging is a risk that can cause mortality. Also do not hedge if the stand is recovering from or is currently under heavy insect attack.

All the examples discussed so far have been in STD configurations (host in separate rows to Sandalwood), however side hedging can be undertaken in MIX configuration as well. This is usually when some LTH have extended their branches across the adjacent rows and are over shading Sandalwood. A hedge, like that shown in Figure 120 can be undertaken. This will remove upper crown branches from the large LTH. In most cases it will also remove some Sandalwood branch tips as well. Operationally this has shown to be beneficial to the plantation as it reduces the shading and assists with letting light into the smaller Sandalwood improving growth and crown health.

NUTRITION

Tropical Australian soils are known to be low in Phosphorous and Nitrogen (Wild, 1958) and many native species have adapted to grow well in these nutritionally poor soils. Soil nutrition analysis across the Quintis estate did confirm that almost all the soils are nutritionally depleted. Moreover, many of the sites that have been established with plantations that have had a fertiliser history. Most did have horticultural crops or improved pasture on, prior to purchase by Quintis. However, in some cases there was a high load of nutrition in the top ten cm of the profile at plantation establishment.

Foliar analysis for the species grown show that on most sites up to age five there are no deficiencies in any of the nutritional elements. Foliar testing across a range of the Sandalwood plantation trees at various ages has shown that there is not generally any deficiency of Nitrogen. Aside from the nutrient requirements of the Sandalwood, having several different host species with potentially differing nutrient requirements means that the targeting of the application of nutrients in plantations is difficult.

When symptoms are observed in long-term hosts species, they may not necessarily be attributed to poor nutrition but other factors such as poor water supply, extreme climate, pathogens and over hosting. However, if those influences can rule out, a foliar analysis and then the application of a targeted fertiliser may be required.

The application of fertiliser is not straight forward. However, if it is decided that fertiliser is required, it is important to consider how it will be applied, foliar, fertigation or granular. Applied liquid fertiliser forms are short lived and for this reason, they tend to be applied at low rates with high frequency. With application through the dripper systems, it is also not possible to target specific species within the plantation. Foliar and fertigation application generally will give an instant result and could be used in cases of extreme nutritional stress but for longer term nutritional supplements the use of granular is more beneficial.

If it is decided to apply a granular fertiliser, then when calculating the application rate, it must be at a level which will be high enough to have an effect. There have been a range of trials investigating the rate to use and these various research trials will be discussed later in this chapter.

Species mix

Sandalwood plantations are mixed species stands, with up to six different tree species, including the Sandalwood. The mix of all these species potentially requiring various levels of nutrition leads to a complicated nutrition program to develop for the stand. To do this it is important to understand the nutritional requirements for each of the species and what the elemental nutrition bank exists in the soil.

The development of a nutrition profile for the host species is relatively straight forward. Once there is an understanding of the minimum requirements for the key nutritional elements, a diagnostic tool can be developed to determine deficiencies and how to rectify them (Reuter & Robinson, 1997)

All the host species used by Quintis were nitrogen-fixing. These are trees that can convert (fix) atmospheric nitrogen into a form that can be utilised in association with symbiotic organisms such as rhizobium bacteria. Inoculation with rhizobium in the nursery is required for most exotic tree species and foliar data for the host species has shown that nitrogen levels are generally high across all planted sites, regardless of initial soil nitrogen levels.

Development and understanding of nutritional requirements of Sandalwood is not straight forward. The nutrient uptake requirements of Sandalwood have produced diverse opinions regarding the nutrients it absorbs from the host through haustoria and from the soil through its root system (Srinivasan, Sivaramakrishnan, Ananthapadmanbha , & Shankaranarayana, 1992) (Barrett, 1998). Despite its parasitic nature, Sandalwood seedling roots demonstrate the capability to absorb nutrients from the soil like other plants (Parthasarathi, Gupta, & Rao, 1974). However, at the three-month stage, seedling growth becomes limited by a deficiency in nitrogen, potassium, and phosphorus, with no such limitation observed for deficiencies in Calcium or Sulfur. Beyond this stage, the establishment of a host connection becomes crucial for Sandalwood's survival and continued growth (Barrett & Fox, 1997). The growth and vigour of seedlings increases following the formation of Sandalwood host connections (Rangaswamy, Jain, & Parthasarthi, 1986).

As discussed in <u>Chapter 2 Species characteristics</u>, the acquisition of nitrogen and phosphorous by Sandalwood from a host was proposed by Iyengar (1960). It was also suggested by a few researchers that a host is also required to support uptake of potassium and magnesium (Rangaswamy, Jain, & Parthasarthi, 1986), calcium and iron (Rao, 1942), copper and zinc (Ramaiah, Parthasarathi, & Rao, 1962). If the host cannot provide enough of these nutrients there are significant effects on the morphological development of the Sandalwood.

All the hosts have rhizobia associations and foliar nitrogen levels across all sites with these species is adequate too high. The exotic Dalbergia requires inoculation in the nursery for this association to produce superior quality seedling. However, even without inoculation, seedlings will eventually gain an association with one of the native soil rhizobia.

The main deficiency that is found in Dalbergia is due to low phosphorous. However, it is uncommon to see visual symptoms. The main symptom is usually reduced stem growth and potentially less root growth as well.

1. NUTRITIONAL REQUIREMENTS

To develop a diagnostic tool to identify nutrient deficiencies before they show as physical symptoms using foliar sampling, it is required to have a knowledge of what the minimum level of each nutrient element is for each species. This methodology is well practiced in horticulture and in forestry to a lesser degree.

References available for horticulture have been developed over a long period of time collecting data (Reuter & Robinson, 1997). The lack of available data for Sandalwood and associated host crops has hindered the use of a valuable diagnostic tool for foresters. In 2016, it was recognised the need to develop a database of elemental requirements for Sandalwood and associated host crops. Quintis started collecting annual foliar data from healthy actively growing trees within their plantations. It was also recognised that the data available was limited when compared to published horticultural data. Quintis continued to build on the database to ensure the data would show as accurately as possible the elemental requirements. Field data results were compared with species in Plant analysis, Chapter 10, Forest plantations (Boardman, Cromer, Lambert, & Webb, 1997) to see if the Sandalwood hosts species foliar elemental levels were like other tropical forestry species.

a) Macro elements

Macro elements are the nutrients that plants need in larger quantities to maintain good growth and health. These are nitrogen (N), phosphorus (P), potassium (K), and then calcium (Ca), sulfur (S), and magnesium (Mg). This discussion will look closer at NPK effects on plantation health.

Nitrogen (N)

Nitrogen is vital as it is a major component of chlorophyll, the compound by which plants use sunlight energy to produce sugars from water and carbon dioxide (i.e., photosynthesis). It is a major component of amino acids, the building blocks of proteins.

Soils

Most of the nitrogen supplied to plants from the soil is organic, which is released by microbes into mineral forms, predominantly ammonium (NH_4^+) and nitrate (NO_3^-). Both exist in exchangeable and nonexchangeable forms. Plants normally use nitrogen in only the ammonium and nitrate forms.

Continual cropping has led to severe declines in nitrogen supply in cereal soil in Australia (Strong & Mason, 1997). Soil nitrogen can be lost in several ways. The denitrification of soil is the loss of nitrogen in soils due to waterlogging in slow draining soils. This is a process that converts nitrate to nitrogen gas, removing available soil nitrogen and returning it to the atmosphere. Leaching into soil sub layers due to heavy rainfall and excessive irrigation can also remove available nitrates.

The amount of organic matter in the soils has a large effect on available nitrogen. By increasing soil carbon, there is an increase in soil nitrogen. As organic matter decays, nitrogen is released as ammonia and subsequently converted into the soluble nitrate form (Albrecht, 1938). There are a lot of benefits to plants by increasing soil organic carbon (SOC). As well as increasing available nitrogen, increasing SOC improves the soil structure, which increases permeability therefore reducing erosion and improving root growth. Quintis soil data shows that soil carbon has significantly increase since plantation establishment. There is a key relationship between the amount of carbon and nitrogen in the soil, the carbon/nitrogen ratio. As carbon increases the amount of available nitrogen also increases. (Lue, 2012)

Table 25: SOC have increased significantly within the plantations since establishment, (Quintis collected soil data. 2023).

Region	Pre estab	Post estab	% increase
Daly	0.37%	0.51%	38%
Katherine	0.35%	0.54%	54%
ORIA	0.40%	0.42%	5%

Rhizobium bacteria biologically fix nitrogen for plants by forming nodules in the roots of legumes and live symbiotically with them. The rhizobium transforms the nitrogen in the soil air into forms that plants can use. The legumes in exchange give the rhizobium a home and glucose. Many of these species live in the rhizosphere (the zone around plant roots) and help plants take up nitrogen from the soil. Although usually associated with legumes, they can occasionally be associated with nonleguminous species (Fernandes & Ravi, 2023). As mentioned in Chapter 2: Species Characteristics, all the host species used are legumes and have rhizobial associations.

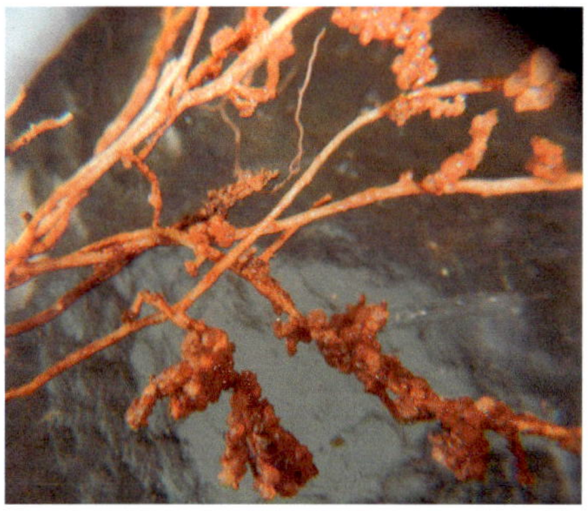

Figure 121: Rhizobium nodules on Cathormion roots, (Robson, 2020).

Crop

Nitrogen deficiency can lead to a variety of symptoms, including decreased photosynthesis and reduced synthesis of amino acids and proteins. This deficiency results in diminished branching and leaf surface area, ultimately leading to stunted growth (Kozlowski & Pallardy, 1997).

The deficiency of nitrogen is seldom observed in Sandalwood plantations, as host plants typically maintain rhizobial associations that ensure nitrogen levels remain within a healthy range. Nitrogen deficiency in Sandalwood primarily occurs in the absence of a suitable host.

Sandalwood

It is well understood that Sandalwood gains its nitrogen from its host (Iyengar A. , 1960) and a host's suitability can be determined by its ability to supply nitrogen to the Sandalwood. Radomiljac, McComb and Pate (1999) stated in their paper that Sandalwood will maintain a higher foliar nitrogen concentration than corresponding host foliage. Moreover, Quintis data from plantations older than four years shows that nitrogen foliar levels for host and Sandalwood are similar and vary considerably over the seasons. However, visual symptoms of nitrogen deficiency only occur when Sandalwood nitrogen foliar levels drop well below associated hosts. If a host is near-by this is a good indicator that hosting is not adequate or there is something inhibiting hosting.

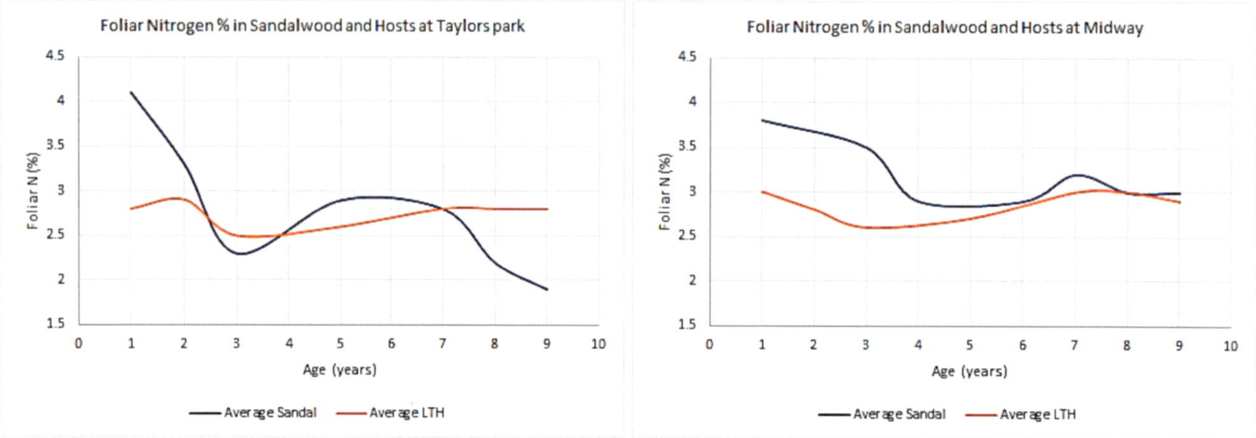

Figure 122: These charts show the mean foliar nitrogen levels over time for both Sandalwood and LTH at Taylors Park and Midway. It shows for most of that time Sandalwood levels have been slightly higher than LTH. However, the levels in Sandalwood at Taylors Park have dropped from age seven. This could indicate that there has been an event that has reduced hosting.

When the average foliar nitrogen percentage for Sandalwood is considered, there is a trend, where foliar nitrogen at establishment decreases from age one to three and then slowly starts to increase again (Figure 123). This is a clear indicator that Sandalwood is predominately obtaining nitrogen from the hosts. The variation is the Sandalwood switching hosts. When hosting the STH the nitrogen levels are high averaging about 4 mg/kg. By the time short-term host dieback occurs, typically around age three, Sandalwood nitrogen levels are at their lowest. Thereafter, nitrogen levels begin to recover as the trees complete their transition to long-term hosts, usually by about age six.

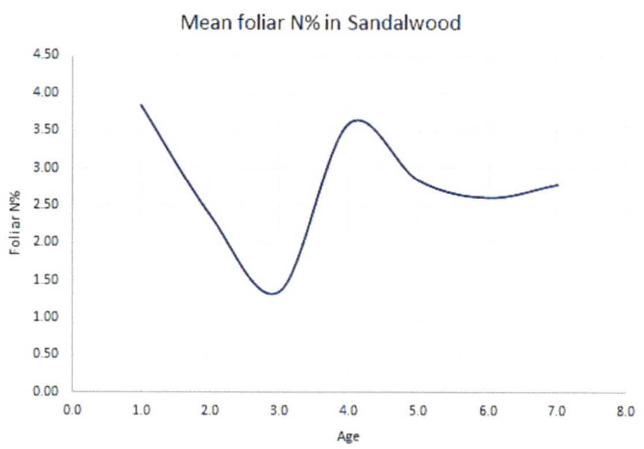

Figure 123: Sandalwood foliar nitrogen levels for Sandalwood at Taylors Park shows a drop off and then an increase as they regain new hosts.

Hosts

The foliar levels in Cathormion and *D. lanceolaria* at age one are on average 3.8% when planted at Midway but lower at Taylors Park for both species with both having a foliar level of 2.7% nitrogen.

There was a slow decline in foliar levels for *D. lanceolaria* for three years at Taylors Park and five years at Midway. After that, their foliar levels have tended to be relatively constant at 3% or just over.

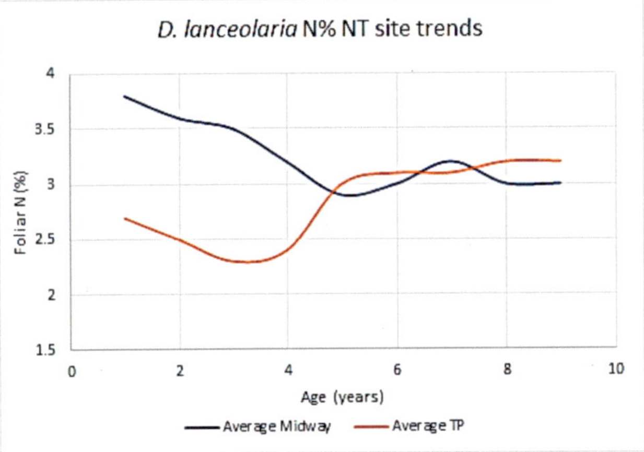

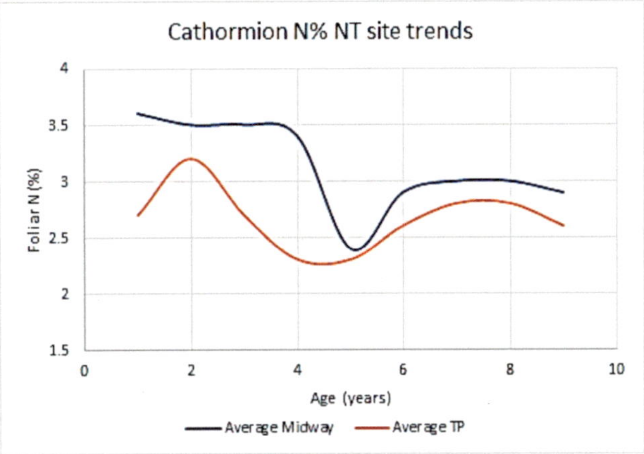

Figure 124: Average foliar nitrogen levels for two key long-term host species at two sites over time.

Phosphorus (P)

Phosphorus is an essential nutrient providing a key component of plant structure compounds and as a catalysis for conversion of a range of biochemical reactions. It is responsible for stimulating growth of roots, increasing stem strength, plays a role in the production of flowers and seeds, and improves resistance to pests and disease.

Soils

Although the nutritional status of the soil under the Quintis plantations have been discussed in Chapter 3, Site Characteristics, it will be beneficial to summarise. As mentioned, the total phosphorus content of most surface soils is low and Quintis sites were no different. In soils with a significant clay content, phosphorus will react with the clay compounds and are bound up or fixed and not available to plants. This reactivity between applied phosphorus and the soil is called 'phosphorus buffering capacity' and is measured by the Phosphorus Buffering Index (PBI) soil test. Soils with a high PBI value will require more phosphorus fertiliser than a soil with a low PBI.

Due to fixation, phosphorus moves very little in most soils and will stay close to its place of origin. Less than 20% of soil phosphorus is absorb in the first season after application (Mosaic, 2023). Leaching does not cause a loss of soil phosphorus. The fixed, residual phosphorus remains in the rooting zone and will be slowly available to trees over several years. Soil erosion and crop removal are the main ways soil phosphorus is lost.

The Colwell P test, used by Quintis, value is independent of the soil's ability to bind phosphorus. Thus, when interpreting the Colwell test it is important to consider the PBI. The Colwell P value required for optimum tree production increases as PBI values increase.

Crop

Phosphorus deficiency is more difficult to diagnose than that of the other macro nutrients. Plants do not usually display obvious symptoms other than a general stunting of the plant during early growth. By the time a visual deficiency is recognised , it may take a long time to correct.

Phosphorus is highly mobile in plants and, when deficient, it may translocate from old plant tissue to young, actively growing areas. So early vegetative responses to phosphorus are often observed. As a plant ages, phosphorus is translocated into the new growth or fruiting areas of the plant, where high-energy requirements are needed.

Sandalwood

Phosphorous is one of the main elements that Sandalwood obtains from its host (Iyengar A. , 1960). When the average foliar phosphorus percentage for Sandalwood is observed, there is a trend, where foliar phosphorus at establishment increases from age one to two. It then rapidly drops off at age three and afterwards slowly starts to increase again (Figure 125). This is a clear indicator that Sandalwood is predominately obtaining phosphorus from the hosts. The variation is the Sandalwood switching hosts as the STH die off from age two to three and the Sandalwood finalising transition to long-term hosts at about age six.

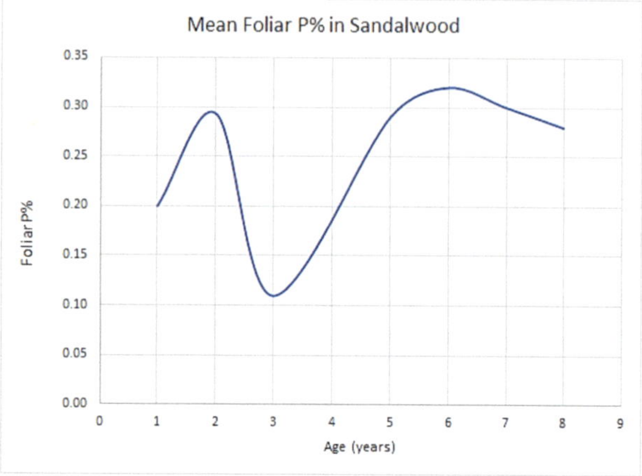

Figure 125: Average foliar P data collected from Sandalwood within the Quintis estate.

Hosts

The phosphorus foliar levels in the exotic host *Dalbergia lanceolaria* have always been lower than the native Cathormion. There was an upward trend in foliar phosphorus for Cathormion until age three and then a decline. It could be theorised that the decline was caused by Sandalwood hosting from that age.

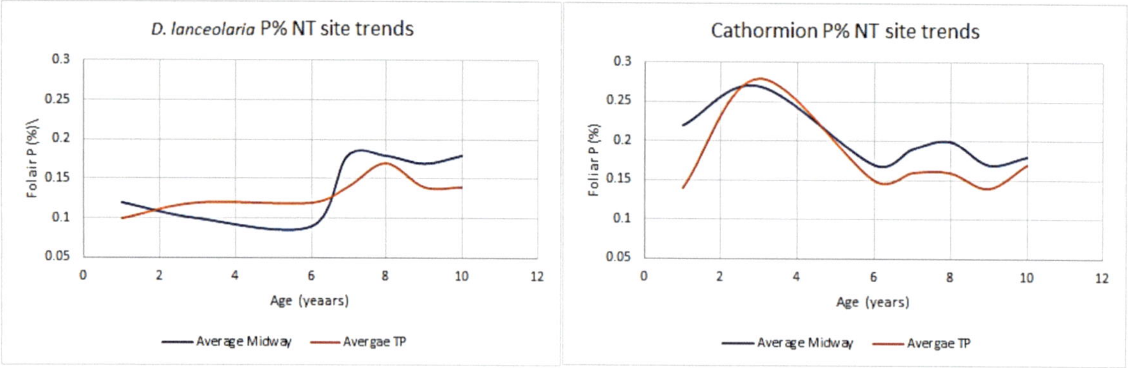

Figure 126: Average foliar phosphorous levels for two key long-term host species at two sites over time.

An operational granular application of 60 kg/ha of elemental phosphorus was administered at age six, resulting in an increase in foliar phosphorus for both species. However, there has been a more pronounced response from Dalbergia. The fertiliser application at Midway has doubled the mean level of foliar phosphorus in Dalbergia.

Potassium (K)

Potassium plays a vital role in the movement of water, nutrients, and carbohydrates within plant tissues. It activates enzymes involved in the synthesis of proteins, starch, and adenosine triphosphate (ATP). Potassium also regulates stomatal opening and closing, thereby controlling carbon dioxide uptake and enhancing photosynthesis. In addition, it is essential for osmoregulation, helping maintain water balance and the concentration of salts within plant cells and tissues.

Soils

Most Australian soils contain adequate levels of potassium (Gourley, 1999). However, most of the soils with Sandalwood plantations on Kandosols, potassium in the top 60 cm is low. Almost all the soils in Northern Territory and Queensland are well below the recommended 0.3 to 0.6 cmol(+)/kg (Gourley 1999).

Some research has shown that potassium deficiency could promote the absorption of Na^+ and Ca^{2+} in species like maize (Du, et al., 2017), but could also inhibit nitrogen absorption in cotton and significantly reduce the content of NO_3 in the leaves (Xu, et al., 2020).

Crop

Potassium is an essential element in regulating water use in plants, particularly the opening and closing of the stomata. Sandalwood relies on mimicking its hosts transpiration and sap flow to facilitate nutrient transfer via the haustoria (Radomiljac, McComb, & Pate, 1999), control of stomata is critical.

Furthermore, potassium is mobile in plants and will move readily from older leaves and root tissue to growing points, deficiency symptoms will first appear in recently matured and older leaves.

Sandalwood

Foliar potassium percentage in Sandalwood should not be below 1.2% and generally, on average it stays above this level.

It is suggested that Sandalwood rely on hosts for a proportion of its potassium (Rangaswamy, Jain, & Parthasarthi, 1986). Like phosphorous, albeit over a longer period, when the foliar potassium trend is observed, there is a drop off during the transition period and then a slow recovery. As the tree ages the level of potassium increases and by age ten, the mean percentage is similar to when it was a seedling.

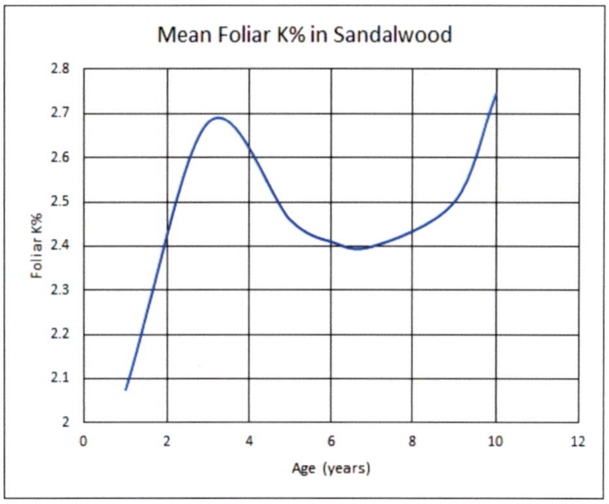

Figure 127: Average foliar K data collected from Sandalwood within the Quintis estate.

Host

Although the levels of potassium in the soils tend to be slightly deficient, generally the levels of foliar potassium in both *Dalbergia lanceolaria* and Cathormion, on average, are well above the minimum levels. The *Dalbergia lanceolaria* would be considered deficient if the foliar levels fell below 0.4%, and 0.16% for the Cathormion.

The foliar trend for both hosts is similar to Sandalwood, with a decline from age three and then increasing from age seven.

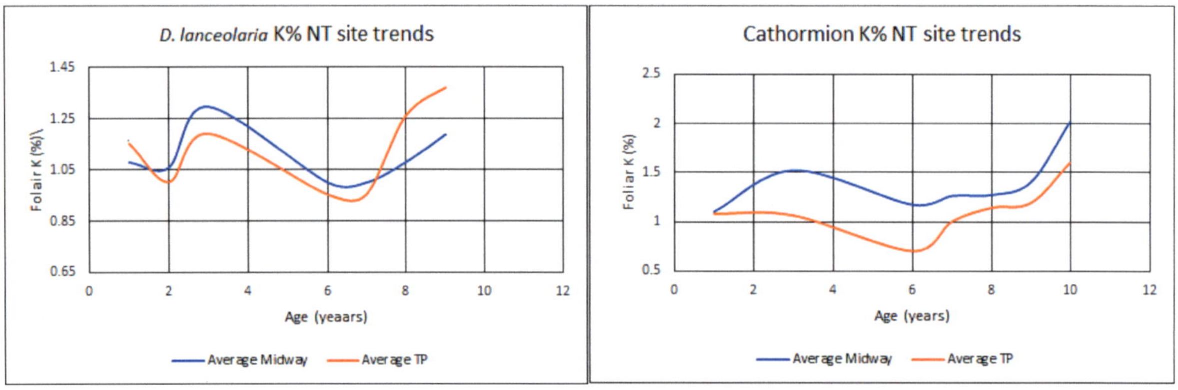

Figure 128: Average foliar potassium levels for two key long-term host species at two sites over time.

Other macro nutrients.

Calcium is an essential element needed for growth and development of plants under both non-stressed and stress conditions. It thereby fulfills a dual function, being not only a key factor for cell wall and membrane stability but also serving as a second messenger in many developmental and physiological processes, including the response of plants to biotic stress.

The Kandosol and Tenosol soils in Northern Territory tend to be low in calcium, whereas the Vertosols in ORIA are high. However, the foliar data from all species in the Northern Territory shows calcium levels similar to the trees in ORIA, well above the recommend 0.3% minimum.

Magnesium fulfills several functions within the plant, as it is a principal component of chlorophyll which is supporting the function to absorb sunlight during photosynthesis. Magnesium acts as a phosphorus carrier in plants and is essential for phosphate metabolism.

Magnesium is low in most of the soils with plantations but is at sufficient foliar levels in all species across all sites. Magnesium percentage in Sandalwood should not be below 0.1%.

Sodium is not an essential element for plants but can be used in small quantities, to aid in metabolism and synthesis of chlorophyll. The amount of exchangeable sodium in soils with low levels of total salts are called sodic soils. Some of the ORIA and Dalbeg soils are sodic. The measure of high exchangeable sodium percentage (ESP) in these soils indicates several soil structure issues that can affect plant growth. High sodium soil levels can also restrict the availability of calcium, magnesium, and potassium for uptake by plant roots. However, this does not seem to be happening in the plantation trees on these sodic soils. The plantation 'Woods' in Dalbeg had ESP levels as high as 12 but the foliar analysis shows there were no deficiencies in calcium, magnesium, or potassium. The sodium percentage in Sandalwood should range between 0.1% to 1%.

Sulfur in plants help form important enzymes and assist in the formation of plant proteins. It is needed in small amounts, but deficiencies can cause serious plant health problems and loss of vitality. Sulfur percentage in Sandalwood should range between 0.07% to 3%.

b) Micro elements

Trees need micro elements in smaller amounts. They should only be added if deficiencies can be found. However, some deficiencies of micro elements are due to deficiencies in macro elements. If the deficient macro elements can be remedied, then, in many cases this can stabilise the micro elements.

Boron plays a key role in a diverse range of plant functions. This includes cell wall formation and stability, maintenance of structural and functional integrity of biological membranes, movement of sugar or energy into growing parts of plants, and pollination and seed set.

Quintis did find boron toxicities in Aquitaine Vertosols in ORIA where it had caused host mortality. Although the Kandosol and Tenosols soils tend to be low in boron, generally the foliar boron levels are in the sufficient range for most species. Boron mg/kg in Sandalwood should range below 10 to 110 mg/kg.

Zinc is a component of the enzymes that control growth. Carbohydrate, protein, and chlorophyll formation is significantly reduced in zinc deficient plants. Zinc deficiency for agricultural crops is found in about one third of worldwide soils due to low total zinc concentrations. Soils containing less than 0.5 mg/kg extractable zinc are potentially zinc deficient (Mertens & Smolders, 2013). Generally, zinc levels are low in the plantation soils with most having less than 0.5 mg/kg extractable zinc. There some evidence of zinc deficiencies in Sandalwood, with 16% of the foliar database showing levels in the deficient range. All of the trees with low levels of zinc are planted on the sandy soils in Northern Territory.

The remaining micro elements are all well within the required foliar range in all species on all sites. Copper is an essential part of reproduction. It aids a tree in absorption and digestion of nutrients through the roots. Chloride, like copper, improves the tree's metabolic processes. Iron is required to create chlorophyll; manganese supports enzymes in digesting carbohydrates and converting nitrogen and molybdenum is utilised for transforming nitrogen.

2. DEFICIENCY SYMPTOMS

a) Nitrogen (N)

A nitrogen shortage causes the loss of green colour from the leaves, resulting in an even paleness (chlorosis). A deficiency of nitrogen makes the leaves pale and small. Old leaves shed early, causing the foliage cover to become thin, and twigs to die back.

Sandalwood can become nitrogen deficient during the transition process or if the host is lost to disease. Growth will slow and trees will lose leaves, the crown becomes thin and the remaining leaves will turn yellow.

However, Sandalwood leaves can have a yellowish tinge to them when exposed to full sunlight when young. This disappears as they age. This yellowing should not be confused with nitrogen deficiency. If in doubt a foliar analysis should be completed.

Figure 129: Sandalwood tree without a host showing classic N deficiencies, (TP8, Robson, 2017).

b) Phosphorus (P)

Phosphorus is essential for the development of flowers, fruit, and roots. It is a mobile nutrient and is moved from older leaves to the newly developing tissue. Older leaves turn a darker green followed by a purplish tint starting from the leaf margins. Leaf tips dry off. The occurrence of red leaves which is not normally found in plants, is caused by anthocyanins. These are natural pigments commonly found in plants that can act as metabolic markers of nutrient deficiency, especially phosphorous deficiency. Researchers have demonstrated that phosphorus deficiency can induce anthocyanin accumulation (Huiqiong , Kunrong , ZhiQiang Zhang,, & Yanru , 2023) .

Figure 130: The leaf symptoms for P, K and Zn deficiency are very comparable with burning of tips or edges and red colouring, (Robson, 2019).

c) Potassium (K)

Potassium deficiency symptoms can vary. They are often not easy to recognise and can be mistaken for other problems. Symptoms include slower tree growth, marginal and interveinal burns on leaves, leaves turning yellow or purple, and smaller than normal leaves, leading to heavy leaf fall.

d) Zinc (Zn)

Zinc deficiency, leaf symptoms include small, narrow leaves, developing interveinal chlorosis. Leaves crowd along short stems (rosetting), and smaller twigs die back. Symptoms are often more pronounced on the sunny side of the tree.

Figure 131: The leaf symptoms for Mn, Fe and Mg and Zn deficiency are very comparable with yellow leaves with green veins, (Robson, 2019).

e) Manganese (Mn)

Manganese deficiency causes a diffuse pale green (chlorosis) mottle between the veins in young and old leaves. The leaf size is normal. A narrow band remains green on each side of major veins. Symptoms are more noticeable on the southern side of the tree.

f) Iron (Fe)

The plant needs iron to produce chlorophyll. Iron deficiency causes a distinct pattern in the leaf caused by a loss of chlorophyll, only the main veins stay green. Iron does not move easily within the plant, so young leaves are the worst affected, showing the symptoms, while older leaves may remain green.

g) Magnesium (Mg)

Magnesium deficiency is most common in leached acidic sandy Tenosols. Magnesium deficiency produces a very distinct yellowing towards the apex of the older leaves, with a triangular area remaining green at the leaf base and affected leaves fall prematurely. The defoliation can be heavy.

Figure 132: The leaf symptoms for Na, Cl and B toxicity, (Llorente, 2020).

h) Sodium (Na) and chloride (Cl) toxicity

Salt toxicity causes leaf burn and yellowing, starting near the tip. Leaf fall is heavy, and dieback follows. Older leaves show the symptoms first.

High sodium levels in the soil upsets tree nutrition, which leads to reduced potassium uptake, so potassium symptoms can sometimes be shown.

i) Boron (B) toxicity

Boron toxicity (like salt toxicity) starts as tip yellowing and tip burn, but subsequent yellowing of the apex tends to be mottled. It progresses into a yellowing of inter-veined areas near the tip. Leaf fall is heavy, leading to dieback.

Foliar baseline

The use of foliar analysis in tree crops to determine elemental nutrient deficiencies is well-practised in horticulture and somewhat in forestry to a lesser degree. This method of diagnostic analysis can be used to great efficiency to determine fertiliser regimes to fix nutrient imbalances and improve growth and tree health. It can also be used to target specific areas of concern.

There are several references available which document the desired levels of both macro and micronutrients for a range of tree crops (Reuter & Robinson, 1997). These tables have been developed over a lengthy period of time collecting data. The lack of available data for Sandalwood and associated host crops has hindered the use of a valuable diagnostic tool for foresters.

In 2016 the Quintis research group recognised the need to develop a database of elemental requirements for Sandalwood and associated host crops and started collecting annual foliar data from healthy actively growing trees within the plantations. Table 26 shows the baseline requirements each of the nutrients tested in a foliar analysis. A healthy tree will have foliar nutrients within the specified levels in the table. For trees exhibiting nutritional symptoms, as described above can be foliar tested and results compared with the optimal data in table 26.

Table 26: The optimal baseline for nutritional elements for Sandalwood and key Quintis LTH species.

	Santalum album	*Senna Saimea*	*Cathormion unbellatum*	*Dalbergia laneolaria*
	optimal	optimal	optimal	optimal
N (%)	2.1 - 5.5	2.1 - 3.4	2.1 - 4.3	2.3 - 4.6
P (%)	0.18 - 0.7	0.16 - 0.45	0.11 - 0.35	0.11 - 0.35
K (%)	0.9 - 5.0	0.9 - 2.4	0.16 - 2.0	0.4 - 2.0
Ca (%)	0.3 to 3	0.35 to 4.0	0.12 to 3.0	0.6 to 3.5
Mg (%)	0.1 to 1	0.1 to 0.4	0.1 to 0.6	0.15 to 1.0
Na (%)	0.01 to 1	0.004 to 0.5	0.004 to 0.6	0.004 to 0.5
S (%)	0.07 to 3	0.12 to 0.3	0.12 to 0.3	0.12 to 0.4
B (mg/kg)	11 - 110	11 - 110	6 - 60	11 - 110
Cu (mg/kg)	2.1 - 10	4.1 - 12	3.1 - 15	5.1 - 30
Zn (mg/kg)	11 - 45	9 - 35	4.1 - 40	4.1 - 40
Mn (mg/kg)	15 to 150	15 to 170	20 to 250	30 to 650
Fe (mg/kg)	20 to 250	20 to 250	30 to 250	40 to 650
Al (mg/kg)	4 to 200	10 to 200	6 to 250	10 to 430
Co (mg/kg)	0.06 to 1	0.06 to .5	0.05 to 0.5	0.06 to 0.2
Mo (mg/kg)	0.2 to 6	0.15 to 2	0.004 to 0.7	0.2 to 1.5
Cl (mg/kg)	0.25 to 2	0.05 to .5	0.1 to 1	0.05 to 0.15

3. QUINTIS NUTRITION RESEARCH

Experience with fertiliser application in the first four years after establishment has shown little benefit and, in most cases, it is not economical to apply. Historically little thought has been given to additions of fertiliser. It had been speculated that once Sandalwood links with the long-term hosts, nutrition supply will be sorted for the rest of the rotation. This may be the case on some sites, particularly if there has been a fertiliser history from previous crops or when there has been early age mortality and stocking is reduced from establishment. The issue of stocking exceeding site capability can affect nutrition. Currently a high percentage of the estate has had incredibly low establishment mortality, and high stocking of both Sandalwood and hosts. High stocking potentially equals inadequate site capacity. Hence, from age five onward there is evidence of what appears to be nutrition stress in some plantations.

Since 2008 there have been over 30 field trials investigating the nutrition of plantations. Most of those trials have looked at granular fertiliser rates, and to a lesser extent liquid fertiliser or biostimulants. This next section will discuss the results for 15 of these trials.

a) Granular fertiliser trials

There were several experiments established between 2005 and 2008, however due to incomplete records of what fertilisers or what rates were applied, they cannot be reported on. The following trial summaries are for experiments that have been well recorded.

Fertiliser trials at establishment

Trial 1: One of the first replicated fertiliser trials to be established was a granular fertiliser trial on newly established sites at Packsaddle Road plantation, ORIA, in 2008. This trial compared the application of a MAP tablet, 50 g of complete fertiliser (NPK+ TE) and a nil control on two soil types, red and black soils. Unfortunately, the elemental rates for these fertilisers were not recorded, so the actual application rate is unknown. However, an analysis of the results after 12 months showed there was no significant difference in the fertiliser over the nil control.

Figure 133: Trial 1 chart showing results for Sandalwood fertiliser treatments after 12 months, (TFS research experiment TSR39).

Trial 2: A second trial established in 2010 was designed to investigate the application rate of a granular fertiliser applied at establishment. This experiment was established on a new site in ORIA. The soil was a high quality Cununurra clay soil that had been previously used for horticultural cropping. The experiment compared a high phosphorous fertiliser with increasing rates.

A mono-ammonium phosphate (MAP N10%: P21.9%) fertiliser, which only has nitrogen and phosphorous elements, was applied at three increasing rates, 100, 200 and 400 kg/ha or 21.9 kg P/ha, 43.8 kg P/ha and 65.7 kg P/ha. A nil control was included in the trial.

The results after 17 months demonstrated no significant differences in treatments over the control for either Sandalwood or Cathormion.

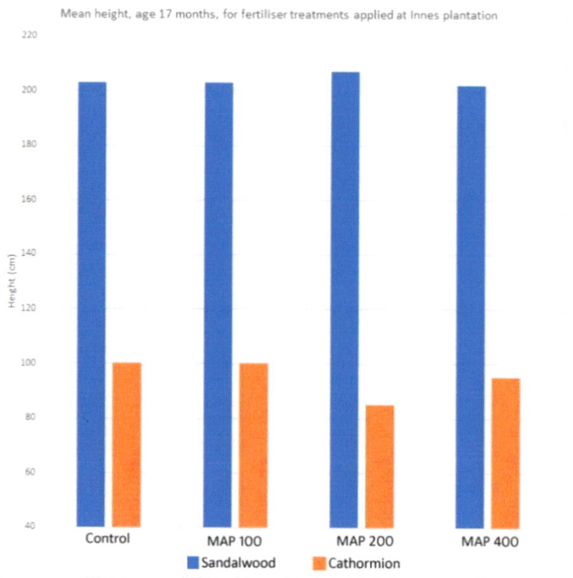

Figure 134: Trial 2 chart showing differences in height growth for 2 species when treated with increasing fertiliser, (ITC research experiment).

Trial 3: The first granular fertiliser experiment in Northern Territory was established at Eagle Park, in 2012. This is a site with sandy, nutritionally low Tenosol soils. A comparison of two establishment fertilisers, Campbells 'Rustica plus' is a general purpose NPK (12% nitrogen, 5.0% phosphorus, 14% potassium, 9.7% sulphur, 3.1% calcium, 1.2% magnesium, 0.1% zinc and 0.02% boron) fertiliser and Campbells 'Platinum Plus', also a general purpose NPK (0% nitrogen, 11% phosphorus, 10% potassium, 6% sulphur, 4.3% calcium, 1.2% magnesium, 0.36% zinc, 0.1% boron) fertiliser. The rates used were 100 kg and 200 kg/ha of product applied around each seedling after planting.

The NPK fertilisers did have significant improvements for Sandalwood but not Cathormion height growth after 12 months. There were no differences between the two fertiliser products, or the rates applied (Figure 135).

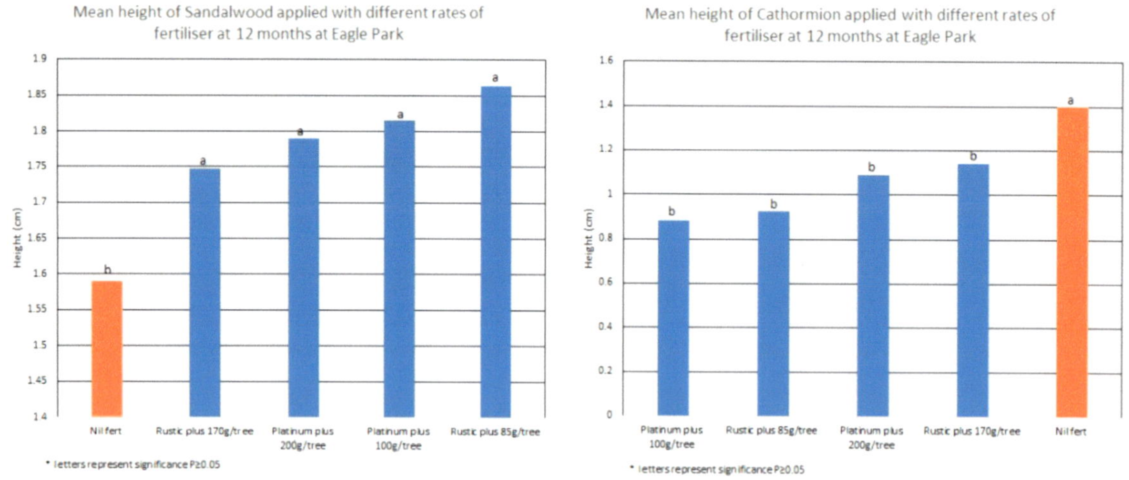

Figure 135: These charts show mean height results for Sandalwood and Cathormion, 12 months after treatments application. The data was analysed with an ANOVA and the letters representing significant differences between the means (P≥0.05), (Quintis research experiment R12-03).

Trial 4: On the same site, a second trial was established investigating increasing rates of specific nutrients, nitrogen, phosphorus and potassium applied as Triple super, Urea and Potash. The rates applied were single plots of nitrogen at 12 kg/ha, phosphorus at 26 and 40 kg/ha, and potassium at 24 and 40 kg/ha. There were also treatments that combined elements (Figure 136).

Trt No	fertiliser
1	20kg/ha Phosphate
2	40kg/ha Phosphate
3	24kg/ha Potash
4	40kg/ha Potash
5	12kg/ha Nitrogen
6	40kg/ha K + 20kg/ha P
7	24kg/ha K + 20kg/ha P
8	40kg/ha K + 40kg/ha p
9	24kg/ha K + 40kg/ha P
10	20kg/ha P + 12kg/ha N
11	24kg/ha K + 12kg/ha N
12	24kg/ha K + 20kg/ha P + 12kg/ha N

Figure 136: treatments applied to Eagle Park experiment.

An assessment of height was taken after 12 months. The results for total tree height were analysed for significant differences between treatments by a one-way ANOVA and compared means with an LSD (0.05) comparison.

There were some differences in the treatments, however, the fertiliser did not improve the height growth of either Sandalwood of Cathormion (Figure 137).

This experiment demonstrated that fertiliser requirements vary considerably among species within the plantation. The relative performance of fertiliser treatments differed for each species, making it difficult to define a single fertiliser regime suitable for all species.

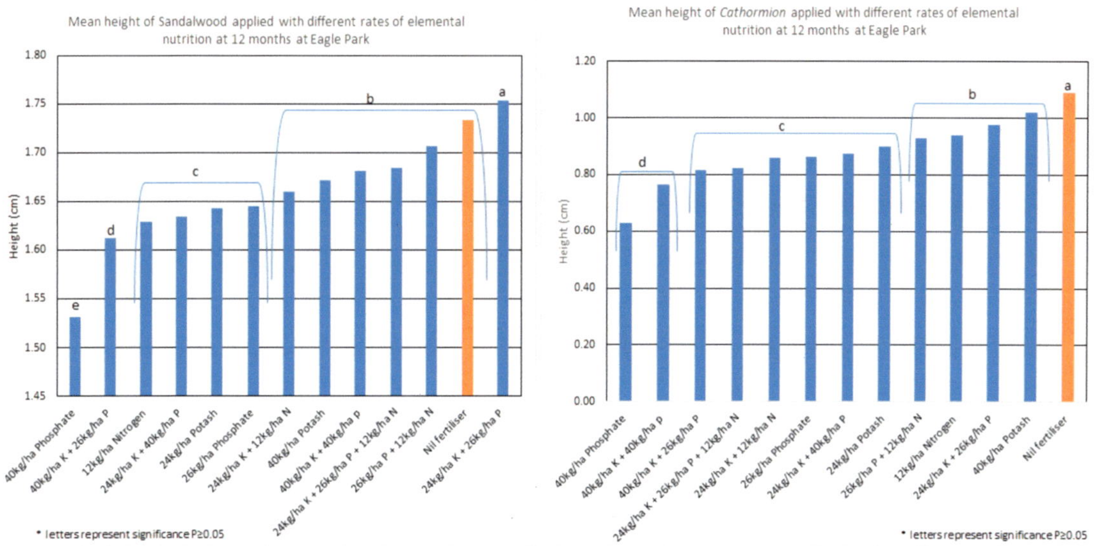

Figure 137: Charts show mean heights for Sandalwood and Cathormion. The ANOVA analysis can be seen with letters representing significant differences between the means (P≥0.05), (Quintis research experiment R12-02).

Trial 5: This experiment was established on an inferior quality one year old Sandalwood plantation at Sandy Creek, planted in 2015. With similar soils to Eagle Park, the soil on this block is a sandy Tenosol. Poor growth and foliar analysis indicated possible nutritional issues as evidenced by all species growing below the expected rates. Many of the short-term hosts had already died. The foliar analysis indicated there were deficiencies in phosphorus, potassium, and copper. The treatments applied were phosphorous and potassium at medium to high rates (Table 27).

Growth data was analysed two years after application with one-way ANOVA and using an All-Pairwise Comparisons Test (LSD P≤0.05). The results showed that two years after application all Sandalwood phosphorus treatments had grown significantly more than those treatments without phosphorus. The more phosphorus applied the better the result. The long-term host, *D. lanceolaria* and Cathormion both did better in the treatments with high phosphorus and potassium.

This trial has shown that additions of phosphorus will increase growth parameters. For most species, the more phosphorus applied the better the response. For the plantations planted on Tenosol (white loam) soils, an application of at least 450kg/ha of Triple Super (90kg/ha P) will achieve significant improvement in growth of all species.

Table 27: Treatments applied in rate trial at Sandy Creek Plantation.

Treatment	elemental P (kg/ha)	elemental K (kg/ha)
Low P	81.3	-
medium P	162	-
High P	243.6	-
Low K	-	8
medium K	-	15.9
High K	-	31.8
Low P&K	81.3	8
Medium P&K	162	15.9
High P&K	243.6	31.8
Nil	-	-

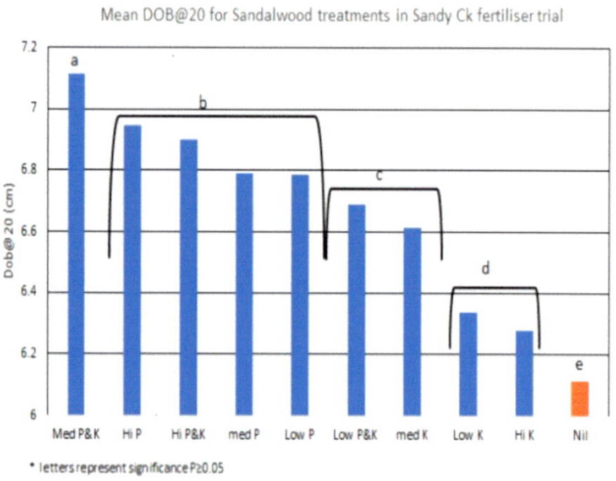

Figure 138:Mean Sandalwood diameter in treatments, two years after application, (Quintis Research experiment R077).

Fertiliser trials at mid rotation

These are trials that were established on plantations that had short-term host mortality and Sandalwood had either transitioned across or were transitioning across to the long-term hosts.

Trial 6: A granular fertiliser rate trial was established in the Top Banana plantation in ORIA in 2010. This was a comparison of two fertilisers at age five. Increasing rates of MAP (10% nitrogen and 21.9% phosphorous) and an OrdCo Special (12% nitrogen, 5.2% phosphorus, 14.1% potassium, 6% sulfur, 4.3% calcium, 1.2% magnesium, 0.1% iron, 0.02% boron) at two rates and a nil control. Results after 12 months showed no significant differences in treatments over the control.

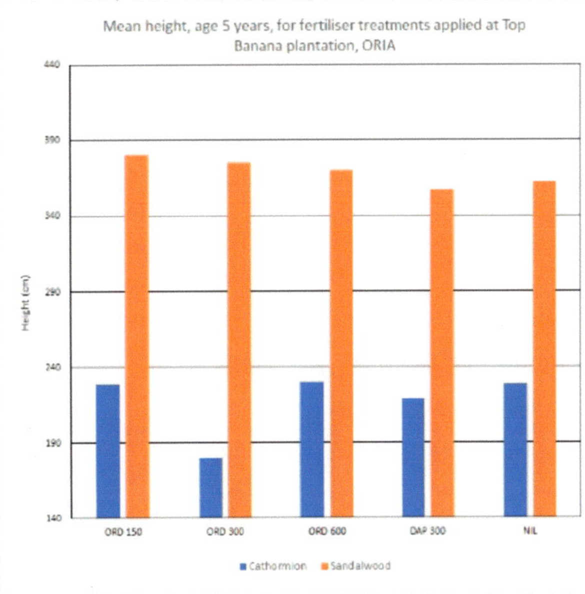

Figure 139: MAP application to five year old plantation in ORIA did not influence tree growth, (Elders Forestry Experiment).

Trial 7: This trial was established at Kingston Rest in 2014, investigated the application of zinc sulphate rates to Senna with declining health. There was also a control and an NPK +TE treatment. The results found that the NPK treatment had a significant effect on the health of the Senna. Application of elemental N at 50kg/ha, P at 63kg/ha and K at 60kg/ha, had a positive effect on Senna crown development. There was no visual difference in the zinc treatment.

Trial 8 and 9: Application of superphosphate on seven year old Senna at PSR 8 in 2013 and ten-year-old Senna at PSR 10 in 2015 investigated the effect of increasing rates of superphosphate (0, 20, 40, 60 & 80 Kg Super/ha) applied by plane. The assessment was foliar leaf analysis and visual health assessment undertaken three months after application.

There were no obvious differences between treatments, but foliar analysis did show an increase in levels of foliar phosphorus. Trial 7 at Kingston Rest showed Senna responded to NPK+TE. It was most likely that the response was to nitrogen and phosphorus, and the results in Trial 9 confirm that. The drop off in Senna nutrition that was observed as they age is most likely due to trees exhausting either the nitrogen or the phosphorus in the soil profile.

Trial 10: A phosphorus application trial across two different sites in the Northern Territory was established to determine the appropriate fertiliser rates and placement/application for trees to obtain the best growth. The trial investigated two moderate rates of phosphorus (60 and 100kg/ha) using monoammonium phosphate (MAP). It also examined the effect of the placement method, applied along the host row near the drip tape (row application) and broad spread across the inter-row (broadcast application) to determine the optimal way for operation application and tree uptake. A liquid fertiliser application treatment high in phosphate using Guano LG was also trialled on top of the granular treatments to compare responses.

An assessment in the third year after application on both sites, showed that the application method was significant between Sandalwood and hosts. Sandalwood had a better response with broad spread and hosts when fertiliser was applied along the row. It is known that a high proportion of the Sandalwood and host connections are in the inter-row (Robson K. , 2019), and the application of fertiliser at this point had better response for the Sandalwood uptake of nutrient.

The Sandalwood demonstrated significant response in the first year after application with 60kgP/ha of granular fertiliser. An additional 40kgP/ha in some treatments to increase rate to 100kg P/ha the following year did not translate to a better diameter increment on Sandalwood trees on these treatment plots. The addition of the liquid fertiliser in the first year also did not significantly improve Sandalwood growth.

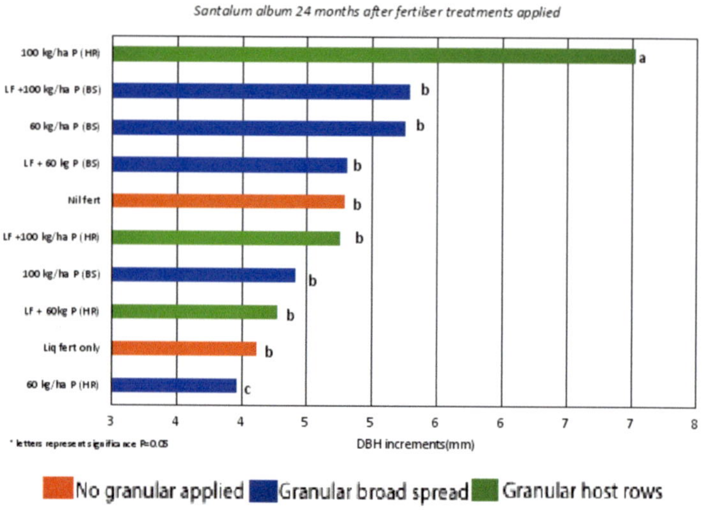

Figure 140: Diameter increments for P rate treatments on Sandalwood after two years, (Quintis research experiment

Although all the hosts improved over the control with the added 60kg P/ha, the along the row application was significantly better. The additional 40kgP/ha in some treatments to increase rate to 100kg P/ha the following year did not significantly translate to a better diameter increment.

Although the application of fertiliser did positively affect diameter growth, some treatments had significant host mortality that reduced stand stocking (tree density). It was found that this reduced stand density had a similar effect to the additions of nutrients.

b) Liquid fertiliser trials

Trial 11: In 2013 an experiment was established in a new plantation at Taylors Park looking at the application of nutrition through the irrigation system. The soluble fertiliser was Campbells 'Diamond special' (8.2% nitrogen, 3.9% phosphorus, 32.2% potassium, 4.2% sulphur, 1.7% magnesium, 0.15% iron, 0.075% manganese, 0.028% boron, 0.023% zinc, 0.017% copper, 0.011% molybdenum) was applied fortnightly at 10L/ha from establishment. The assessment investigated total height growth after three months. The results indicated that Sesbania was the only species that had a significant response to the fertigation. In fact, most of the species were doing better without the added fertiliser.

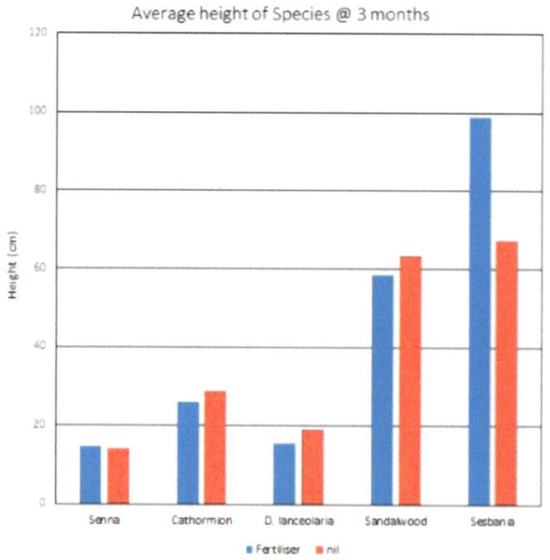

Figure 141: The results from the fertigation trial at establishment indicated that only Sesbania was benefiting from the addition, (Quintis research experiment OPS19).

Trial 12: The ability for Sandalwood and host to quickly develop good root systems is critical for good early survival and continued growth. There are a number of liquid fertilisers on the market that claim to increase root growth when used. This experiment tested the claims for two of these products, Super Stan Phos (Agrichem) and N-Fix (Sabroll). The application of Super Stan Phos to Sandalwood and Dalbergia seedlings applied as a 30 second dunk prior to planting significantly increased height growth over both the control and the N-fix treatments. There was no response to the fertilised treatments for the Acacia, Senna or Cathormion.

Trial 13: There are many products on the market that claim to enhance and stimulate early root growth that result in improved early survival and establishment. These are not fertilisers, but bio stimulates, and their ingredients usually include seaweed products that have amino and humic acids. Their role is not to increase the soil nutrition but improve the soil and micro flora so that the trees have more opportunity to take up what nutrients are available. Quintis operationally used the Agrichem product "Super Stand Phos" at 3L/ha soon after establishment.

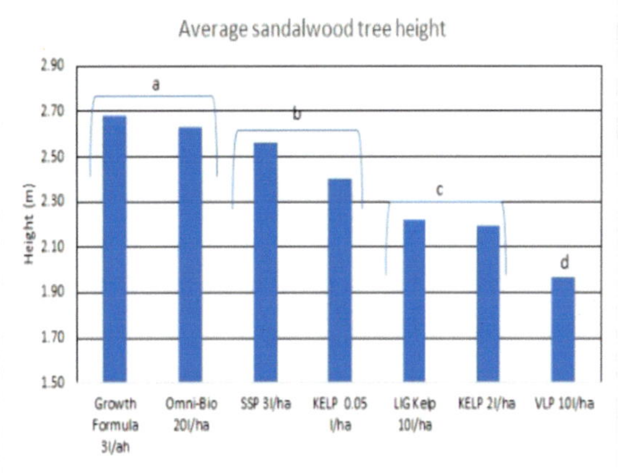

Figure 142: Average Sandalwood tree height for each treatment after two years, (Quintis research experiment OPS 34).

An experiment was established to test the benefit (or not) of Super Stand Phos in a controlled experiment by comparing it with several other comparable products. This trial was established at Roper Plains in 2016 and tested six available bio stimulate products in a one year old plantation on nutritionally poor Kandosol soil.

An assessment was undertaken two years after treatment application. The assessment looked at total tree height and a subjective tree health score. The results were analysed with a one-way ANOVA and a comparison of means (LSD≤ 0.05).

With mean tree height there were four significant groups, the best performing treatments were Omni bio, Growth Formula and Super Stand Phos (SSP) all with a mean height growth of over 2.5 metres. Omni bio and Growth Formula with an average significantly taller trees. In addition, trees were subjectively assessed with a five point health score (1 dead to 5 excellent health). There was no significant difference between treatments.

Trial 14: The health of Sandalwood in plantations on sandy soil can be significantly reduced when short-term hosts die off and trees are transitioning to the long-term hosts. Bio stimulates like Seasol advertise that they can improve root growth. If there is an increase in root activity in Sandalwood and hosts, this could potentially shorten the transition phase as trees will connect sooner. An experiment to examine this was established comparing two bio-stimulate products, Seasol and Rhizovator.

Results indicated that the application of liquid Seasol and Rhizovator to the growth of Sandalwood trees (in transition during the trial duration) improved significantly compared to those without it. Recurring

Seasol and Rhizovator applications over ten months improved Sandalwood diameter increment by 18 to 20% and 7%, respectively, 17 months after the first treatment application.

Moreover, this study suggests that Seasol would provide good remedial action for Sandalwood stands undergoing stress, such as stands in transition. In this case, the application in liquid form of the readily available nutrients and growth hormones present in the product have shown to be beneficial.

Trial 15: On some sites there was a slow depletion of phosphate from the soil bank. This correlated with there being a slow drop in tree foliar phosphorus. In the plantations, the species where this is most common is Dalbergia. A trial was established at Mustang Hill to see if the application of phosphoric acid (Phos acid) can improve foliar phosphorus. A formulation of phos acid and trace was developed in consultation with Agrisol fertiliser consultancy.

The initial experimental design was originally designed to apply 60 kg/ha of phosphorus, 20 as granular and 40 as liquid Phos acid. However, a total of only 44 kg/ha was applied (24kg/ha granular and 20kg/ha in four irrigated applications). The reduction was based on the theory that the total reduced application of P/ha will still see a significant response in tree growth and health, even though the amount of phosphorus is lower than the recommended. Liquid phosphorus in this form is available for plant uptake immediately, where granular can take much longer for plant uptake (Holloway , et al., 2001). There is a loss of phosphorus by leaching and uptake by grasses in granular applications over time. The sandy soils have a low soil Phosphorus Buffering Index (PBI) and the soils across the Northern Territory plantation estate indicate the likelihood of phosphorus loss due to leaching is high. This loss, the actual available phosphorus would be lower than the applied 60 units.

In this trial there was no nil control to gauge the response, however there was a significant increase in foliar phosphorus in the treated stands. Prior to application, in Sandalwood all treatments had foliar levels less than 0.15% P, after ten months this has increased to almost 3%. The foliar analysis of the hosts confirms a significant increase in foliar phosphorus levels. The considered minimum level for foliar phosphorus in Dalbergia and Cathormion is 0.11%. Prior to treatment application the mean levels in Dalbergia were 0.12% and in Cathormion 0.11%. After ten months of treatment application, all treatments had significantly increased foliar phosphorus. For the Cathormion, there was an increase in foliar phosphorus of 60% plus, and for the Dalbergia treatments there were up to a 77% increase.

c) Biochar trials

Biochar is a charcoal-like substance that is made by burning organic material from agricultural and forestry wastes (also called biomass) in a controlled process called pyrolysis. Although it looks a lot like common charcoal, biochar is produced using a specific process to reduce contamination and safely store carbon. During pyrolysis organic materials, such as wood chips, leaf litter or dead plants, are burned in a container with little oxygen. As the materials burn, they release little to no contaminating fumes. During the pyrolysis process, the organic material is converted into biochar, a stable form of carbon that cannot easily escape into the atmosphere (Spears, 2023).

Biochar can enhance soil structure, increase water capacity, improve porosity, regulate nitrogen leaching and improve microbial properties. As mentioned in Chapter 3 Site characteristics, all the soils planted within the Northern Territory plantations have poor soil structure, water holding capabilities and low organic content. The application of biochar to these soils would improve them and consequently improve the carry crop.

As Sandalwood is a mixed species crop where only Sandalwood is utilised, there is a substantial amount of left over biomass. This biomass can be turned into biochar. Quintis research did make and test biochar from excess biomass and it was of high quality (Gonzales, 2021).

Trial 16: A field experiment in collaboration with a Kununurra agricultural company established a trial using Quintis biochar by incorporating it into rows planted with melons and cotton. Although the biochar did not improve the productivity of the melon crop, it did significantly improve the cotton crop.

Trial 17: An experiment was established in inferior quality plantations at Kingston Rest and Drovers in Western Australia. This spread one tonne of biochar/ha onto the soil surface in selected plots. Twelve months after application, there was no significant difference in the treated plots over the control.

d) Summary of trial results

The trials established before 2010 were on the heavy clay Vertosols in ORIA, Kununurra. These soils have a medium to high PBI and therefore a bank of phosphorus available for uptake by the trees. Conversely, the high PBI could have potentially restricted available phosphorus because it was bound to the clay colloids in the soil. There were trials at establishment and at age five, with NPK and various phosphorus rates applied. All these experiments failed to show an improvement in the application of fertiliser. Due to the high PBI, the rates applied possibly were not high enough to see a response.

The development of the foliar nutritional database and baseline nutritional status has played a significant role in pointing where the fertiliser trials need to focus. The decline of foliar phosphorous, particularly in the long-term host has led to the establishment of most of the fertiliser trials.

The experiments on the sandy soils in Northern Territory with rates of phosphorus less than 60 kg/ha also failed to produce significant benefit to the trees. Trial 5 at Sandy Creek with medium to exceedingly high rates of phosphorus applied, demonstrated significant benefits to all species. This was also the case in trial 7 on red soil at Kingston Rest.

The large phosphorus rate and placement experiment, trial 10, show that the rate of 60kg/ha of elemental phosphorus are adequate to increase foliar phosphorus levels in all species. The trial also indicated that the best method of application for granular nutrition is by broad spread and not along the row.

The application of liquid phosphorous fertiliser has shown that an increase in foliar phosphorus levels can be achieved with lower applications of phosphorus per hectare. Additionally, the application of bio stimulant products have also shown to be beneficial particularly at establishment and during the transition phase of the plantation.

The application of biochar, although not specifically fertiliser trials, can assist with improving soil conditions and nutrient uptake.

4. OPERATIONAL PRESCRIPTIONS

The operational fertiliser procedure has been developed with input from the trials describes in the previous section in partnership with best practice fertiliser application. Which includes:

a) Nutrient matching withs soil and plant tissue analysis

Periodic soil testing is utilised to gauge nutrient sustainability for Sandalwood production. Soil test results indicate whether soil fertility is increasing, decreasing, or remaining constant and leads to responsible nutrient management decisions.

Plant foliar analysis provides a quantitative assessment of nutrient concentrations in plant tissue. Used alongside soil testing, it helps evaluate soil fertility and overall nutrient availability. Conducted during the growing season, foliar analysis can identify nutrient deficiencies and guide corrective action for current tree health. It is also a valuable tool in nutrient management planning. Sampling foliage from both healthy and poorly performing areas improves understanding of optimal nutritional requirements and informs management practices.

b) Nutrient application

Based on research results, fertiliser application principles have been developed that consider type, rates, frequencies, and site of applications.

Placement & Timing

An important part of optimising tree response to a fertiliser nutrient is placing the nutrient in such a way that it provides rapid uptake by the tree and reduces potential losses. The mobility of a nutrient in the soil is one of the major considerations in its placement. For example, low mobility of phosphorus in calcareous soils leads to short-term utilisation of the phosphorus that can generally be improved when it is placed close to the tree. However, due to the parasitic relationship with Sandalwood and host, and haustorial connections tending to be in the inter-row, there is some uncertainty as to optimal fertiliser placement for this crop. That said, placement can be a powerful management tool to minimise nutrient losses. Under ideal conditions, the goal is to apply nutrients in the plant-available form in reasonable proximity to the plant roots.

The demand for a nutrient by a growing tree generally varies through the growing season, with the highest uptake associated with the period of most rapid growth. Timing fertiliser applications to provide a plant-available supply of nutrients when the tree needs them is a desirable goal. Trees subject to a deficiency during specific periods of growth may not recover to achieve full yield potential.

5. PLANT NUTRITIONAL REQUIREMENTS

a) Observed deficiency symptoms.

With observations of what appears to be a nutrient deficiency, these tend to be in two groups:

- long-term hosts only with yellowing and poor crown development,
- Sandalwood only with yellowing and poor crown development.

Long-term hosts with yellowing and poor crown development.

Soil nutrition analysis confirms almost all the soils with plantation on, tend to be nutritionally low. However, foliar analysis for the species grown indicate that on most sites up to age five they are not deficient in any elements and any added fertiliser would not improve growth.

The symptoms observed that in some cases long-term hosts may not necessarily be attributed to poor nutrition but other factors such as poor water supply, extreme climate, pathogens and over hosting. However, if these reasons can be rule out, a foliar analysis and the application of a basic fertiliser may help.

If fertiliser is required, the method of application must be considered, foliar, fertigation, or granular. Liquid fertilisers generally have a short residual effect and are therefore applied at lower rates but more frequently. When delivered through the irrigation system, however, they cannot be targeted to individual species within the plantation.

Foliar and fertigation application generally will give an instant result and could be used in cases of extreme nutritional stress but for long-term nutritional supplements the use of granular would be more beneficial.

If it is decided to apply a granular fertiliser, then when calculating the application rate, it must be at a level which will be high enough to have an effect. The research trials have shown that higher applications primarily phosphorus (P) are required to see a response. Research Trial 4 applied elemental phosphorus at rates of up to 40 kg/ha and observed no response. In contrast, Trials 8 and 13, which applied 81 kg/ha and 60 to 100 kg/ha respectively, produced measurable responses. These results, together with findings from other species and sites, indicate that higher phosphorus application rates may be required to achieve a significant growth response (Keenan, Hambleton, Robson, & Webb, 1998).

Tropical Australian soils are known to be low in phosphorus (Wild, 1958) and nitrogen, and many native species have adapted to grow well in these nutritionally poor soils. Although most host species used by Quintis were not native, except for *C. umbellatum* and *A. trachycarpa*, all are nitrogen-fixing.

These are trees that can convert (fix) atmospheric nitrogen into a form that can be utilised in association with symbiotic organisms such as rhizobium bacteria. Foliar testing across a range of the Sandalwood plantation trees at various ages has shown that there is not generally any deficiency of nitrogen.

A range of "off the shelf" commercial NPK fertilisers, when tested in plantation field trials, have shown only moderate increases in growth and have not usually been consistent for the range of species grown within the plantation. The most likely reason for this has been that actual rates of elemental Phosphorus have been too low to see a long-term response.

Sandalwood with yellowing and poor crown development

It is rare to see Sandalwood in this condition while there are health active short-term hosts (STH). However it is quite common to see Sandalwood in this state when STH start to die off. This is put down to the transition stage, a period of poor growth and health whilst the Sandalwood find another host to attach to. All plantations go through this stage and will last a varying amount of time dependent on the site and growth of the trees. The transition period tends to be shorter in "Mixed" configurations compared to "Standard" configurations.

Fertiliser applied through drippers or as a foliar spray during this stage may be beneficial, and previous applications have produced favourable responses. Various combinations that have been used to beneficial effect include APP (3/ha), BioMax (10l/ha) with BioTrace (1l/ha), Supa Stand Phos (3L/ha), Kelpak (2L/ha), Seasol (10L/ha) and Phos Acid (10L/ha).

Within irrigation plantations, a considerable proportion of any fertiliser application would be lost due to leaching or runoff (particularly in the flood system). This may mean that any nutritional benefits gained with granular fertiliser application may not be long-lived and require regular "top-up" applications. Moreover, it could mean that rather than applying a high once-off rate, lower biennial or triennial application may be more beneficial.

The timing of the fertiliser application needs to be carefully considered. In dripper systems, the application should be timed so that rain can wash in the fertiliser. In flood systems, incorrect timing (and water speed) can see a high proportion of the spread fertiliser washed through into tail drains.

b) Granular fertiliser

For mid rotation (ages five to 12), in sites showing phosphorus deficiency symptoms, an application of a granular fertiliser high in phosphorus applied as broad acre is currently the best option. As mentioned previously, due to tree size, a sufficiently high rate may be required to see any result.

For a forestry crop, mid rotation fertiliser applications are rare, but on poorer soils it can be beneficial. However, due mostly to economic reasons, any fertiliser application is generally high but a once off.

MAP (mono-ammonium phosphate) has been an important granular fertiliser for many years. It is water-soluble and dissolves rapidly in adequately moist soil. Upon dissolution, the two basic components of the fertiliser separate again to release ammonium (NH_4^+) and phosphate ($H_2PO_4^-$). Both of which plants rely on for healthy and sustainable growth. The pH of the solution surrounding the granule is moderately acidic, making MAP an especially desirable fertiliser in neutral- and high-pH soils. Quintis did use two variants of MAP (Table 28). Triple super has been used but is difficult to get in Australia.

Table 28: Granular fertiliser Products and rates used by Quintis.

Product	Soluble	kg/ha	Elemental rate
MAP	yes	300	10% Nitrogen, 21.9% Phosphorus
MAP + Zn	No	300	10% Nitrogen, 21.9% Phosphorus, 1% Zinc
Triple Superphosphate	No	300	20% Phosphorus, 17% Calcium, 1.5% Sulphur

c) Liquid nutrition

Fertigation via irrigation system

Application of the previously mentioned fertigation can be used to give tree health an instant boost but longer-term applications of a granular fertiliser high in Phosphorus will be necessary for a longer-term response.

There are a range of options available for liquid nutritional application (Table 29). Fertilisers such as Campbells range which are fully water soluble, chloride free fertilisers containing the required nutrients for consistent growth and development of a range of plants providing immediately available nitrogen, potassium, phosphorus, sulphur, magnesium and trace elements. Applied at a fertigation rate for horticultural crops between 30 to 70kg/ha. It must be noted that care should be considered as all 'off the shelf' liquid general fertilisers do have nitrogen rates of between 10 to 15%.

As phosphorous is the main element required, a phosphorus specific fertiliser such as Agrisol Phos Acid Quintis mix, could be used, or one of the tech grade MAP fertilisers. Research trial 15, demonstrated that the Agrisol Phos Acid Quintis mix when applied in five applications prior to and after the wet season significantly increase Phosphorus foliar levels in all species.

Table 29: Soluble fertilises used by Quintis as fertigation.

Product	Soluble	L/ha	Elemental rate
Diamond 19	yes	30 to 70	19% Nitrogen, 8.0% Phosphorus, 16.0% Potassium, 1.7% Sulphur, 1.2% Magnesium, 0.04% Iron, 0.02% Manganese, 0.01% Boron, 0.020% Zinc, 0.011% Copper, 0.001% Molybdenum
Campbells Diamond P	yes	35 to 70	12% Nitrogen, 18.8% Phosphorus, 10% Potassium, 1% Sulphur, 0.1% Iron, 0.05% Manganese, 0.02% Boron, 0.015% Zinc, 0.011% Copper, 0.007% Molybdenum
Campbells Solu-MAP	yes	35 to 70	12.1% Nitrogen, 27.1% Phosphorus
Agrichem SUPA APP	yes	10 to 30	15% Nitrogen, 23% Phosphorus
Agrisol Phos Acid Quintis mix	Yes	15 to 20	26.4 Phosphorus, 2.6% Potassium, 1.78% Sulphur, 0.53% Magnesium 0.53% Iron, 0.05% Manganese, 0.55% Zinc, 0.25% Copper, 0.04% Molybdenum

Foliar application

It is well demonstrated that Sandalwood will respond to additions of nutrients when applied through a through a foliar application. However, the response is short lived. The use of foliar application of nutrients can be used in stands that are struggling through transition and for areas where there has been mass host mortality and infills have been planted and the Sandalwood need a nutrient boost.

Bio stimulants

There are natural low levels of organic matter in the soil across the estate. Soil conditioners improve soil structure by enhancing water retention and cationic exchange, adding organic matter, increasing nutrient uptake and suppling or stimulating microbial growth. As shown in trial 7 and trial 14, these products can show improvement in early the growth of young plantations on poor sandy soils. Without these products the addition of a single application of foliar NPK plus trace fertiliser did not improve the growth of the Sandalwood. Table 30 list products trialled with success in young Sandalwood plantations.

Table 30: Bio stimulant products used by Quintis.

Product	Soluble	L/ha	Elemental rate
Agrichem Kelpack	yes	2 to 5	Organic Liquid Seaweed Concentrate
Agrichem Supa stand phos	yes	3 to 6	5% Nitrogen, 10% Phosphorus, 3% Potassium, seaweed
Omnia Rhizovator	yes	15 to 30	Biological stimulants with chelated trace element
Omnia Bacstim	yes	1 to 3	2 strains of *Brevibacillus laterosporus* 1 strain of *Bacillus amyloliquefaciens* 2 strains of *Bacillus licheniformis*
Growth Formula	yes	3 to 6	IAA, NAA root stimulates
Seasol	yes	5 to 10	0.2% Nitrogen, 0.02% Phosphorus, 3.7% Potassium, seaweed

d) Methods of Application

The application method will depend on the age of the plantation and whether specific species need added nutrition. They are five systems that can be used for fertiliser application. The calibration of each system to ensure the correct amount is applied is important.

Irrigation

The application of nutrition through the dripper irrigation system is the most efficient method of getting nutrition to the roots of the tree. However, care needs to be taken to ensure that the fertiliser is well dissolved and drippers are not blocked. All new fertiliser formulations to by injected into the system, need to be approved by tape manufacturer prior to use.

Orchard sprayer

Application of foliar nutrition to small areas of trees showing specified deficiencies, can be done with an orchard sprayer. This system is time consuming and expensive but is a good option for small areas within a stand.

Individual tree Application (ITA)

The individual tree application (ITA) method is suited when only one species in a mixed configuration requires fertilising with granular fertiliser. This methodology is the application of a measured amount of fertiliser applied by hand to selected individuals. It is an efficient method to use in a mix configuration where the trees that need to be fertilised run diagonally across the rows.

Tree row

Granular fertiliser may need to be applied specifically to the host rows. This can be done efficiently using a tractor-mounted fertiliser hopper calibrated to deliver fertiliser to a single row only.

Broad spread

Research has shown that a broad spread application of granular fertiliser across the entire stand is the most beneficial for growth response. Quintis used an aerial application methodology for broad spread fertilising. This system is very efficient for large areas that require granular fertiliser. However, there can be some burning of the leaf tips if the application is done when leaves are moist. There is also a restriction on the amount that can be applied per hectare at any one time (max 300kg/ha pf product).

WEED COMPETITION

Weeds pose a significant challenge in Sandalwood plantations as they compete for environmental resources such as light, water, and nutrients, which are often limited in supply (Fround-Williams, 2008). This competition can lead to reduced biomass for the intended crop. Consequently, weed control is a primary expenditure item during the establishment phase of Sandalwood plantations.

The land designated for Sandalwood plantations would typically have had a previous land use, which influences the weed load and species present. In the Northern Territory, many sites were previously utilised for improved pasture or sown with hay species. While the land in ORIA and Dalbeg was previously used for horticulture. This prior land use plays a significant role in determining the types of weeds that can affect tree growth if not controlled.

The tropical environment, as detailed in Chapter 3: Site Characteristics, provides an ideal environment for weed proliferation. Given the high rate of weed growth, the timing and appropriate methodology for weed control is crucial.

1. COMMON WEEDS

The problematic weeds commonly encountered in Sandalwood plantations fall into four categories: grasses, herbs, shrubs, and vines. The significance of these weed groups varies depending on the plantation stage. During the establishment phase, grasses, herbs, and shrubs are the primary weed concerns. As the Sandalwood plantation transitions and develop a canopy, shrubs and vines become the predominant weed issues.

Significant weed species

There are many weed species that are found in the plantation. The following is a description of the most significant of the weeds found.

Figure 143: Sickle pod can grow in dense thickets and shade out small seedlings, (unknown).

Senna obtusifolia (Sicklepod) is found in Northern Territory and Queensland plantations. Sicklepod (*Senna obtusifolia* and *Senna tora*) are closely related annual weeds. Sicklepod is a vigorously growing, extremely competitive woody shrub which grows from 1.5 metres to 2.5 metres high by one metre wide. Senna prefers well drained, fertile soils. Soil seed reserves of 2000 seeds per metre^2 of soil have been recorded in dense stands. It can germinate at any time of the year under favourable conditions and seed can remain viable in the soil for up to ten years (Mackey, Miller, & Palmer, 1997).

Figure 144: Hyptis in plantation, (Mustang Hill. Robson, 2024).

Hyptis suaveolens (Horehound) has been and on some site continues to be a major weed species in young plantations in Northern Territory. Hypsis is a Latin American exotic erect annual herb which can grow to three metres high with a woody base and long hairy stems. The leaves have a strong mint aroma when crushed. It favours disturbed areas such as plantation edges and firebreaks. Hyptis will grow on most soil types, except those which become waterlogged. It can take over plantation inter-rows, forming dense thickets. Hypsis is known for its resilience to slashing and mowing and its prolific seeding will ensure that it will remain troublesome.

Sida cordifolia (Flannel) can be found in Northern Territory and Queensland plantations. Sida is a tough erect perennial shrub, growth from one to three metres high with a strong branching taproot. It is an invasive unpalatable shrub that competes with and excludes native plants as well as pasture and crop species. Prolific seeding has contributed to Sida becoming a common weed in areas where soils have been disturbed and or degraded. It is found on most soil types except some of those derived from limestone and seasonally flooded clays.

Figure 145: Wild Sesbania flowering, (unknown).

Sesbania cannabina (Wild Sesbania) found in West Australian, Northern Territory and Queensland plantations. Sesbania is an annual herb that will grow on heavy soils along watercourses and low-lying areas. It grows to approximately three metres in height. It is the most widespread of the Sesbania species and commonly grows in areas subject to inundation, where it can withstand waterlogged conditions.

Vines

Passiflora foetida. (Wild passionfruit) is a problem vine in both Northern Territory and West Australian plantations. Native to the West Indies and South America, it is now widespread in tropical regions. It is a creeping, scrambling vine, often found growing over and around other plants and trees but can also be found as a ground cover. It is the most consistent weed species in mid rotation Sandalwood plantations. It can cause mortality when the Sandalwood tree crown become covered in this vine.

Figure 146: Passionfruit is a big problem vine in Sandalwood plantations, (TP15.Robson, 2022).

Macroptilium atropurpureum (Siratro), **Convolvulus & Ipomea spp** (Bell vine), ***Neonotonia wightii*** (Glycine) are pasture vines and are an issue in young to mid rotation plantations in Queensland.

Siratro and Glycine were introduced into coastal Queensland as a cattle pasture plant. Both are legumes, Siratro native to tropical America and Glycine native to tropical Africa, can form dense infestations along forest edges, and will gr ow completely over young trees, effectively smothering them. The Bell vines consist of thirteen species that are vigorous, rapidly growing, perennial vines that can climb up to 12 metres in trees or spreading across the ground. They are opportunistic colonisers of open, disturbed habitats. Under favourable conditions of full sun, ample moisture and fertile soil it can grow very rapidly, smothering all other vegetation. Its climbing habit enables it to compete successfully with trees and shrubs. They will reproduce primarily from broken fragments of stems that produce new roots at the nodes.

The thirteen invasive congeners recorded in Queensland are *I. alba, I. batatas, I. cairica, I. carnea subsp. fistulosa, I. hederaceae, I. hederifolia, I. nil, I. ochracea, I. pestigridis, I. purpurea, I. quamoclit, I. triloba, I. alba* and *I. quamoclit*.

Operculina codonantha (Potato vine) is a problem over the wet season in newly established plantations in ORIA. Potato vine occurs in all three states but is only a problem species in young plantation on the flood irrigated ORIA. It can grow tendrils up to ten metres long and pull over young seedlings during the wet season.

Figure 147: These vines can very quickly completely cover trees and pull them over (unmanaged Millaroo stand, (Robson, 2024).

Centrosema pascuorum (Cavalcade) is an exotic hay crop species from central America. It is an issue in young to mid rotation plantations in Northern Territory as it can cover small to medium size trees. The twining stems extend up to two metres from the crown of the plant. It is most common on sites that were planted with improved pasture or used for commercial production of hay prior to plantation establishment.

2. TYPES OF WEED CONTROL.

It is essential for the successful establishment of Sandalwood and hosts that weeds are kept to a minimum. To do this, a range of methods, physical, environmental, biological and chemical control practices in an integrated weed management approach, to control the weeds within the plantations is used.

Environmental

Sandalwood plantations are in the monsoonal tropics, with wet summers and long dry winters. Application of water by irrigation systems is essential for good survival and growth of tree species. However, wet soils also provide unseasonal growing conditions for weeds. Irrigation scheduling, discussed later in this chapter, is a system where water is applied as it is needed according to soil type and weather conditions. Overwatering will encourage weed growth and spreading of weed seeds. An initial high stocking of plantation seedlings can very quickly compete and shade out many weed species. The initial tree stocking within Sandalwood plantations, dependant on the configuration, can be as high as 1400 stems per hectare. These seedlings can quickly create a canopy along the row and shade out weeds and grasses.

Biological

The selection of tree species is important for the success of the plantation. A range of both fast growing and slower growing species is used. When placed in the appropriate planting design, they complement each other, and the plantation system rapidly develops. This development of the trees can inhibit growth of most detrimental weed species. These fast-growing species require regular pruning. The pruning offcuts remains within the plantation as mulch, reducing weed growth. On some sites, the pot host Alternanthera is planted between trees to form a natural weed mat that inhibits weed growth. The recommended sowing of a cover crop between the tree rows can also out compete the undesirable weed species.

The use of grazing animals in an agroforestry system to reduce weed competition has been trialled with some success. Goats and young cattle have been utilised during mid rotation.

Physical

Stubble management is the management of the land between tree rows. If left unmanaged, the inter-row will grow weeds, particularly during the wet season, which will provide a seed bank for irrigated rows. The inter-rows cover can be managed by several methods, slashing, cover crop retention, mulching and cultivation.

During the first two years, occasional shallow cultivation within the inter-row can be used. This will remove any unwanted weeds and can assist with irrigation infiltration in crusting to hard setting soils.

In the northern tropics, strong afternoon winds can prevent the use of spray herbicides close to small seedlings. During the establishment phase, hand labour can be utilised to pull or chip weeds away from small seedlings. Historically, hand weeding was the predominate method of weed control. In the early 2000's there was little understanding of the effect, if any, of applying herbicide on the weeds that the Sandalwood could potentially be hosted to. As the plantations age, vine became more of problem weed and hand removal and cutting vine stems was utilised.

The use of cardboard tree guards sheltered seedlings from chewing animals, wind and herbicides, and helped to compete with weeds. With the use of tree guards protecting seedlings, cheaper methods of weed control such as chemical can be used.

Chemical

With a large-scale annual planting program, the weed growth in the tropics cannot be controlled by physical methods alone. The careful use of herbicides is essential for a successful established plantation. This includes employing a range of herbicides and application methods, such as spot spraying with shielded wands which only spray selected weeds; the use of residual herbicide applied prior to planting and side direct application with the use of a tractor mounted spray.

The parasitic relationship Sandalwood has with nearby plants means that care must be undertaken when using any herbicide near Sandalwood. There are some herbicides that have clearly shown that when used, can have a serious detrimental effect to Sandalwood. Each herbicide belongs to a mode of action group. Herbicides within the MOA group 5 (PS II inhibitors) such as atrazine and Gesatop, Group 4 (Synthetic Auxins) such as clopyralid, fluroxypr, picloram and triclopyr, if used will have serious phytotoxic effects on Sandalwood.

Herbicides in the MOA Groups 1 (grass specific), Group 3 (Inhibitors of microtubules), Group 14 (inhibitors of PPOs), Group 22 (inhibitors of photosynthesis at photosystem), Group 9 (Inhibitors of EPSP synthase) and Group 10 (Inhibitors of glutamine synthetase) have shown to have herbicides that are safer to use with Sandalwood.

a) Herbicide trials

Prior to 2012, herbicides were sparingly used as there were several incidences of phytotoxicity of chemical sprayed near the Sandalwood seedlings. The main form of weed control was hand chipping from around the seedlings and planting mound with Alternanthera plugs. However, the scale of the establishment program increased significantly from 2012, and a concerted effort was undertaken to examine if certain herbicides could be used to control weeds from establishment.

Pre-emergent herbicides

Trial 1. Several herbicide trials were undertaken by Elders Forestry at Goldtyne plantation. This trial was investigating the use of Atrazine and Simazine, both Group 5 herbicides, widely used by the agricultural industry in the 2000's. This trial examined increase rates (0, 1, 2, 4, 8, & 10 litres per hectare) over Sandalwood. All treatments, except zero had a phytotoxic effect on Sandalwood. This trial also tested a treatment of Clomac (Clopyralid), Group 4 herbicide, at 1000 grams per hectare. Sandalwood had a strong negative reaction to this herbicide.

Trial 2. This experiment investigated the use of Oryzalin (Prolan), a Group 3 herbicide. This herbicide action inhibits cell elongation in the newly emerging root shoots of those plants with fibrous root systems (grasses). The objectives of the trial were to test the weed control effectiveness of Oryzalin at two rates. It investigated any phytotoxic effects of the herbicide on the establishment growth and survival of Sandalwood and host species.

The trail was established in stand 39 at the Dalbeg plantation, a flood irrigated stand. There were three treatments, a control (200 litres per hectare water only) and two rates of Oryzalin, 4 litres per hectare and 6 litres per hectare. Treatments were applied with a broad acre spray unit.

A total height measurement was completed 80 days after planting, indicating there was no difference between treatments for the Sandalwood, Acacia and Sesbania. The Cathormion in the 6 litres per hectare treatments were significantly taller than both other treatments. The Senna in the 6 litres per hectare was also significantly taller than the control. The subjective health assessment indicated there was no difference between the treatments.

The trial ran for 80 days, and the results indicated Oryzalin treatments did not affect the health or growth any of the planted seedlings. The herbicide did slightly affect the growth and establishment of the alternanthera, but after 40 days, newly infilled seedlings displayed no ill effects. After 80 days the Oryzalin pre-emergent plots had a tenth of the germinated weed on the tree beds compared to the control.

Trial 3. This trial was established to test the performance of a range of pre-emergent herbicides across two sites, one in Northen Territory, the other in Western Australia. The two sites were, Taylors Park, NT (stand 128, TP10) and Pioneer 37, ORIA. The experiment was a randomised complete block with three replications.

There were three preplant treatments, a control (200 litres per hectare water only), Pendimethalin (Group 3) at 4 litres per hectare and a mix of Oryzalin at 4 litres per hectare and Oxyfluorfen at 2 litres per hectare. In addition, there were also three post plant treatments, Oryzalin at 4 litres per hectare, Sulfometuron methyl (Group 2) at 60 grams per hectare, and the mix of Oryzalin at 4 litres per hectare and Oxyfluorfen at 2 litres per hectare.

The trial was assessed five months after treatment application. At the Taylors Park site, the herbicides, Pendimethalin and Sulfometuron methyl, had some effect on the growth of Sandalwood, Sesbania and Cathormion. Unlike the Vertosol clay site at Pioneer, where only the Sulfometuron methyl treatment had a detrimental growth on seedlings.

The mix of Oryzalin and Oxyfluorfen pre-emergent herbicides was effective with a wide weed spectrum and incurred no or little phytotoxic effects to the tree species used. While the Pendimethalin was as effective as the Oryzalin at inhibiting weed germination, on the sandy Taylors Park site, there was a slight effect on Sandalwood height growth and a major effect on Sesbania height growth. However, this was not the case on the clay site in the ORIA with no detrimental growth effect by Pendimethalin.

Although the Pendimethalin treatment reduced the weed and grass cover the most on both sites, all of herbicides controlled the weeds and grasses better in the Pioneer site than the Taylors Park site, reducing weed competition by 80% compared to the control.

Trial 4. In 2013, a trial was established to investigate weed control and seedling growth by testing three treatments with the two herbicides Oryzalin and Oxyfluorfen. The treatments were 1 - Oryzalin at 6 litres per hectare as a preplant, 2 - Oryzalin at 6 litres per hectare and Oxyfluorfen at 3 litres per hectare as a post plant pre-emergent, and 3 - Oxyfluorfen only at 3 litres per hectare as a post plant pre-emergent. There were two post plant applications in the first ten months after establishment.

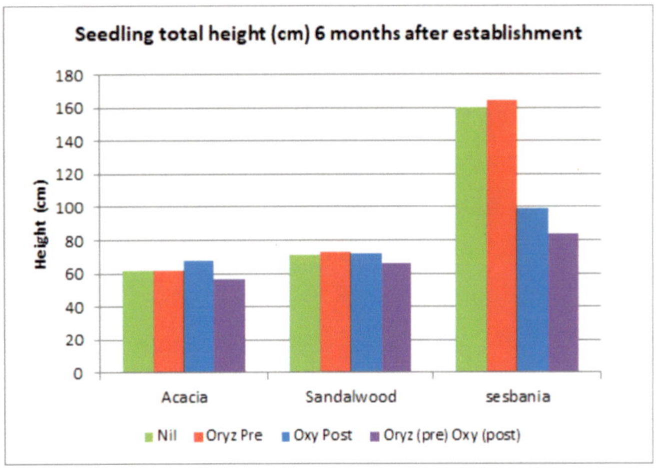

Figure 148: Seedling height growths when treated with pre plant and post pant pre-emergent herbicides.

The results after six months found that the Oryzalin at pre plant had no effect on growth of any tree species. Only the Sesbania was significantly affected by the addition of Oxyfluorfen. The Sandalwood and Acacia did show a slight reduce growth but not significant.

Operational use: With research trials indicating that both Oryzalin and Oxyfluorfen were safe to use as a pre plant pre-emergent, the operational use of Oryzalin at 4 litres per hectare, in mix with Oxyfluorfen at 2 litres per hectare as a pre-plant pre-emergent was recommended and implemented in new Quintis plantations between 2013 and 2014. This mix worked quite well on the sandy clay loam soils in Northern territory with no phytotoxic effect to young plantation seedling growth.

Oxyfluorfen does have a knockdown effect on weed germinates when used at rates of 4 litres per hectare and above. At some point in 2014 an operational decision was made to apply Oxyfluorfen at 4 litres per hectare and above to young established plantations. This proved to be detrimental to Sandalwood seedling growth. Sandalwood established in sandy soils in Taylors Park 10 and 11 did not grow at the expected rate. Inspection of extracted seedlings revealed little new root growth. Soil testing from around the Sandalwood indicated Oxyfluorfen was potentially the cause.

Systemic and contact herbicides

Trial 5. Testing of Halosulfuron (Sempra), a Group 2 herbicide and Haloxyfop (Verdict), a Group 1 herbicide to treat nut grass (*Cyperus rotundus*) infected sites. This experiment was established in Dalbeg stand 19 in 2012. The Halosulfuron was applied at 120 grams per hectare and the Verdict at 200 millimetres per hectare to a four month old plantation.

The Sempra was effective at killing off the nut grass on the site. The Sempra treatment had a severe negative effect on the Sesbania. The Sandalwood, Acacia and Cathormion were also affected by the chemical, although to a lesser degree. All the species grew more in the control untreated plots.

Trial 6. This was a pot trial to see the effect of spraying the alternanthera at the base of the Sandalwood with increasing levels of glyphosate. Three rates of glyphosate (1, 2 and 4 litres per hectare) were applied to the alternanthera pot host. These were left over Sandalwood seedlings that had been in the pot for three months after the rest were field planted.

Assessments over a 45 day period showed Sandalwood declining health in three treatments. The control treatment which had the pot host physically removed showed no decline in health. It was clear that even lesser amounts of glyphosate when applied to the pot host had a phytotoxic effect on Sandalwood seedlings. This phytotoxic effect of the herbicide was phenotypically significantly greater than the physical removal of the pot host. The effect of the higher rates (2 and 4 litres per hectare) on the Sandalwood seedlings after one month were obvious with leaf drop and overall decline in plant health. Although the lowest rate (1 litre per hectare) did not appear to have a pronounced effect on the appearance and health of the seedling there was a clear effect on the root development.

Trt 1 Remove althernathera @ 30 days   Trt 2 1L/ha Glyphosate@ 30 days   Trt 3 2L/ha Glyphosate @ 30 days   Trt4 4L/ha Glyphosate @ 30 days

Figure 149: The effect on the above ground foliage and the roots when host is sprayed with increasing levels of glyphosate.

<u>Trial 7</u>. This large pot experiment focused on the effect of pre-emergent and knockdown on Sandalwood growth and health. The treatments were the following: herbicides sprayed over the Alternanthera and soil around the base of the Sandalwood with 1 - Oryzalin at 10 litres per hectare + Glufosinate at 3 litres per hectare, 2 - Oryzalin at 10 litres per hectare + Oxyfluorfen at 4 litres per hectare + Glufosinate at 3 litres per hectare, 3 - Glufosinate at 4 litres per hectare and 4 - control (water only). Sandalwood was established in 30 litre plastic pots each treatment was applied three times over a ten week period.

After 12 weeks, the seedlings applied with Glufosinate only had significantly higher increment growth compared to those applied with Oryzalin and Glufosinate; Oryzalin, Glufosinate and Oxyfluorfen; and water (control). Similarly, seedlings applied with Glufosinate only had largest above ground dry mass and dry root mass, the largest basal diameter and the best root-shoot ratio compared to those seedlings applied with all other treatments. However, when subjected to a statistical test, they did not significantly differ with control seedlings in terms above ground dry weight and root dry weight, basal diameters and root-shoot ratio.

<u>Trial 8.</u> This was an operational trial that investigated the possibility of applying a Glufosinate herbicide with side directed tractor mounted spray rig over alternanthera onto young four month old Sandalwood. The rate was 3 litres per hectare with a water rate of 200 litres per hectare.

The trial was assessed three weeks after the second treatment application. The assessment consisted of visual health assessment of the Sandalwood and alternanthera. There were no obvious phytotoxic effects in the sprayed rows. The attached photos (Figure 150) are typical of the condition of the Sandalwood and alternanthera in both treatments.

Figure 150: Photo 1 shows the herbicide treatments three weeks after application, photo 2 is the control treatment, (TP6, Robson, 2015).

Herbicide incident. There have been several incidents with herbicide over-sprays that have damaged plantations. In all cases this has been due to the use of MOA Group 4 herbicides. The herbicides in this group typically have an ester or amine formulation and this make them more volatile. With this formulation, particularly esters, under certain climatic conditions they can volatilise and drift away from target species (Delvalle, 2022). In 2012 this occurred to six month old plantation at Eagle Park in Northern Territory.

Figure 151; Spray drift affected Sesbania at Eagle Park, (Robson, 2012).

A site preparation aerial spray of Garlon (triclopyr) and Comet (fluroxypyr) had been completed at Taylors Park 3 and 4, which is directly six kilometres from Eagle Park. Two days after that spray was completed all the Sesbania at Eagle Park turned yellow.

Initial effect was the leaf yellowing and stem curl typically associated with poisoning by auxin growth regulators. After the initial yellowing, all affected leaves fell off the tree and a high proportion of the branches started to go black from the primary growing point and this continued to spread slowly downwards. However, most of these trees still had a green trunk and started to reshoot after a week.

Weed control comparisons. A cost comparison of the various weed control options was undertaken in 2014. The options available to managers were hand weeding, tree guards and weed matting, pre plant pre-emergent and post plant herbicides.

Hand weeding was the original weed control method that was used by managers in newly established plantations since the start of the plantation program in 2001. This consisted of a mix of hand weeding that involved using casual labour hire to physically remove by hand pulling and chipping, with occasional spot spraying of herbicide. The documented cost associated with this method at Florina Road plantation in 2012 was $3700 per hectare.

Figure 152: hand weeding Sandalwood at establishment, (Robson, 2012).

The use of plastic ground cover was thought to be a possible efficient method to control weeds and reduce the control costs. It was a significantly cheaper method than hand weeding coming in at $1700 per hectare. Costs included:

- Plastic mulch: $600.
- Labour installing: $400.
- Additional cost in planting: $400.
- Additional land prep cost: $300.

Although it did significantly reduce weeds along the rows, the plastic ground cover caused issues with Sesbania growing roots between the soil and the matting. This caused roots to crimp the irrigation tape, stopping irrigation. As it was under the matting, it was difficult to determine where the blockage was. It also created a safety hazard, as the environment beneath the matting became hot, humid, and moist conditions favourable for frogs. The resulting increase in frog populations also led to more snakes, thereby increasing the risk to workers.

Figure 153: Plastic weed mat along the rows to prevent weed growth, (Goldtyne, Robson, 2010).

The application of one pre-emergent and knockdown before planting and one post plant herbicide application as a side spray on hosts and a knapsack spray on Sandalwood rows in Taylors Park in 2014 was $1360 per hectare.

Costs included:

1. 2 x pre-emergent & application cost: $470.
2. Knapsack spray: $750.
3. Side direct spray: $140.

When tree guards are installed onto all host trees, and with herbicide applied to Sandalwood rows by knapsack, the per hectare cost at Taylors Park in 2014 was $1244. Costs included.

- Tree guards & stakes: $204.
- Labour install: $150.
- Knapsack spray: $750.
- Side direct spray: $140.

Figure 154; Use of tree guards on hosts at Dalbeg, (Robson, 2016).

Tree guards on Sandalwood was also trialled, but it was found that the alternanthera would grow inside the tree guard and smother the Sandalwood, so this idea was discarded.

Summary of trials

The use of herbicides is essential to control the tropical weeds from across the estate. However, some types of herbicides have shown to be detrimental to both Sandalwood and host trees.

The soil type is important in determining the detriment when using a preemergent. The heavy clay soils show less phytotoxicity at higher herbicide rates than the sandy soils.

Herbicides in the mode of action (MOA) Groups 1, 3, 9, 10 and 14 have indicated that when used with care and at rates recommended by the label, they can be used safely within Sandalwood plantations. Specifically, the herbicides, Haloxyfop, Fluazifop, Oryzalin, Pendimethalin, Flumioxazin, Carfentrazone, Glyphosate and Glufosinate. In 2022, the production Oryzalin was discontinued. News reports suggest this was due to damage in the manufacturing facility (Danial, 2022).

The most used non-specific herbicides within plantations are Glyphosate and Glufosinate. Trials have shown that Glyphosate can cause damage to Sandalwood if the alternanthera foliage is sprayed. There is also anecdotal evidence that spraying trees stems is also detrimental to growth and health.
Herbicides in the MOA Group 22, the non-selective contact herbicide Diquat and Paraquat have also shown to not be phytotoxic to Sandalwood plantations when sprayed on nearby weeds. However, these herbicides are no longer in use as they have shown to be toxic to humans by the oral route and moderately toxic through the skin (EPA, 1997).

Herbicides in the mode of action (MOA) Groups 2, 4, 5, and 6 have shown to cause phytotoxic effects in Sandalwood and host. The following herbicides have proven to have detrimental effects when used within, or in some case near, Sandalwood plantations include, Halosulfuron, Atrazine, Simazine, Duiron, Dicamba, Clopyralid, Fluroxypyrr, Picloram, and Triclopyr.

The exception to this is 2,4-D and 2,4-DB, MOA Group 4. Quintis did use these herbicides to control vine in the inter-row of the plantation in Queensland. The main issue with this group of herbicides is the ester and amine formulation that can move in the air, as described in the over spray incident. However, Quintis used this herbicide effectively in a low volume shrouded sprayer that sprayed only the weeds between the tree rows.

Figure 155: Shrouded sprayer of application of 2,4-D in inter-row, (Dalbeg. Robson, 2014).

The current operational weed control method from planting, is the establishment of tree guards onto hosts trees and the use of a non- specific herbicide (Glufosinate) to control weeds and grasses between the trees. The cost comparisons (Table 31) shown below indicate that the tree guard option is the most cost effective. However, over a three-year period, the use of the plastic mulch would be the most cost-effective method.

Table 31: Weed control options cost comparisons for the first year of establishment (per ha).

Method	Labour	Chemical and application	Plastic Mulch	Tree Guards & bamboo stakes	Site prep extra	Total costs
Hand weeding	$3,700					$3,700
Plastic mulch	$800		$600		$300	$1,700
Herbicide only	$750	$610				$ 1,360
Tree guard + Chemical	$150	$890		$204		$1,244

Due to the overuse of Oxyfluorfen in the latter half of 2014, the use of pre-emergent was banned by Quintis and this has increased the cost and overuse of non- specific herbicides.

Legal requirements

The use of chemicals in Australia is limited by regulations for use as determined by the label for that particular chemical. Most labels have the following statement, "not to be used for any purpose, or in any manner, contrary to this label unless authorised under appropriate legislation".

Australian law requires all agricultural and veterinary chemical products sold in Australia to be registered by the Australian Pesticides and Veterinary Medicines Authority (APVMA).

Once a product is registered, it is approved for the purposes and uses stated on the product's label. The APVMA are an independent statutory authority responsible for assessing and registering pesticides and veterinary medicines proposed

Situation/ Crop	Weeds Controlled	State	Rate/Treated ha	Critical Comments
Eucalypt forestry plantations	Amaranth *(Amaranthus spp.)*, Annual Ryegrass, *(Lolium rigidum)*, Asthma Plant *(Euphorbia hirta)*, Barnyard Grass, *(Echinochloa spp.)*, Chickweed *(Stellaria media)*, Crowsfoot Grass *(Eleusine indica)*, Deadnettle *(Lamium*	All states	6.75 to 9 L/ha	**PRE-PLANTING:** Apply 7 -2 days prior to transplanting. Use the higher rate on heavier textured soils or soils with high organic matter. Mechanical disturbance should be minimised during planting to optimise performance. **POST-PLANTING:** Apply immediately following planting. Before emergence of weeds. Repeat applications may be made as directed sprays to provide

Figure 156: The pendimethalin label is an example of restricted use only in eucalypt plantations, (TITAN AG Pty Ltd , 2024).

for supply in Australia. It is generally understood that if the label does not mention a specific pest or crop, it cannot be used without an approved permit. Sandalwood plantations are not mentioned on any agricultural label in Australia.

A minor use permit application is made to APVMA and are issued to legalise the use of a chemical product or active constituent in a use situation considered to be minor. This would be used on a minor crop situation, where no registered products exist for the proposed use. In addition, the use of the product would not produce sufficient economic return to register the product, or there will be limited use in a major crop where no registered products exist for the proposed use.

Unfortunately, as Sandalwood is reasonably new to the forest industry and minor in the Australian forestry landscape, there are almost no commercial chemicals available for use in Sandalwood plantations. Almost all chemicals used within forestry in Australia specify either Pine or eucalypt plantations or both.

B. COMMERCIAL, INDUSTRIAL, NON-AGRICULTURAL AREAS and FORESTRY PLANTATIONS

CROP/ SITUATION	WEEDS	STATE	RATE	WHP	CRITICAL COMMENTS
Commercial & industrial areas, forest plantations, rights-of-way and other non-agricultural areas	See list of weeds controlled in Table 1.	All States	1.0 to 5.0 L/ha	-	Determine the recommended rate of use by considering the criteria WEED SPECIES, WEED STAGE OF GROWTH, WEED DENSITY and CLIMATIC CONDITIONS as described above in **Part A** of the Directions for Use table. **Warnings:** Do not allow spray or spray drift to contact desirable plants. To avoid potential crop damage, refer to the label sections on **Application** and **PROTECTION OF CROPS, NATIVE AND OTHER NON-TARGET PLANTS**.

Figure 157; The label on Glufosinate does not specify tree species and is available for use within Sandalwood plantations, (BASF Australia Ltd , 2024).

Prior to 2012, Quintis did not use significant quantities of chemical. However, there was several permits that were applied for but have since expired (figure 158). It is important to note that most of these expired permits were applied for the use in Spicatum plantations, not Album. However, as the permit states for Sandalwood, it does also cover Album plantations.

Permit No.	Type	Description	Status	Issued date	Expiry date
PER5982	AG	Oxyfluorfen / sandalwood / weeds	EXPIRED	11-Oct-2002	30-Sep-2007
PER5980	AG	Paraquat and Diquat / sandalwood / weeds	EXPIRED	30-Sep-2002	30-Sep-2007
PER5979	AG	Glyphosate / sandalwood / weeds	EXPIRED	30-Sep-2002	30-Sep-2007
PER5462	AG	Trichlorfon / Sandalwood plantations / Browsing ripand insects and caterpillars	EXPIRED	01-Sep-2002	30-Sep-2005
PER3765	AG	Oxyfluorfen / sandalwood / weeds	EXPIRED	19-Jul-2001	30-Sep-2002
PER3764	AG	Simazine flowable / Sandalwood / Weeds	EXPIRED	19-Jul-2001	30-Sep-2002
PER3763	AG	Paraquat and Diquat / sandalwood / weeds	EXPIRED	19-Jul-2001	30-Sep-2002
PER3762	AG	Glyphosate / sandalwood / weeds	EXPIRED	19-Jul-2001	30-Sep-2002

Figure 158: List of expired APVMA minor use permits for weed control chemicals available for Sandalwood plantations, (APVMA, 2024)

Post 2012, there has been minor use permits approved for Album plantations, mostly for use of specific insecticides to treat termites, fig leaf beetle and rats. Of the herbicides discussed in the chapter, only Glyphosate and Glufosinate have commercial forestry use on their label. In 2024, there were no current minor use permits of any herbicide for use in Sandalwood plantations.

There is a special APVMA permit (PER 7250) that allows use for research trials but not operationally. In Quintis's experience the process applying for a minor use permit is convoluted, time consuming and has a high chance of being rejected.

The other option is to collaborate with the chemical company in trials to get Sandalwood onto the label. But experience has shown that because Sandalwood is only a small industry, most chemical companies are not interested.

b) Slashing and cultivating

Physically cutting of weeds and grasses by slashing in the inter-row has a number of control benefits. Slashing the weed/grass load maintains a green cover in the interrow which helps stop overland erosion flow. It lowers the soil temperature as bare earth can reach over 50°C. Sandalwood roots are predominately in the top 20 cm and elevated temperatures could be detrimental to root growth. As soil temperature rises, ammonium (NH4+) toxicity increases. There is also an increase in soil denitrification and availability of multiple nitrogen forms, such as NH4+ and nitrate ions (NO3−) (Fonseca de Lima, Kleine-Vehn,, De Smet, & Feraru, 2021).

Slashed material will sit on top of the remnant weeds and grasses and act as a mulch, slowing weed recovery. Until this mulch breaks down, it will sit on top of stubble but will eventually be in direct contact with the soil contributing organic matter. Although slashing will not remove weed species, if timing is right, it will prevent tall weeds from flowering and adding to the seed bank.

c) Silvopastoral systems

As Sandalwood plantations mature, a primary management objective is to reduce the reliance on chemical applications. Despite this goal, ongoing vine emergence have created a persistent challenge, necessitating continual use of chemicals for effective weed control.

<u>Agroforestry and Silvopastoral Systems</u>

Agroforestry presents an alternative strategy to address weed management issues. Silvopastoral systems, a form of agroforestry, integrate tree crops with grazing pasture. Typically, these systems feature pasture areas with widely spaced or clustered trees, allowing livestock to graze among them. In grazing silvopastoral systems, cattle or other animals graze on pasture located beneath scattered or widely spaced

Figure 159: AFM goat herd being relocated between stands, (Robson, 2025).

trees. This promotes a balanced ecosystem and potentially reduces the need for chemical interventions (El-Hage Scialabba, 2022).

Although silvopastoral systems are not a novel concept, they have traditionally been considered unsuitable for the silvicultural structure of Sandalwood plantations. There have been proposals to cultivate crops such as sorghum between the rows, but a 3.6 metre spacing is insufficient for current harvesting equipment. Concerns have been raised that grazing hoofed animals could cause severe damage to irrigation infrastructure, such as drip lines. Given that Sandalwood possesses a shallow root system, there is potential for animal activity to disrupt root connections and host relationships. Research in agroforestry has demonstrated that cattle may harm trees by rubbing against them or causing them to fall (Silveira Pontes, Rodrigues, Souza Carneiro, Chiezi Mendes, & Porfírio-da-Silva, 2024). Indeed, wild cattle that enter plantations can inflict substantial damage on both irrigation infrastructure and trees.

Case Studies

During the mid-2010s, Santanol adopted a grazing silvopastoral system utilising goats to manage weed issues in mid-rotation Sandalwood plantations in Kununurra. This method required fencing all grazed areas and employing a herdsman to oversee the livestock and ensure their removal from the site each night. The goats proved highly effective in controlling weeds located beneath the trees. Although there is limited documented evidence regarding their efficacy, visual assessment suggested that this could potentially be a cost-effective alternative to chemical weed control.

Figure 160: Dalbergia lanceolaria with patches of back chewed off by foraging goats with the plantation, (Robson, 2025).

Figure 161: A plantation Sandalwood with extensive bark chewed off by foraging goats, (Robson, 2025).

In 2024, Apical Forestry Management (AFM) introduced a small herd of goats into a plantation within the Northern Territory. This plantation had experienced persistent vine weed problems and reliance on chemical control measures. Initially, the goats targeted the weeds without impacting the trees, leading to an increase in their numbers. However, the additional goats caused considerable damage to some tree patches in the plantation. Consequently, the goats were removed from the site and are no longer utilised for weed management.

In 2025, AFM initiated a trial using young cattle for weed control, achieving positive results. The cattle successfully managed problematic weeds without causing harm to the trees or irrigation infrastructure. Nonetheless, close monitoring is ongoing to ensure the continued effectiveness and sustainability of this practice. Special thanks to AFM management for providing this information,

WATER

Santalum album thrives naturally on the Indian peninsular in areas which receive rainfall of between 650-1600mm (Radomiljac A. M., 1998). Successful cultivation has been observed for centuries in semi-arid and arid zones in India, sometimes with only initial establishment irrigation or none at all. A study by Jain, Ranajan, & Hemany (2007) demonstrated economic yields of heartwood from Santalum album trees. Some trees yielded up to 102.8 kg of heartwood after 23 years, even in arid regions of India receiving an average of only 300mm of rainfall annually.

The presence of naturally occurring S. album wildlings in native forests in and around the ORIA and BRIA, where annual rainfall is less than 900 mm, suggests that the species can persist under relatively dry conditions. This indicates that Sandalwood plantations in tropical and subtropical regions of Australia receiving more than 650 mm of annual rainfall present minimal risk and have the potential to produce economically significant quantities of heartwood. However, supplemental water during the dry season generally enhances growth and survival, prompting all Australian plantations to implement permanent irrigation systems directly supplying water to the trees.

This water section is designed to give an overview of the irrigation systems, the scheduling methodology and the research used to help develop those methodologies. It is not intended to be a technical description of the two systems utilised to deliver water to the plantations.

1. IRRIGATION SYSTEMS

Sandalwood plantations use two types of irrigation methods to irrigate trees, this being Flood Irrigation and Drip Irrigation. The type utilised in a particular plantation is driven by the location and the presence of communal irrigation infrastructure. Plantations that exist within the Ord River Irrigation Scheme (WA) and the Burdekin Haughton Water Supply Scheme (QLD) are flood irrigation systems with water provided by local authorities. All other plantations operate on drip irrigation systems utilising water from underground aquifers and supply infrastructure.

a) Flood

Flood irrigation provides deep watering which replenishes subsoil moisture levels through the entire soil profile to the point of saturation. Subsequent drainage and water use encourages the replenishment of consumed and displaced oxygen in the soil profile. As a result, intervals between flood irrigation events can be quite long (depending on other factors such as tree age, stand density and soil type).

The site is prepared to ensure that there is a fall from one end of the row to the other. Irrigation channels or flume piping run along the end which supplies water. This water is siphoned (sucked) from channels and flows down furrows or between raised plant beds to the lower end of the field.

Figure 162: A- pipes syphon water from the channel to run down beside the rows in ORIA (Melon farm, Robson, 2008). B – In BRIA, water is released from a flume pipe laid out at the end of the tree rows to release water, (Dalbeg, Robson, 2020).

While the water is traveling down the row it is soaking into the soil profile and across into plant beds. Excess water is drained off into main drains at the end of the rows. The less water (hence more infiltration) that drains off the field after all furrows have flowed, the more efficient the irrigation.

Fields are prepared by laser levelling to a uniform grade (slope) of at least 1 in 2000 (> 0.05%) in the direction of irrigation flow. The field can have a slight grade across the direction of flow depending on the field's natural layout and some management priorities.

A flood setup can use both small (19mm) or large (37mm or 50mm) diameter siphons. The size of the syphon changes the amount of irrigation inflow. The head pressure is determined by the height of water level in the channel compared to height of the level in the tumble area. The maintained water level in the channel will depend on the amount and size of active siphons, compared to the amount of water flowing into the head channel. In a typical situation, an example flow of 24 megalitres per day for 37 mm siphons sucking on a 12 to 20 cm head pressure (about eight siphons per megalitre) would be expected.

The key to successful flood irrigation is getting the water on and off the field as quickly as possible. This will reduce evaporation, deep drainage and tail water. Field length, slope, head pressure, siphon size and irrigation frequency determine how long a field section needs to be flooded to sufficiently soak the bottom end of the field. A section is finished when all or most furrows flow into the drain. Most annual field crops are watered in eight or 12 hour shifts. However, established Sandalwood plantations may take 24 to 48 hrs per section. When a section is finished all the siphons must be stopped and the next section can be started.

The advantages to this system are there are no ongoing delivery costs as the water is driven by gravity. The entire site receives water, the tree row and inter-row and profile wet down significantly. The disadvantages are that it is a reasonably inefficient method of irrigating, although the application is cheap, the water use efficiency is poor. If the site leveling is not correct the water can 'pond' and cause waterlogging or, conversely, run off too quickly and the trees do not get enough water. This can cause uneven watering across the stand, if the soil changes across the stand the irrigation flow through can change.

b) Dripper

The dripper system for irrigation is water efficient but complex. The water comes from deep aquifers and is pumped to the surface and distributed by a network of mains, submains, risers, and drip tape.

The water is moved using either diesel or electric motors to drive the pumps. A computer program opens and closes valves at prescribed times. The water is delivered to the base of all the trees along the row.

Figure 163: Drip irrigation along Sandalwood a row at Taylors Park, (Robson, 2014).

The advantage of this system is water efficiency, applying only the required prescribed amount of water. Although all rows have drip lines, with some effort, lines can be turned off and only specific rows or areas can be irrigated. The disadvantages are the system is expensive to install and run. At times, the large motors and pumps are required to be running continuously. Due to the complexity of the system, the maintenance required to ensure efficient water delivery is high. The drip lines can slip down the tree mound and move away from the tree line, they can be chewed by various animals and water is not delivered to where it is needed.

2. FACTORS AFFECTING WATER DEMAND.

There are four factors that can affect the amount of water required to be applied to provide adequate and effective irrigation (Van Beuzekom, 2023), being water requirement changes as the plantation ages, soil texture, individual species requirements and climate.

As a plantation grows and the trees age, lateral roots push deeper into the soil profile allowing trees to source water from deeper in that profile. However, soil probes across the estate have indicated that it is uncommon to see extraction lower than 700 mm below the surface in a Sandalwood plantation. As trees get larger, and their canopy increases in size, there is a higher demand for more water to sustain good growth.

The soil texture plays an influential role in determining water availability. The lighter sandy soils are usually free draining. This allows water to move though the profile quickly, meaning they dry out much faster than heavy clay soils. Conversely, in lighter soils, available soil moisture (moisture trees can extract) can be obtained when total soil water content is from 10% and above. On heavy soils, due to tight chemical bonds between water and soil particles, total soil water content needs to be around 25% before any soil moisture is available (Rai, Singh, & Upadhyay, 2017).

Different tree species have varying water requirements and tolerances to overly wet or dry soils. For example, Sandalwood is a drought tolerant species and can be one of the last species in a stand to show signs of drought stress. However, it cannot cope with excessively prolonged periods of soil saturation. On the other hand, *Sesbania formosa* thrives in wet conditions but is quick to lose vigour if soil moisture levels are low for prolonged periods of time. When *Cathormion umbellatum* is young, it is drought sensitive but as it ages it becomes more drought tolerant. The table below outlines tolerances of commonly grown species within a Sandalwood plantation.

Table 32: Drought and moisture tolerances for plantation species.

Species	**Drought Tolerance**	**Tolerance to Waterlogging**
Sandalwood	High Moderate (will become dormant)	Low
Sesbania formosa	Low	Very high
Acacia species	High	Low
Senna siamea	Moderate to Low	Low
Dalbergia species	Moderate – High (when dormant)	Moderate
Cathormion umbellatum	Moderate (will become dormant)	High
Albizia lebbek	High	Moderate

Due to different water requirements for each species, prioritisation is required when determining when to irrigate a stand. The purpose is to irrigate drought intolerant species such as *Sesbania formosa* and young *Cathormion umbellatum* at a frequency that prevents them from sustaining severe drought stress, while not over-watering drought tolerant species such as Sandalwood and Acacia species (Van Beuzekom, 2023).

Evapotranspiration from trees and evaporation of soil moisture varies greatly depending on climatic factors such as temperature, radiation, humidity, and wind. In the cooler months of the year, water loss from plants and soil is much lower than during hotter months with water requirements varying accordingly. Prolonged high winds have the ability to increase water loss significantly. so moisture levels and signs of drought stress on trees should be monitored closely during these periods.

3. IRRIGATION SCHEDULING

All irrigated agricultural crops in Australia have developed irrigation scheduling programs. Scheduling irrigation is about planning when and how much water to be applied to a crop to achieve crop production or a particular quality. The principles of scheduling water are the same for all crops, however the critical timing, methods and techniques will vary with the crop. There are three key components to successful efficient irrigation scheduling.

The soil water balance calculations required for scheduling must have climatic data, in particular Evapotranspiration rates. It is important to have knowledge of the readily available water (RAW) content, data on soil moisture and dry out rates across a range of sites, and knowledge of crop water use. Data for these three components will allow accurate and efficient irrigation to plantations.

a) Evapotranspiration

Evapotranspiration (ETo) is the combination of two separate processes; evaporation is when water is lost from the soil surface, and transpiration which is the water lost from the crop itself.

Evaporation is the process whereby liquid water is converted to water vapour (vaporisation) and removed from the evaporating surface (vapour removal). Water evaporates from a variety of surfaces, such as lakes, rivers, pavements, soils, and wet vegetation.

Transpiration consists of the vaporisation of liquid water contained in plant tissues and the vapour removal to the atmosphere. Crops predominantly lose their water through stomata. Nearly all water taken up by the trees is lost by transpiration and only a tiny fraction is used within the trees.

The United Nations Food & Agriculture Organisation (FAO) standard methods for modelling Evapotranspiration (ETo) use a Penman-Monteith equation (Allen, Pereira, Raes, & Smith, 1998). This is a complicated equation which is automatically worked out by modern weather stations and Bureau of Meteorology (http://www.bom.gov.au).

There are several factors that affect evapotranspiration. The key ones include solar radiation, air temperature, humidity and wind speed. ETo data for regions can be taken from BOM extrapolated data or collected on site by weather stations. Typical ETo levels across the estate are discussed in Chapter 3.

b) Readily available water

Not all water held in the soil is readily available to plants. Some water is bound so tightly to soil particles it cannot be used by plants. For irrigators it is useful to know how much water is readily available for plant use. Readily available water (RAW) is the water that a plant can easily extract from the soil. RAW is the soil moisture held between field capacity (maximum water) and a nominated refill point for unrestricted growth. In this range of soil moisture, plants are neither waterlogged nor water-stressed. Plant roots can continue to take water from the soil after the refill point is reached, but this water is not as readily available, and the crop finds it difficult to extract. If the soil dries to a permanent wilting point, the plant can no longer remove any water from it, although some water may still be present it is completely unavailable. There are two ways used to commonly measure RAW, soil moisture percentages taken from soil moisture probes and soil texture tables.

Moisture probes

A capacitance probe is a device for measuring soil water content. It consists of a number of sensors mounted on a vertical probe column which is installed down the soil profile. The sensors are connected to a central data logger and record sensor readings at pre-determined intervals. Each sensor consists of an oscillator and several electrodes, an array of parallel metal spikes that form a capacitor. The soil adjacent forms the dielectric of the capacitor, which completes an oscillating circuit. A capacitance field is generated between the two rings of each sensor. This field extends into the soil adjacent to the sensor and is affected by the soil's dielectric. The soil dielectric is affected by water content. Changes in water content cause a change in the frequency of the oscillating circuit which can be related to actual soil water content via a calibration curve.

To be useful for irrigation scheduling the moisture probes must record water use in the root zone. In young Sandalwood plantations (two to six years) the observed host rootzone depth range is from 70 to 100 cm and averaged 87 cm. While Sandalwood rootzone depth ranged from 30 to 85 cm and averaged 47 cm (Kew, 2010) .

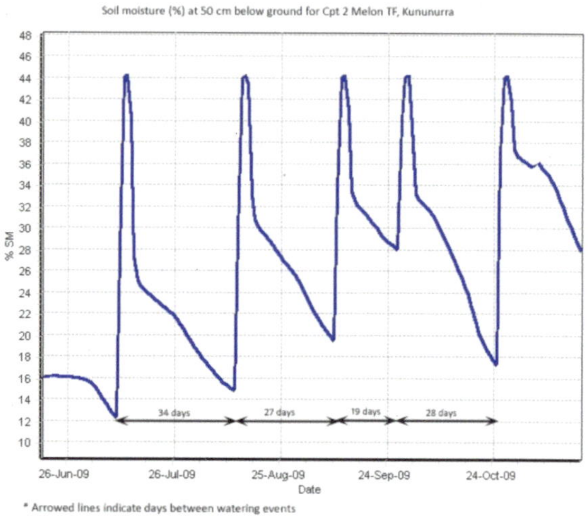

Figure 164: Soil probe chart showing soil moisture % in top 50cm of the profile in two year old Sandalwood planation in ORIA and the flood irrigation events over a five month period.

Figure 164 shows the water profile reduction in the root zone of a two year old Sandalwood plantation. The chart shows the stacked water in the soil profile in the root zone after an irrigation event, refilling the profile to 44% and the gradual drawdown over 34 to 19 days to a wilt point of 14% soil moisture. It is important that the moisture profile does not reach the wilt point. It is much better for the trees if the profile is refilled before it reaches that point.

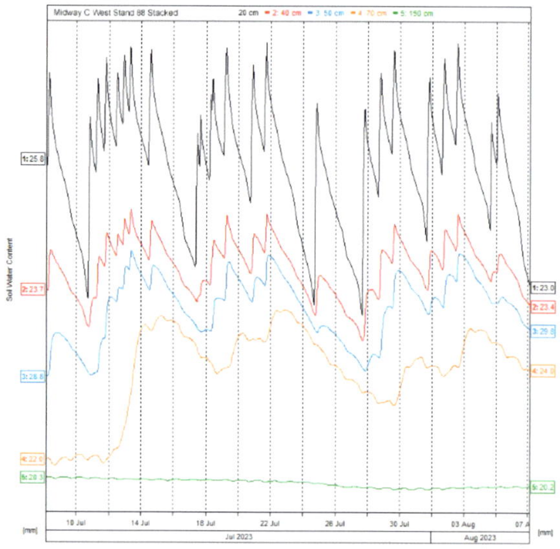

Figure 165; Stacked soil moisture probe chart showing soil moisture content down the profile. Note this is from a drip irrigated site. The chart shows the daily additions of water to the profile.

Soil probes are also necessary for monitoring the effectiveness of scheduled irrigation. Probe data can be viewed as a summed or stacked chart. Figure 164 is a summed chart which shows the overall moisture content for the profile. Whereas figure 165 shows the stacked data, the moisture content for the various levels down the profile. This is useful to determine how far down the profile the irrigation is reaching and to determine how far down the profile water is being used by the crop. The data displayed in figure 165 shows the daily addition of irrigation water by the drip system and the water is reaching down to 70 cm but not to 150 cm.

RAW Soil texture tables

The second method commonly used to work out RAW is to use the soil texture method which uses a table developed for use in agricultural crops. The tables deduce soil RAW by comparing increasing soil textures against soil water pressure (taken at wilt point). Tensiometers can be used to determine this soil water pressure as a measure of how dry the soil is. The table converts the tensiometer pressure reading into volume of available water for the particular soil texture at particular compaction and level of course fragments (Table 33) (McKenzie D. , 2011).

Table 33: RAW soil texture table. This table was provided by SoilMgt (McKenzie D. , 2011) and based on the data from (Moore 1998) and (Wetherby 2003).

RAW (-8 to -60 kPa), mm per cm of soil profile

Sand	Coarse	0.20	0.20	0.16	0.16	0.12	0.12	0.08	0.08	0.04	0.04
	Medium	0.40	0.40	0.32	0.32	0.24	0.24	0.16	0.16	0.08	0.08
	Fine	0.44	0.44	0.35	0.35	0.26	0.26	0.18	0.18	0.09	0.09
Loamy Sand / Clayey Sand	Coarse	0.43	0.43	0.34	0.34	0.26	0.26	0.17	0.17	0.09	0.09
	Medium	0.48	0.48	0.38	0.38	0.29	0.29	0.19	0.19	0.10	0.10
	Fine	0.51	0.51	0.41	0.41	0.31	0.31	0.21	0.21	0.10	0.10
Sandy Loam	Coarse	0.43	0.63	0.34	0.51	0.26	0.38	0.17	0.25	0.09	0.13
	Medium	0.49	0.63	0.39	0.51	0.29	0.38	0.20	0.25	0.10	0.13
	Fine	0.63	0.77	0.51	0.61	0.38	0.46	0.25	0.31	0.13	0.15
Loam		0.57	0.77	0.46	0.61	0.34	0.46	0.23	0.31	0.11	0.15
Sandy Clay Loam		0.57	0.68	0.46	0.55	0.34	0.41	0.23	0.27	0.11	0.14
Clay loam		0.54	0.69	0.43	0.55	0.32	0.42	0.21	0.28	0.11	0.14
Clay		0.51	0.80	0.41	0.64	0.31	0.48	0.21	0.32	0.10	0.16

c) Crop Factors (Cf)

The crop factor is the ratio between evapotranspiration observed in a specific crop and that of a reference crop under the same conditions. While evaporation is relatively easy to measure, transpiration is more difficult to quantify. For this reason, crop evapotranspiration is more practically estimated by relating it to daily evaporation using a crop factor

The amount of water a plant uses depends on the weather. For example, it cannot be assumed that this plant uses 1.5 litres per day. Instead, a ratio called a crop factor is used. It is the ratio of the water used by the plant to evaporation. The proportion of daily evaporation that must be replaced by irrigation is referred to as the 'crop factor' and this varies with crop type, vigour and growth stage.

A list of standardised crop factors has been worked out for many crops and can be found in Allen's "FAO 56" (1998). The crop factors for other tree horticultural crops from this document are around the 0.7.

Studies have shown that in summer, Sandalwood (in Kununurra) only transpire for a few hours per day (5 am to 8 am). This indicates a low crop factor determined for Sandalwood of 0.3 by Cater et al (2010). However, Lewis et al (2006) points out that a Sandalwood plantation has other species which do transpire for much longer periods.

Flood scheduling

The application of irrigation water as a flood provides a 'feast and famine' scenario to the plantation. Just after application, the site is effectively waterlogged until gradual infiltration and run off reduce water levels. If the levelling is not correct or if there has been 'slumping' across the site, water can lay in the stand causing mortality due to waterlogging. Over a period, which varies dependent on the climate, the site dries down. Usually, the top of the mound dries first and the inter-row last.

Research trials and a lot of trial and error by irrigators has determined that in ORIA and BRIA, a 40-day flood schedule is suitable to provide sufficient water to the trees without causing detrimental health effect due to waterlogging or drought.

Drip Scheduling

The scheduling of irrigation used for the drip system can be tailored for individual groups of plantation stands. This is somewhat limited to the valves and the stands within that valve. Several valves are connected to a subterranean sub-main water supply. The limit to supply the water is how much the bore can provide and how much can be pumped by the system. Valves supply this amount of water to several stands at a time. Once a valve has been opened and the stands have received their allocated amount of water, the valve then closes, and the next valve opens to irrigate the next group of stands.

The scheduling calculations for the start of the irrigation season are from a nominated end of the wet season. In most cases this is the end of April when soil profile moisture is determined in the root zone of 800mm. The soil water calculations are the same used in Trial 2 described in the Research section below and utilise ETo, profile RAW in the root zone and a crop coefficient. This varies dependent on the sites, and takes into consideration stand age, species composition and stand stocking.

A critical component of the drip system is the maintenance required to keep the system running effectively. The maintenance required for the engines and pumps, the filtration systems and the dripper lines are constant.

4. RESEARCH TRIALS.

Trial 1: There are several instruments that can be utilised to monitor soil and water use. As discussed, the main method used operationally for determining when to irrigate is by monitoring the gradual reducing soil water profile, using capacitance probes. However, the research team has also investigated several other instruments to assist with monitoring soil water use by the plantation.

Tensiometers

Soil water potential is the most direct measure of plant available water. A tensiometer is an instrument that directly measures soil moisture suction. It consists of a porous ceramic tip, a sealed water-filled plastic tube and a vacuum gauge. The porous cup is buried in the soil and allows water to freely move between the water-filled tensiometer and the soil (figure 166). As the soil around the cup dries, the suction increases and water moves out of the tensiometer until the suction within the tensiometer is the same as that of the soil water. The soil suction reading relates directly to plant water tension, and hence it is a more meaningful measure of plant stress than the soil water content.

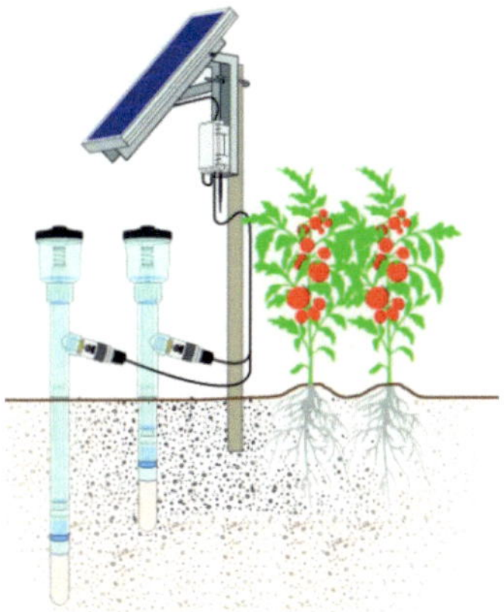

Figure 166: Typical soil tensiometer, (ICT International).

Dendrometers

A dendrometer is a sensor used to measure the trees circumference or diameter. The effectiveness of logging dendrometers in recognising water stress and drought resilience in trees has been identified in several tree crop species (Wheeler, Black, & Bugbee, 2023) (Cuevas, et al., 2010). Under soil water deficit situations, transpiration during most of the day exceeds root water supply. Greater water potential develops between the xylem and the living tissue in the stem than under well-watered conditions. The rate of water flow between these tissues is increased in dryer soil conditions, causing greater trunk shrinkage. Trunk diameter fluctuations occur in response to the potential gradients within the trunk, which reflect the balance between tree water supply and demand. Therefore, there is a close association between daily evolution of tree water status and trunk diameter oscillations.

Sap Flow Meters

Sap flow meters are instruments for the measurement of sap flow or transpiration in plants. The most effective sap flow sensors use the Heat Ratio Method (HRM) principle and can measure high, low and reverse flow rates in small woody stems and roots as well as large trees. The sensor consists of 35mm needles installed in the stem of the tree, the middle needles pulses heat into the sap and the outer needles detect the rate of heat transfer to the outer needles (Figure 167).

The concept behind using sap flow meters is the fact that under water stress, the magnitude of the sap flow velocity in a monitored tree crop is reduced. Information about the sap flow trends can then be used to determine when irrigation is required (Figure 168).

Figure 167: Sap flow meter used by Quintis.

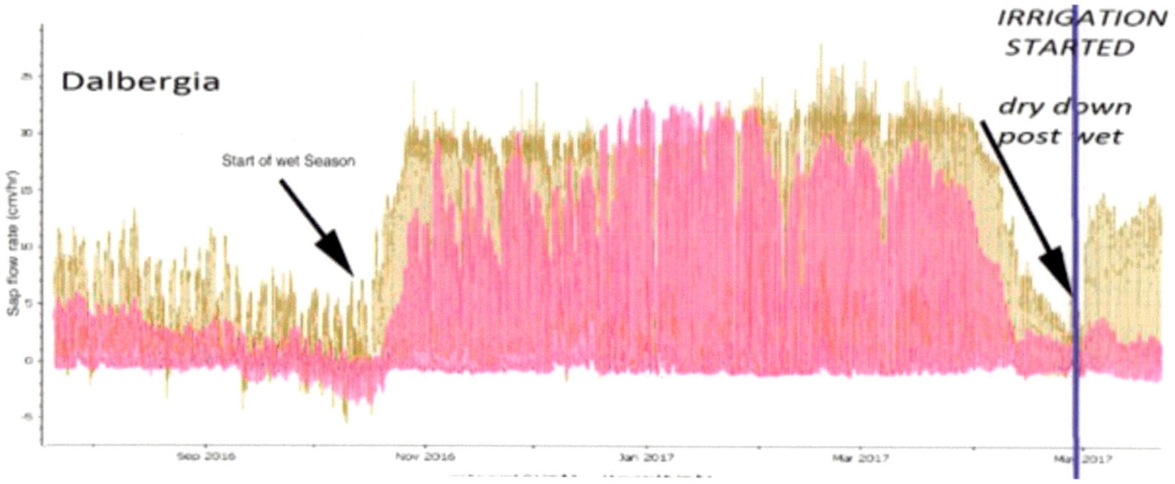

Figure 168: Data from a Sap Flow Meter on a long-term host tree. The data clearly shows the tree's water use before, during and after the seasonal wet season, as well as indicating when the tree needs irrigation. The chart shows the sap flow rate during September which has not been watered for several weeks. Note the sap flow rapidly increase with the onset of the wet season and conversely the sap flow slowly declining each day after the wet season finished. A point in late March indicting much reduced sap flow and the need for irrigation.

<u>Trial 2</u>: A study was conducted across seven properties for Elders Forestry in July 2010 to understand this period of dry down time across flood irrigated stands in ORIA (Robson K. , 2011). Although the soils across the ORIA are generally heavy cracking clays there is some variation in clay content between various plantations and this could potentially affect the RAW. The rootzone RAW values were calculated from water retention data for clay soil samples from Kununurra (Table 34). A soil moisture deficit of -8 to -60 kPa was used. This is based on *Sesbania formosa* as it is most susceptible to water stress. The results indicate that there is significant variation in RAW between sites, up to 48% between the highest RAW at Kneebone to the lowest at Scott's and Melon (Table 34).

Table 34: Average RAW for Elders Sandalwood plantations in ORIA.

Property	Sandalwood Rootzone (mm)	RAW (mm)
Kneebone	450	221
Doc Wally	450	191
Innes	400	184
Top Banana	525	177
Mock Nth	500	152
Scott's	525	149
Melon	510	148

The study tested several crop factors to find the best suited, using the climatic data in 2009, the dry time in Figure 164 and RAW in Table 33. In July 2009 Compartment 2 Melon took 33 days to go from field capacity to just above a considered wilt point.

To calculate the water use, a crop factor was multiplied by the daily evapotranspiration.

- Crop water use (mm/day) = crop factor x evaporation (mm/day)
- In this case if we use a Cf of 0.7 : 0.7 x 6.5 = 4.5mm/day

The RAW was divided by water use to estimate how long that water will last.

- Maximum irrigation interval (in days) = RAW ÷ Crop water use (mm/day)
- In this case: 148 / 4.5 = 32.5 days

The soil moisture probes showed that it was 33 days between watering, and using a crop factor of 0.7, with a RAW of 148 calculated an estimate of 32 days.

To estimate the drawdown period for scheduling, if the August data was used for the same compartment, with an ETo of 8.3 and an actual irrigation interval of 26 days, the Cf of 0.7 is closest to the mark (Table 35).

This trial demonstrated that using the known RAW values for each property, the climatic ETo values and a Cf of 0.7 for Sandalwood plantations between three and six years of age accurately predicted the dry down time between field capacity and near wilt point (Table 35).

This study showed that a flood irrigation schedule of four to six weeks for Sandalwood plantations across ORIA, dependent on the Eto, is satisfactory for retaining good soil moisture.

Table 35: Soil water calculation data to determine the correct Cf and irrigation interval.

Month	Evapotranspiration per day (mm)	Actual days between Field Cap & wilt point	Crop factor	Daily water mm	Approximate Irrigation Interval (days)
					148
July	6.5	33	0.35	2.3	65.1
July	6.5	33	0.5	3.3	45.5
July	6.5	33	0.7	4.6	32.5
August	8.3	26	0.35	2.9	50.9
August	8.3	26	0.5	4.2	35.7
August	8.3	26	0.7	5.81	25.5
Sept	9.2	20	0.35	3.2	46.0
Sept	9.2	20	0.5	4.6	32.2
Sept	9.2	20	0.7	6.4	23.0

Trial 3. Several low irrigation trials were established between 2015 and 2017. In 2016, an experiment was conducted at two sites with different soil types. One site was in a three year old planting at TP11, a sandy topsoil site. The other at Sandy ck, a two year old planting, with a loam topsoil site. Each site had two 0.078-hectare plots, one had irrigation removed and the other was the irrigated control.

The plots were assessed each year for diameter and survival of all species. The results found that for Sandalwood although initially the irrigated plots significantly grew more than the non-irrigated, both treatments grew similar amounts in the subsequent years. The irrigated trees at the Sandy creek (SC) site did grow better than the irrigated trees at the TP11 and both non irrigated plots.

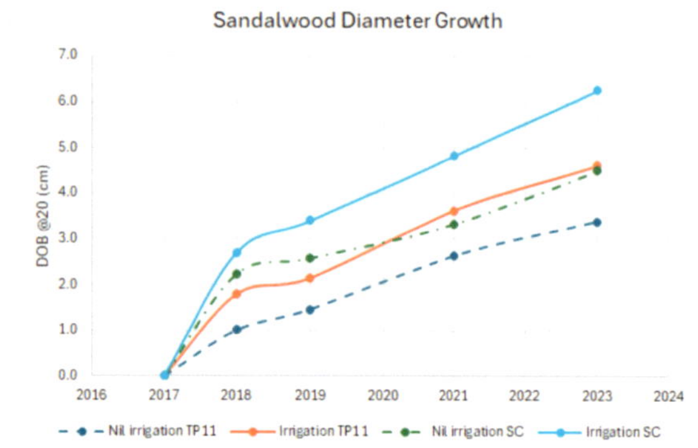

Figure 169: Annual Sandalwood diameter growth for irrigation treatments at TP11 and Sandy creek.

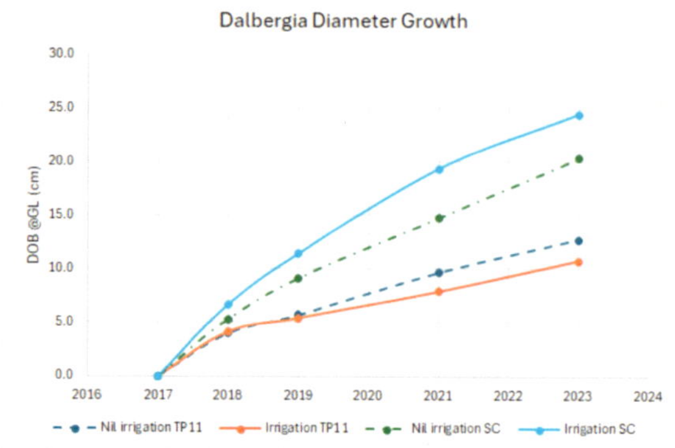

Figure 170: Annual Dalbergia diameter growth for irrigation treatments at TP11 and Sandy creek.

Although the Sandalwood mean DOB at 20cm was similar for each treatment, the percentage of larger trees, greater DOB at 20cm than 13cm, was bigger in the irrigated plots at both sites. However, the percentage of mid-size trees between 9 and 13cm was larger in the non-irrigated plots.

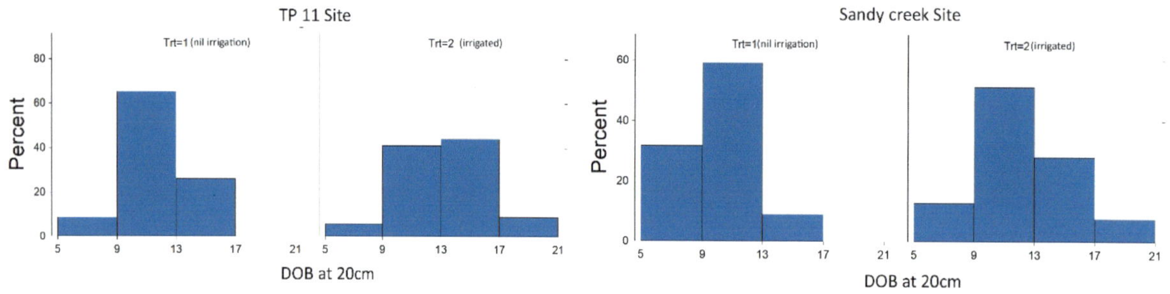

Figure 171: Histograms of diameter mover bark at 20cm for Sandalwood trees at the 2023 assessment, at the sandy Creek and TP11 sites in irrigated and nil irrigated treatments.

The results for *D. lanceolaria* were similar. The irrigated trees grew more than the non-irrigated at sandy creek but not at TP11 where they grew at a similar rate to the non-irrigated. The addition of water did not significantly increase tree growth for either species.

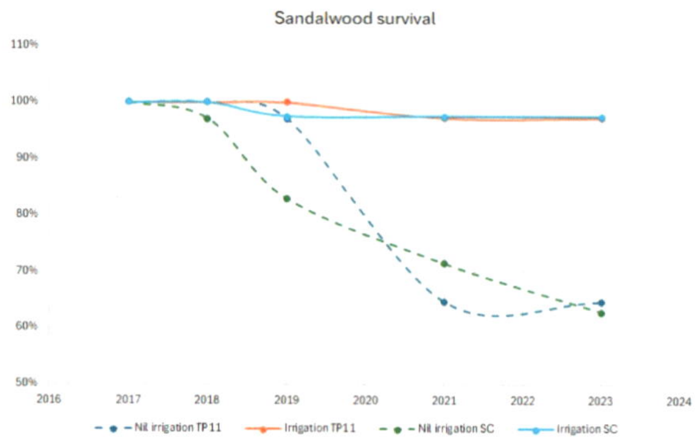

Figure 172: The removal of irrigation significantly affected Sandalwood survival on both sites.

However, the addition of water did increase Sandalwood survival. The irrigated treatment survival of hosts was similar in both treatments, the addition of water to the Sandalwood made a significant difference. On both sites the non-irrigated plots had a 35% to 37% mortality rate over the irrigated plots which had only lost 2% of the trees after six years.

At establishment, Sandalwood trees in both irrigated and non-irrigated treatments had Sap Flow meters installed. As expected during the first post wet season the sap flow in the non-irrigated was lower than the irrigated. However, it was not until the beginning of the buildup season where the Eto is highest that the sapflow in the non-irrigated trees significantly slowed (figure 173). This drop in sap flow could be visually observed in the trees. The non irrigated trees had less leaves and appeared to be drought stressed.

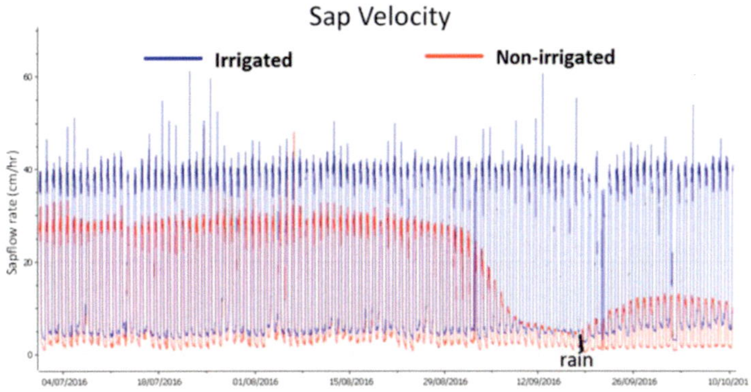

Figure 173: Sandalwood sap flow from July onwards in a non-irrigated plantation (red) at Taylors Park, compared to the irrigated plantation (blue). Trees exhibited similar growth up till the end of August.

The experiment demonstrated that removal of irrigation may not reduce mean tree size but will reduce tree density, but what effect did the treatment have on heartwood development? A small number of individuals (n=2) from each treatment in TP11 were destructively sampled to review heartwood development at age ten.

Care was taken to sample trees of similar diameter. The results found that the two trees sampled in the non-irrigated treatments had significantly defined heartwood at 20cm above ground. Whereas neither tree in the irrigated treatment had any obvious signs of oil-bearing heartwood. The irrigated trees had more transitional heartwood than the non-irrigated trees but no dark-coloured oil bearing heartwood.

Destructively sampled Sandalwood in irrigated treatment 12cm DOB at 20cm.

Destructively sampled Sandalwood in non-irrigated treatment 12cm DOB at 20cm .

The cut stem at 20cm above ground showing significant development of transition heartwood but no oil-bearing heartwood could be visually determined in either tree.

The cut stem at 20cm above ground showing defined oil-bearing heartwood development and surrounding transition heartwood in the sampled trees in the non-irrigated treatment.

Figure 174: The trees and stem biscuits from two typical trees destructively sampled in TP11 irrigated and non-irrigated plots, (TP11 Robson, 2024).

It was surprising that the non-irrigated trees had more oil-bearing heartwood than the irrigated. Although, the sample number was not high enough to say that irrigated stands will have less heartwood, it does create questions as to why this is the case on this site.

Trial 4. A trial was established at Kingston Rest in 2016, investigating the effect of reduction of irrigation on Sandalwood. In this experiment taps were installed on the Sandalwood drip line and water was turned off on the Sandalwood rows only, the host rows were irrigated are normal. The irrigation was removed after the wet season and there was no discernible difference in the sapflow between the treatments. No growth reduction was observed in non-irrigated Sandalwood when compared to irrigated trees until mid-September, more than four months after irrigation started in early May. Sandalwood taps were turned back on in early October and the Sandalwood sap flow quickly recovered over the following week (Figure 175).

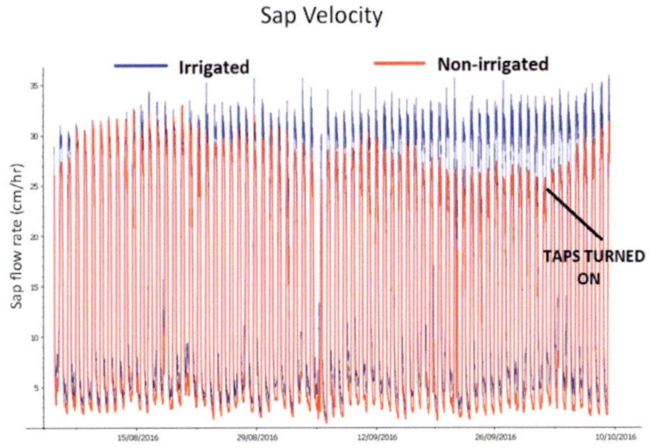

Figure 175: Sandalwood sap flow (cm/hr) in KR Farm 1 from August onwards.

The low irrigations trials have shown that Sandalwood, after they have established, can be grown with reduced water. The low term effect of irrigation is the mortality of the smaller, weaker individuals.

5. DETRIMENTAL EFFECTS

Plantation growth rates and survival are impacted by a broad range of issues related to irrigation and water use. This includes a poor understanding of, ideal soil moisture levels, watering depths which may lead to sub-optimum plantation growth and health, inefficient control, monitoring mechanisms that may allow plantation to overwater or drought with subsequent tree mortality, poor health and growth.

a) Water deficit

Drought is the most important abiotic factor adversely limiting growth and crop production. Drought stress results in non-normal physiological processes, which will influence one or a combination of biological and environmental factors. The damage to the normal metabolism may reduce growth, or plant death (Fathi & Barari, 2016).

Sandalwood is known for its drought tolerance and typically does not succumb directly to reduced soil moisture. However, decreased soil moisture can trigger mortality in host species, which, in turn, may lead to Sandalwood death due to physiological shock from sudden host loss and susceptibility to biotic factors. Water supply to plantations is meticulously monitored and scheduled, considering the varying water needs of different tree species. For instance, Sesbania requires more water compared to Sandalwood and Cathormion. Maintaining the right balance of water is crucial, as both overwatering and underwatering can have adverse effects.

Water stress can affect trees in several ways, with its impact depending on factors such as climate, soil water holding capacity, water availability, and tree characteristics like age and crown size. Water availability significantly influences nutrient uptake and metabolite transportation, especially critical for

Sandalwood due to its parasitic nature. The maintenance of a certain level of water availability is essential for Sandalwood's parasitic relationship with its hosts.

Plants typically respond to water deficit by either dropping leaves or closing stomata. Sandalwood belongs to the isohydric category,. It is capable of adjusting stomatal conductance to maintain a constant water potential despite changes in soil water potential and evapotranspiration. Prolonged water stress can lead to stomatal closure, reducing transpiration and photosynthesis, eventually causing carbon starvation and dieback. The strategy of stoma modification to open or close all or some is essential for Sandalwood due to the requirement to maintain sap flow potential constant with the host (Radomiljac, McComb, & Pate, 1999).

McDowell et al (2008) explains water stress results in two basic physiological events, short-term water stress leads to hydraulic and symplastic failure (tree runs out of water), or to a carbon starvation as a result of long less drastic water stress.

With long-term reduced water availability, Sandalwood will close stomata to maintain sap water potential. Reduction in transformation will eventually lead to carbon starvation. With the stomata closed transpiration does not occur and the tree has to rely on stored carbon. But if this is prolonged then that stored carbon runs out and the plant becomes carbon negative, this in turn causes die back.

The uptake of nutrients is difficult when soils are dry and to increase the uptake requires extra energy. This is difficult as photosynthetic activity has slowed due to the partial or complete stomata closure. However, for trees that are not water stressed for too long and have significant starch (carbon) reserves stored in the leaves and trunk, this is not an issue (Moore, 2012).

Short-term water stress can result in hydraulic and symplastic failure, prompting plants to shed leaves as a defence mechanism. The sudden onset of water stress triggers the release of stress hormones like ethylene and ABA, leading to leaf and fruit abscission. Foliage symptoms of water stress include shedding, scorching, chlorosis, and browning, while water-stressed trees become more susceptible to pathogen attacks.

Root size significantly influences hydraulic uptake, and water deficit affects fine root growth and haustorial connections, further impacting water uptake by plants. Thus, understanding and managing water stress are crucial aspects of plantation management, particularly in tropical environments where it can be challenging to avoid.

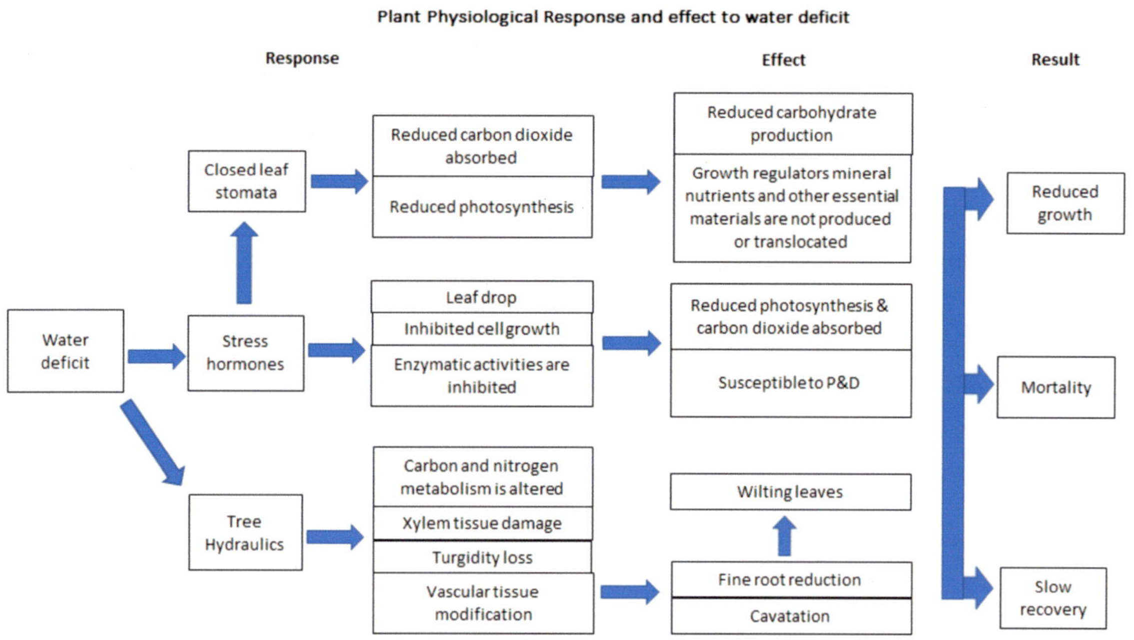

Figure 176: The response to a water deficit in plants can vary the effect and ultimately the result, which can range from a slowdown in growth to tree death, (modified from (Coder, 2018).

As mentioned, Sandalwood is a reasonably drought tolerant species. However, while connected to the short-term host species sesbania, which is not a drought tolerant species, Sandalwood is not drought tolerant. Host shock is when there is the sudden removal of the key host. Usually, the short-term host die back over a period, and this gives the Sandalwood time to detach and start developing foraging roots. The sudden removal of the main host does not allow for this to happen. This can be sometime observed when sesbania are heavily topped and due to this it quickly dies. The Sandalwood leaves would turn brown and eventually drop. In most cases the Sandalwood would recover when they find a new host, but if this takes too long, they will die of water stress or a biotic factor.

b) Water logging

Sandalwood does not like to have its roots in anaerobic conditions for long. Sandalwood naturally occurs on sloping sites that drain well and occasionally outside mangroves on sandy well-draining soils. Evidence from 20 years of plantation development shows that when Sandalwood is subjected in inundation for more than ten days (approximately), they are susceptible to a range of pathogens that will cause mortality.

ENVIRONMENTAL STRESS

Sandalwood plantations are a living system, albeit an artificial system. Unlike most forestry plantations, these plantations are mixed species plantings, which are known but poorly understood to be more structurally complex than monocultures (Zemp, et al., 2019). Due to diversity, mixed species planting often show a higher resilience to some natural events that can cause crop stresses. When establishing large green spaces on new sites there is always the possibility for unforeseen detrimental events.

It is well understood that a key reason for some individual mortality is due to site capacity and competition. Where sub dominate trees just cannot compete for the necessary resources over a more dominate nearby individual. As tree mortality is due to competition, lack of nutrition, poor soils and over hosting has already been discussed in other sections of the book, this section will focus on the other main stress factors that can detrimentally affect Sandalwood growth and pose a significant risk to the plantation commercially reaching full rotation.

Plants are subjected to a wide range of environmental stresses which reduces and limits the productivity. Two types of environmental stresses are encountered by plants and can be categorised as abiotic stress and biotic stress. The stress from abiotic factors have caused major crop losses worldwide and these factors include extremes in climate, inferior quality soils and drainage and anthropogenic factors. On the other hand, attacks by various pathogens such as fungi, bacteria, oomycetes, nematodes and herbivores are included in biotic stresses. As plants are sessile in nature, they have no choice to escape from these environmental cues. Many native species tend to have developed various mechanisms to overcome these threats. However, exotic species usually do not have the same resilience, and this makes them vulnerable to a range of onsite abiotic and biotic stress factors (Gull , Lone, & Islam Wani, 2019).

1. STRESS DIAGNOSIS.

Over the period of the rotation, the stocking levels in some areas can be observed to significantly reduce. Occasionally tree mortality or decline in health is observed within the plantation. In some cases, the reason for this is not obvious as there are several vectors that can cause tree decline. These are primarily, abiotic and biotic factors or poor and unintentional silvicultural management.

It is beneficial to understand the stress factors and the threat they pose to plantation health. In ecology, biotic and abiotic factors encompass all the living and non-living parts of an ecosystem. Biotic factors pertain to living organisms and their relationships. Abiotic factors are the non-living components of the ecosystem, including sunlight, water, temperature, wind, and nutrients.

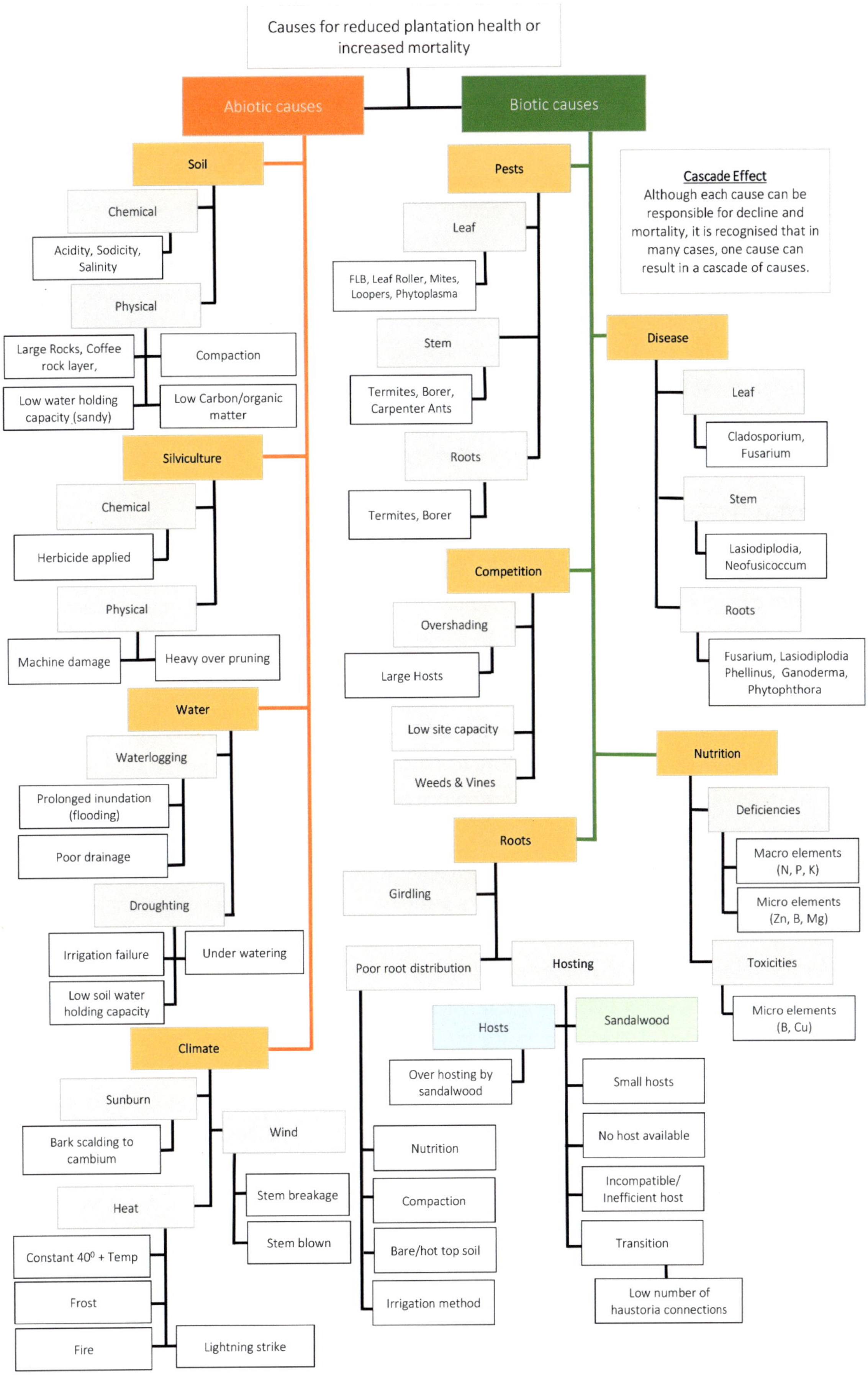

Figure 177: Risk matrix for Sandalwood and hosts in plantation, (developed by Robson, 2019).

These factors can be broken into two categories, primary and secondary stressors. Primary factors are those that are the first to apply stress to the tree. While secondary factors only become effective due to the tree being weakened by the primary factor.

An example of this is the mortality has been observed in the plantation Senna in the Northern Territory. Mass mortality only occurred once the Sandalwood transition occurred. The stress of hosting the Sandalwood weakened the Senna and it became susceptible to the pathogen Lasiodiplodia, which eventually killed the tree.

Certain factors may independently result in mortality, such as fire or specific anthropogenic influences like herbicide poisoning.

Experience has shown that it is usually a combination of two or more stress factors that causes mortality. Mortality is caused by cascading effect of these factors. The matrix shown in Figure 177 indicates most of the biotic and abiotic vectors that can cause stress.

2. BIOTIC STRESS FACTORS

Pest and diseases can affect plant health and compromise the ability of the plantation to meet their management objectives if unabated. As the plantations age, risks to plant health from pest and disease increase.

In 2014, there was a major defoliation event from Hyposidra looper in several Sandalwood stands in Florina Road. In 2015, Figleaf beetle heavily infested Sandalwood, while crusader bugs and leaf rollers heavily infested Cathormion across the Northern Territory and Queensland plantations. In the same year, Crexa moth and borers also heavily infested stands in Kununurra. From 2016 to 2018, there were more frequent heavy defoliation events. These were caused by Figleaf beetle, looper and leaf miner or a combination of two or three of these insects simultaneously in Sandalwood and leaf rollers and twig girdlers in Cathormion in the Northern Territory plantations. In 2018, bean pod borers also infested sesbania in the Early Storms plantation. Other than the existing insect pests, some pests considered minor for several years flared up and severely damaged Sandalwood in the Northern Territory in 2020, such as mealybugs and spider mites. In 2021, grasshopper and locusts started to affect Sandalwood across all the Sandalwood plantations in Australia.

Equally damaging with these herbivores are the northern giant termites (*Mastotermes darwiniensis*) and different small termites' species feeding on live Sandalwood and host trees. Occasionally, feral pests such as rats and native wallabies chew drip tapes, causing droughting stress to trees. Stray cattle from neighbouring properties sometimes encroached the plantation to feed and break Sandalwood and host trees.

Since 2016, incidences of plant pathogen diseases increased in host trees, predominantly Senna and occasionally in *Dalbergia lanceolaria* and Sandalwood. Identified diseases affecting Sandalwood plantations in Kununurra, were Phytophthora spp and Lasiodiplodia spp. In 2020, more

Botryosphaeriaceae, Fusarium, Cylindrocarpon, and Pythium were observed in plantations affecting Dalbergia, Cathormion and Sandalwood.

With losses of hosts considerable numbers of Sandalwood in the Northern Territory plantations had sunburn or sunscald. This can render a large tree non-economically viable because they may either die before end of the rotation or produce little heartwood.

This section will discuss the most apparent risks to plantation health, including pathogen dieback, insect pests, waterlogging and droughting, fire, cyclone and wind, sun scorch, and heartwood rot.

a) Pathogen dieback

PHYTOPHTHORA

The first identified signs of tree dieback were observed in Kununurra Sandalwood plantations in 2004. In a 2005 report, Dr Giles Hardy (Murdoch University) observed the disease symptoms within the plantation were very typical of a biotic factor (an infectious agent such as fungi, or bacterial) rather than abiotic (such as waterlogging, salinity) factors. This was because the pattern of deaths observed though widespread were not continuous but in mosaics. An abiotic cause would tend to result in more even deaths across each compartment. Subsequently Dr Hardy isolated a Phytophthora species from the roots and from the soil baits.

The source of Phytophthora in Ord valley Sandalwood plantations is the irrigation water supply. In 2005 two species of Phytophthora were identified that have been isolated from several locations in the irrigation channels of the Ord Irrigation Area (Davidson & Cunnington, 2005). The two species are *Phytophthora cinnamomi* var. *Parvispora* and a Phytophthora spp. that was an undescribed species in 2005.

Phytophthora has also been identified in four year old drought stressed Sandalwood in Taylors Park in Northern Territory. Pathology identified Phytophthora in soil from around the root system in small numbers.

Phytophthora cinnamomi

Phytophthora cinnamomi is an introduced water mould. Based on historical evidence of agricultural impacts in the 1800s, the pathogen is believed to have been brought into Australia by early European settlers, presumably within infected plants. First detected in Australia in 1935 (Burgess, et al., 2021), *P. cinnamomi* has since been introduced to many locations across the Australian continent. It has contaminated thousands of hectares in Western Australia, Victoria, Tasmania and South Australia, as well as wet coastal forests in Queensland. Once an area is infected with *P. cinnamomi*, the organism cannot be eradicated, only managed.

Infection of plants by *P. cinnamomi* is favoured by free water in the soil or ponding on the soil surface. Warm wet soils, especially those with impeded drainage, favour the germination of *P. cinnamomi* chlamydospores (spores that enable the pathogen to survive in an area during harsh environmental conditions). As the roots of the infected plant die, long-lived resistant chlamydospores are produced again. This provides a source for future re-infection of seedlings in the area. Soil conditions that affect *P. cinnamomi* include moisture, temperature, structure, chemistry and organic matter content. Soil moisture levels and temperatures significantly influence the activity of *P. cinnamomi.* High organic matter content provides for increased micro-organisms which may compete with and prey upon *P. cinnamomi.*

Management Practices

An integrated approach to the management of *P. cinnamomi* within the Sandalwood plantations is required. Although the disease cannot be removed, several strategies can be put in place to ensure minimal impact.

This would include irrigation management, as the use of spore contaminated irrigation water is unavoidable. If the amount of water on each plantation compartment can be managed to reduce the risk of infection. *P. cinnamomi* prefers damp soil conditions, to ensure soil moisture conditions are optimal for tree growth, and over watering is avoided, soil moisture should be closely monitored. Soil moisture probes that are strategically placed around the plantation to monitor soil moisture levels can be used. In addition, a site-specific watering regime should be developed for each plantation to consider localised soil and micro-climate factors. In flood irrigated site, during site preparation, careful levelling is undertaken to ensure water does not pond but will flow constantly through the plantation. Monitoring of flow rates is undertaken to measure amount of water flowing in and out of plantation compartments.

The retention of organic matter within the plantation will assist with the beneficial micro-organisms. Rows are slashed regularly to reduce weed and grass competition. Slashing reduces weed and grass competition, but the organic waste is retained along the rows. This slash waste contributes to the organic matter which will eventually decompose back into the soil. Higher levels of organic matter make conditions suitable for micro flora and fauna antagonistic to *P. cinnamomi.*

Well known horticulture crops susceptible to *P. cinnamomi* such as the Avocado industry have been managing the disease by using regular applications of phosphoric acid, applied as Agriphos 10L/ha. Additions of Phosphoric acid to trees strengthens root cell walls inhibiting infection by the pathogen.

Research experimental work has shown that the aerial application of Agriphos at 10L/ha applied twice prior to the wet season can significantly increase Sandalwood root cell wall thickness making it more resilient to *P. cinnamomic* infection (Robson & Tapscott, 2010).

Impact on plantation

The time of the year when symptoms are most obvious is after the wet season. During this time of the year, the Ord valley has had a lot of rain, soils tend to be waterlogged, ponding can occur, heat and humidity combine for optimal conditions for the disease. Some of the worst affected areas have had up to 10% of the trees showing some symptoms. Trees that show infection symptoms by *P. cinnamomi*, wilt and drop most of their leaves. A high percent of these will recover and within four months have replaced all their leaves. This loss of leaves could however influence the growth rates.

BOTRYOSPHAERIA

The Botryosphaeriaceae fungi represent a significant disease affecting various trees and shrubs in forests and managed landscapes. Currently, this classification encompasses over two dozen genera, commonly associated with cankers on woody trees. They are non-mobile, cellular organisms, whose structure is composed by threads called hyphae. They produce their compounds by absorption or exchange (heterotrophy) and reproduce by spores (Wijayawardene , et al., 2017).

Among the common tropical species in the Botryosphaeriaceae family are Lasiodiplodia spp, *Neoscytalidium dimidiatum*, and *Pseudofusicoccum adansonia*. Within Sandalwood plantations, species like *Lasiodiplodia theobromae* and *L. crassipara* have been identified, isolated from boabs and mangoes in the Kimberley region, posing a threat to stressed trees (Barbour, Norris, & Burgess, 2010).

Typical symptoms include sudden leaf browning, failure or reduced new leaf shooting in the spring, although manifestations can vary depending on the host and underlying stresses. Weakened trees are susceptible to extensive cankers in the cambium and sapwood, disrupting water and mineral transport. Infections may range from small necrotic areas to extensive branch dieback and trunk cankering. Severely affected plants often exhibit water sprouts, stained bark, and sunken lesions with oozing sap (Brazee, 2023). Drought-stressed plants are particularly vulnerable to this disease, with fungi commonly entering through wounds.

In Sandalwood plantations, affected trees typically show a black powdery appearance on the stem, with branches or stems dying from above the affected point (Figure 178A). When observed from a distance it looks as if the tree has died from bottom up, but on closer examination the symptoms start from the top down always from a wound. Cross section of trunks and branches reveal a blue/black staining of the wood (Figure 178B).

Observations from mangoes in Kununurra have shown these symptoms were a result of infection by Botryosphaeria after mechanical pruning (Sakalidis, Ray, Lanoiselet, Hardy, & Burgess, 2011). The same pruning was undertaken on the Senna and Dalbergia and the same symptoms were observed (Figure 178C).

Management Practices

Management practices primarily focus on prevention, as there are no known effective treatments in Australia. The disease is latent across all the plantation estate, however, observation of infected trees in the plantation indicates although mortality occurs, some individuals will recover.

Some species tend to be more resistant to the infection than others. *Dalbergia latifolia* seems to be resistant to infection with very few observations of individuals affected with this disease. Cathormion and Sandalwood are also resilient, however weak poorly hosted Sandalwood trees can succumb to the disease. The most susceptible species are Senna and *Dalbergia lanceolaria*. Although recovery has been observed in both species.

Figure 178: A - Senna with typical blackened stem symptoms indicating Lasiodiplodia. B – cross section of main stem showing blue/black stain symptoms and C- hedging wounds in Senna are the main source of infection, (photos taken from Burgess report (2014).

Stress, particularly water stress, increases susceptibility, emphasising the importance of crown management and hedging of long-term hosts to prevent disease spread. As discussed in the pruning and competition sections of this chapter, the crown management of long-term hosts is essential for a balanced plantation. The hedging of the long-term hosts has also been responsible for the spread of Lasiodiplodia along the host row. Only unstressed trees not affected by the disease, and during a period with good soil moisture must be include in any hedging program.

Impact on plantation

The impact on the plantation due to Lasiodiplodia die-back in some cases has been catastrophic, with the loss of all long-term hosts. This has a major detrimental effect on the health and growth of the Sandalwood. The loss of the hosts will stop Sandalwood growth and expose the Sandalwood stems to afternoon sun which will cause sun scald.

Figure 179: Complete die off of the Senna row due to Lasiodiplodia has increased the risk of sun scald for Sandalwood, (Florina Road, Robson, 2023).

Careful monitoring and management of suspected sites must be implemented. Although by the time a major disease spread has been noticed it is too late, and remedial action must be undertaken. This remedial action is for no further hedging and an increase in irrigation must occur particularly during the build-up season (August to November).

At the sites where there has been a major loss of hosts on both sides of the Sandalwood, this has been due to the loss of a susceptible species like Senna. In addition, due to the extra hosting pressure on the remaining host, they become less resistant, and mortality occurs. Major host loss can occur gradually over three to four years, emphasizing the need for proactive monitoring and management programs to mitigate disease spread and its detrimental effects on plantation health.

HEARTWOOD ROT

The occurrence of Heartwood rot is a considerable loss to the value of the Sandalwood. Heartwood rot was first noticed in trial harvests in 2010 from trees at Frank Wise research Station in ORIA Western Australia.

Growing plantations in tropical climates create favourable conditions for fungal diseases due to the moist environment. Despite the long dry period between May and October outside of the monsoonal seasons, all plantations rely on irrigation to maintain moisture essential for tree transpiration, photosynthesis, and growth. In areas like ORIA, irrigation also prevents deep cracking of clay soils, which could damage tree roots.

Sandalwood trees establish a connection to their host through root haustoria, facilitating the exchange of assimilates and water. Although it is unclear if fungal infections can spread through this connection, the proximity of host and Sandalwood trees presents a potential pathway for fungi to move between species. Some host species might be more susceptible to fungal attack, serving as reservoirs for disease and enabling cross-infection into Sandalwood. Disease management in a parasitic plantation system consequently becomes significantly complex despite the biodiversity benefits.

Observations during a destructive sampling process of Sandalwood trees at the Frank Wise Research Station in Kununurra revealed wood rot within the bole wood, even in actively growing trees. Bark damage often signalled, but not always, the presence of fungus, with the loss of bark exposing the cambium to fungal infection (Barbour, Norris, & Burgess, 2010). Where wounding and fungal entry occurred, the tree responded by encapsulating the infection with a Sandalwood-scented darker ring of wood, attempting to isolate fungal spread. Later harvest confirmed Sandalwood would attempt to isolate fungal infection by depositing Sandalwood oil around the wood rot.

Usually, the inner area within this ring is degraded to such an extent that a hollow forms and this can allow entry by either ants, termites or other pathogens into the core of the tree. The primary wood rot infection does not kill the tree but rather it is the secondary entry of other pests and pathogens.

It is now known that stressed Sandalwood are susceptible to various pathogens, especially phytophthora (Done , Kimber, & Underwood, 2005) and Lasiodiplodia (Burgess T. , 2014). Reports of fungi attacking Sandalwood were mainly limited to nurseries, where Phytophthora and Fusarium have been observed. Several studies have sought to identify fungi responsible for heartwood rot, with candidates including Ceripora, Irpex, Fomes, Ganoderma, Phellinus, or Peniophora, although none have been confirmed.

Figure 180: A - A ten year old tree with evidence of heartwood rot. B – the same ten year old stand with a tree with heartwood rot that has progressed to a void. HW = heartwood, TH= transitional heartwood, SW = sapwood, (Kingston Rest, Haverkorn, 2023).

During the eight year old Sandalwood harvest in 2010, it was evident that bark and cambium damage resulted in lower sesquiterpene synthesis compared to infections in the core of the tree. Cambium damage potentially impaired the tree's ability to produce sesquiterpenes, compromising its capacity to effectively isolate wood rot (Barbour, Norris, & Burgess, 2010).

Management

The cause of heartwood is more than likely due to damage of the bark of the tree. Burgess et al (2014) determined the pruning wounds are an entry pathway for canker and rot fungi in Sandalwood.

The Burgess et al (2014) project very clearly indicated that a mechanism by which fungal pathogens enter Sandalwood and potentially compromise the heartwood as the tree grows. Heartwood fungal rot can take several years to develop before they are externally visible. The presence of these pathogens has a strong economic impact on Sandalwood by reducing tree growth and subsequently heartwood development and oil production. Therefore, the key to reducing heartwood rot is to reduce any form of wounding. This includes pruning of large branches that leave open scars. This has been discussed in the Pruning section.

The other potential entry point for pathogens is via delving. Delving is running a shallow V-shaped board down the row to scrape of the top layer of detritus and mulch (Figure 181). Unfortunately, Sandalwood has very shallow roots and a high proportion of its roots just below the surface. Figure 181 C shows Sandalwood coppice in the middle of the row. This coppice is from roots damaged by the delving operation. Although there are no studies specifically indicating delving as a source of pathogen entry, the study by Burgess et al. (2014) certainly indicated this is a strong possibility.

Figure 181: A- Un delved inter-row. B- Tractor delving row. C- Coppice shoot from root damaged by devlving, (Chapmans, Robson, 2022).

b) Insect pests

GIANT NORTHERN TERMITES

Subterranean termites have a superficial resemblance to ants but are in fact more closely related to cockroaches. While many termite species feed on grass or decaying wood, around ten native subterranean termite species can cause damage to timber in Australia. The most destructive is the giant northern termite (*Mastotermies darwiniensis*) (QLD Govt, 2024).

The giant northern termites (*Mastotermies darwiniensis*, colloquially known 'Masto'), pose a significant threat to tropical Australian plantations, affecting both Sandalwood and its host trees. Infected trees typically display symptoms such as yellowing leaves, the wilting and drying of leaves, and the death of shoot tips or whole branches on an otherwise healthy-looking tree (Thisleton, Neal, & Smith, 2013). Due to their subterranean nature, detecting termites early is challenging, as they often only become apparent when trees show visible signs of damage, which may already be too late.

The lifecycle of Masto begins with eggs that hatch into workers, soldiers, or nymphs. Nymphs can develop into winged adults called alates, capable of dispersing and establishing new colonies. However, the establishment of new colonies by this method is uncommon. The workers can moult into wingless secondary reproductives, called neotenics, and they are the main egg-laying caste in giant northern termite colonies. Colonies increase in size by producing these neotenics that will bud off from the main colony. This is an important consideration for control since a termite-free area is at the greatest risk of colonisation underground along its borders with existing colonies.

In the native vegetation, colonies tend to be small in extent and numbers. However, in areas developed for horticultural and forestry use, colonies can be huge, covering several hectares underground and consisting of large numbers of individuals (Thisleton, Neal, & Smith, 2013). These colonies often have many nests with secondary reproductives and nurseries of immature termites. It has been shown that nests are connected by underground tunnels and that the termites move freely between them.

When trees become infested with termites, they do not often show any external signs. Worker termites enter the tree through the root system or the base of the tree below the ground. They then move up through the heartwood and start to hollow out the upper branches weakening the structure of the tree. Occasionally they will move to the outside of the tree and run several deep grooves that look like rings around the tree.

Termites have been found in all the species used in Sandalwood plantations. Although it is unclear if the termites have a species preference, the Sesbania is a good indicator species. Usually, the first sign of an infection is 'folding' branches in the sesbania (Figure 182A). Other symptoms can include epicormic shoots emerging anywhere along the stem (Figure 182B), and circular cracking in the back around the stem that is occasionally filled with mud (Figure 182C), yellowing, wilting and drying of leaves, and the death of shoot tips or whole branches on an otherwise healthy-looking tree.

Figure 182; A – folding branches in Sesbania, B- coppice along the stem in Sandalwood, C- circular ringbarking in Sandalwood, (Kingston Rest, Ong, 2016).

Once detected, treatment timing and a good understanding of their behaviour are essential to minimise or control them. Prior to plantation development, a survey for signs of colonies and then treatment is completed during the dry season.

Treatment strategies

Treatment strategies for plantations have evolved since 2012. Initially, it involved injecting trees and soil with Fipronil, a group 2B insecticide, through to setting trap bins across sites with positive identification and treating the bins once infected, to searching for infected trees and directly injecting them.

<u>Irrigation injection</u>. Termites were detected across significantly large areas of Kingston Rest in 2013. There was urgency to quickly treat a large area and Fipronil was injected through the irrigation tape in

KR1. Although advice from Brian Thistleton (2013) was that this method was not suitable for Masto control, Fipronil was injected three times over a three-month period. Initially surveys did show a reduction in Masto activity, but this was only temporary.

Soil injection. In 2016 to control giant northern termites in established plantations, the Quintis developed a soil injection system. This system injected Fipronil into the base of every tree in every fourth row. The procedure was to drive the probe at the base of the tree by pushing it deeply according to the length of the probe and at an angle using the foot. When the probe was inserted at the base of the tree, a 250 ml Fipronil solution was released.

Figure 183: Soil injection probes use to insert insecticide solution near tree, (Robson, 2016).

Although early trial work indicated injection every fourth row to be an effective method to control Masto's, observations several weeks after injections, have shown giant termites were still actively infesting treated trees in treated areas. These observations were even noted in stands where every tree in every row was treated using the hand injection machine.

Sentricon. In 2018, Quintis was provide with Sentricon AlwaysActive™ (SAA) (Hexaflumuron) and Recruit AlwaysActive (RAA) (Noviflumuron) products for field evaluation against Mastotermies. A trial investigated the efficacy of SAA and RAA as preventative and remedial measures against Mastotermies infestation in Sandalwood and host trees. It also determined the optimum number of stations per hectare that would effectively protect trees and control termites.

The results showed that none of the Hexaflumuron or the Noviflumuron baits attracted significant Mastotermies feeding but did attract the smaller grass feeding termites. However, it was observed that when Mastotermies termites did feed on the baits there was some evidence of efficacy. Primary indicator bins with rods positively fed on, did not have Mastotermies activity after 11 months. However, due to Mastotermies minimal feeding in the trial it was not possible to determine the optimum number of stations per hectare to protect trees and control Masto colonies effectively.

Aggregation drums. A system developed by Dr Brian Thistleton at Northern Territory DPI&F was implemented across the plantations in 2016. This involves installation of aggregation drums in identified hotspots. This method involved locating active galleries in trees and placing drums, seventy-five litre plastic rubbish bins with the bottom removed. The drums are placed over an active stump and the bottom outside edges covered in soil. The drum is filled with sesbania chips which is moistened.

The lid is screwed down so it cannot be opened. After four to six weeks, the drums are potentially highly active with termites and are treated by knocking the bin over spreading chips and termites over the ground. Then a diluted mix of chemical at 1.5 mls of Fipronil (200g/L a.i) per one litre of water is tipped over the termites and chips. The low dose is not to kill the sprayed termites but so they can take the chemical back to the colony. The low dose coating makes them sick, they clean each other and spread the chemical, then stop feeding and eventually the colony dies out. After treatment, a number of bait drums are set to monitor future incursions.

Figure 184: A set aggregation drum in the plantation boundary, (Roper Plains, Mende, 2019).

Stump treatment. This method is like the aggregation drum method, except that instead of setting a drum, chemical is poured directly into galleries of infested trees. As the chemical runs down into the galleries, it coats the termites and has the same effect as the drum method. Although the drum method, with a fully infected drum will coat many more individuals than a single stump.

DEFOLIATING INSECTS

There are many defoliating insects that can cause damage to plantation trees. Some of the defoliators of Sandalwood include Fig Leaf Beetle (Poneridia sp.), Hyposidra loopers (*Hyposidra janiaria*), Wattle Blue caterpillars (*Theclinesthes miskini*), Sauris loopers (Sauris malaca), Painted Pine Moth caterpillars (*Orgyia australis*), Crexa moth caterpillars (*Genduara punctigera*) and Coastal Valanga grasshopper. *(Valanga meleager*). These vary in their threat posed but all, without treatments can cause significant stress to Sandalwood trees.

Defoliators of host trees include, Sheesham looper (*Plecoptera reflexa*), Lemon Emigrant caterpillars (*Catopsilia pomona*), Brown leaf rollers (Pyralidae), Crusader Bugs (*Mictis profana*) and their nymphs, Cleora loopers (Cleora sp.), and Red-Shouldered Leaf Beetles and other leaf beetles *(Calomela testacea*) and their larvae.

Additionally, several species of mealybug that are known to occur in plantations, including Spherical Mealybug (*Nipaecoccus viridis*), Pink Hibiscus Mealybug (*Maconellicoccus hirsutus*), and others. Also found in the plantations and affecting trees are Seychelles Scale (*Icerya seychellarum*) and an unknown cushion scale (Icerya sp.) are wax-covered, sap-sucking pests.

Furthermore, there are also a number of borers, stem borer moth (*Gonioma sp.*) larvae bore into Sandalwood, Dalbergia, Senna and Cathormion timber and the large Auger Beetles *(Bostrychopsis jesuita*) consume and burrow into dead and untreated timber from Sandalwood and other trees.

Figure 185: Pest poster produced to inform plantation staff about the major Sandalwood pests, (Robson & Ong, 2016).

Fig leaf beetle

The Fig Leaf Beetle has been the most persistent and damaging of the defoliating insects and is worth looking at closely in this section.

Two closely related species of Figleaf Beetle (*Poneridia australis, P. semipullata*) larvae and adults are severe defoliators of Sandalwood. Successive waves of defoliation can cause Sandalwood deaths as the adult beetle defoliates, lay their eggs which hatch and then starving larvae consume the emerging buds and prevent trees from replacing lost foliage.

Figure 187: Defoliation of Sandalwood by FLB in Taylors Park 9 in 2016, (Mende, 2016).

Figure 186: Figleaf Beetle (Poneridia australis), (Mende, 2016).

Fig leaf beetles were first sighted in 2012 in Sandalwood plantations in the Northern Territory, with only the odd individual tree defoliated. By 2015, infestation had significantly increase to whole rows in some stands repeatedly completely defoliated (Figure 187).

Initially they infect plantation edge trees and then spread along Sandalwood rows. Adults fly to adjacent trees, consuming foliage and breeding as they go. Experience has shown that once an area as had an infection, that site is prone to re-infestation even after chemical control.

Both egg-laying and pupation occurs on the ground at the base of a defoliated trees, while egg laying may also occur on foliage. Trees that have been defoliated by Fig leaf beetle larvae are also likely to be damaged again when the same insects mature and emerge as adults. Multiple overlapping generations can occur in Sandalwood plantations.

The Fig leaf beetle is less active during the cool, dry northern Australian winter and outbreaks worsen with the arrival of warm weather, although individual beetles are present year-round. However, extreme weather may prompt adult beetles to spend less time feeding and seek shelter under loose bark.

Control

Aerial application of a nonspecific insecticide will control Fig leaf beetle, as will application of the systemic Clothianidin. Significant trial work was undertaken to evaluate the effectiveness of Clothianidin applied in Sandalwood plantation so that the chemical company Sumitomo would register the chemical for label use in Sandalwood plantations. Application can be either by irrigation or by coulter wheel, although irrigation application was cheaper and has a quicker response than coulter wheel. The treatment is effective for at least two seasons but starts to wane in the third year and should be reapplied (Mende, Robson, & Llorente, 2019).

OTHER BIOTIC STRESSORS

The plantation hosts numerous minor pest species, typically existing at low densities but can occasionally proliferate rapidly, leading to outbreaks causing minor to moderate damage. However, sporadically, in time or space, some species may grow rapidly to damaging numbers, developing outbreaks which may persist for a variable length of time before subsiding.

Outbreaks are often associated with extreme environmental conditions, such as drought, flooding, or hot or cold weather. Outbreaks tend to occur at sites within the overall geographic range of the plant, such as in stands of uniform age structure, on soils with similar nutrient or water content. In the case of insect herbivores, but not plant pathogens, outbreaks have been observed to occur following a decline in the abundance of predators (Jones C. G., 2001).

Sandalwood plantations have experienced two major outbreaks involving entirely different species, resulting in considerable damage over large areas. Understanding these outbreaks and their causes is crucial for effective pest management in Sandalwood plantations, to ensure the health and productivity of the trees.

Sapsuckers

Sapsuckers include aphids, mealybugs and scale insects. Major outbreaks occurred in 2009 and 2020, predominantly affecting stands aged five to eight years. The damage in uncontrolled stands was considerable. It is still unclear why there were such large outbreaks in these years.

Aphids, mealybugs and scales appear in clusters and feed on the sap by inserting a needle-like sucking tube into the plant and drawing out the juice. In high numbers, the removal of large quantities of sap will cause the tree to lose leaves and the shoot tips to die off. After the sap has been used by the insects, it is excreted as honeydew, which forms the base on which a black fungus grows. This fungus is known as sooty mould and its presence can reduce photosynthesis.

Seeding the plantation with beneficial insects such as ladybird beetles (Cryptolaemus spp) and lacewings (Chrysoperla spp) will generally control natural populations of aphids and mealybugs.

However, during these plague type outbreaks control must be with a specific insecticide, usually applied with a boom sprayer, not by plane.

Figure 188: A - a heavily infected shoot on a sesbania. B- Sandalwood shoot completely coated in mealy bug, Kununurra (Robson, 2009).

Rats

As shown in the section on climate in chapter 3, the wet seasons can significantly vary and during those dry years, the only source of water for animals is within the plantation. The wet season in 2017 and 2018 were exceptional to above average, whereas the 2019 wet season failed and was one of the lowest on record. Large rat populations that had built up during the good wet seasons moved into the plantation in 2017 to 2020.

They chewed the irrigation tape and bark off trees and burrowed under the trees. The damage to the irrigation system was not only economically significant but caused severe drought stress, sunscald and significant mortality to many trees in the worst affected stands.

Figure 189: A- droughted plantation effect due to rat damage to irrigation system. B – consistent damage to irrigation tape by rats, (D. Firth, 2019).

Phytoplasma

Phytoplasmas are insect-vectored bacteria that cause disease in a wide range of plants species. Specifically, Spike disease is a phytoplasma bacteria that is a significant disease of *S. album* in India.

Phytoplasmas are phloem-limited plant pathogens that have been identified in over 1000 plant species worldwide. Outbreaks of the phytoplasma-related disease, papaya dieback, has resulted between 10% to 100% crop losses in South-East Queensland and Western Australia. Strawberry lethal yellows and green petal disease outbreaks in Queensland have led to 10% to 50% of strawberry runners being destroyed. Lucerne yellows disease has been reported to cause an annual loss of AU$7 million to the lucerne seed industry (Streten & Gibb, 2006).

The majority of the Australian phytoplasmas are assigned to the 16SrII group. However, a member of the 16SrXII group is more commonly associated with economically important diseases in Australia such as strawberry lethal yellows, papaya dieback and grapevine yellows (Streten & Gibb, 2006).

A 2023 discovery of suspected phytoplasma affected Sandalwood trees in a Quintis plantation, identified a phytoplasma similar to the Buckland Valley grapevine yellows phytoplasma (BVGYp) (Manners, 2023). This disease has previously been detected in Victoria in grape vines and also from *Exocarpus latifolius* (scrub Sandalwood) in the Torres Strait. This is the first time this disease has been found in Australian *S. album*.

The symptoms found in these affected trees are similar to the Indian disease. As the disease progresses, the new leaves become smaller, narrower or more pointed and fewer in number with each successive year until the new shoots give an appearance of fine spike 'Witches broom.' At the advanced stage, inter nodal distance on twigs becomes small, with the haustorial connection between the host and sandal breaking. With the Indian disease the affected trees die within two to three years after the onset of spike disease (Swaminathan, Hosmath, & Mallesha , 1998). There is evidence of Sandalwood mortality in this outbreak but not enough evidence to determine if it is as catastrophic as the Indian Spike disease.

The disease is naturally spread through leafhoppers or other sap sucking insects and root-grafting. With the Indian Spike, the infected plants can take a long time to express disease symptoms and eventually die. According to Weintraub and Beanland (2006), phytoplasmas are restricted to the phloem. Therefore only insects that feed on phloem sap are responsible for the transmission of phytoplasmas. These include aphids, mealybugs, scale insects and flatid planthopper. Although a collection of planthoppers from the site have been sent for analysis for evidence of the phytoplasma, so far all have assessed as negative (Dodd, 2023).

Figure 190: A- an affected S. album with rosetted coppice shoots along the stem. B – a closer view of the fine leafing Witch's broom look, (Robson, 2023).

Impact and control

The spike disease in India has reduced viability of Sandalwood growing stock by 25% over 20 years (Selvaraj, Suryanarayana, Venkataravanappa, Rao, & Mandal, 2022). Although the severity of this disease here is still unknown, if not managed, the potential impact is significant.

Figure 191; Flatid planthopper is the vector most like responsible for the spread of phytoplasma, (Ong, 2017).

Quintis did initiated a management program that includes, quarantine of all vegetative material, including seed from these plantations. An application of systemic insecticide to reduce the numbers of vectors and a survey of the plantation to map trees with symptoms and removal of trees that are showing symptoms.

3. ABIOTIC STRESS FACTORS

Abiotic stress factors are nonbiological events that can be detrimental to plantation health. This section will discuss the detrimental effect of extreme water, fire, wind, heat and the sun.

a) Waterlogging and droughting

The stress imposed by water, either the lack of or over abundance are discussed in the section on Water in Chapter 5.

b) Fire

Although Sandalwood trees are not resistant to intense heat from fire, they can recover from mild scorching. There have been instances of small fire outbreaks within the Sandalwood plantations, often caused by lightning strikes or sparks from machinery, such as slashers striking rocks, which ignite dry grass within the plantation.

Most fires within the plantations have been relatively mild, thanks to well-managed fuel loads and additional moisture from irrigation. The damage typically involves slight to severe scorching of the lower branches. Observations suggest that Sandalwood and Dalbergia can recover from moderate scorching, whereas Cathormion and *Acacia trachycarpa* cannot.

However, if significant fuel loads accumulate and environmental conditions are favourable, intense wildfires that pass beneath the trees and heat both the stems and branches can cause Sandalwood mortality. While the above-ground stems may die, the base of the trees can coppice. It is often common to see new growth sprouting from root coppice in the middle of the rows.

Figure 192: Plantation fire in 2-year-old Midway A in 2015. The fire was in the inter-rows and did not burn under the tress but still caused significant scorch, (Robson, 2015).

Fire control.

Managing the fuel load both within and surrounding the plantation is crucial to prevent wildfires from entering. Establishing and maintaining external and internal firebreaks can significantly mitigate the risk of uncontrolled wildfires.

The plantations are typically bordered by native vegetation woodlands, which are prone to annual wildfires. Grass fuels, if not reduced through burning or slashing in the late wet season and early dry season, have a high potential to ignite from lightning, arson, or escaped burns in the late dry season. These fires tend to burn hotter and more intensely in the late dry season, with greater rates of spread, posing the greatest annual risk to the plantation estate over its 15-year rotation.

In the Northern Territory, the late dry season is characterised by strong south or southeast afternoon winds and low humidity, which further complicates fire control efforts once ignition occurs.

The risk can be mitigated through scheduled and timely fuel reduction burns around the plantation, regular maintenance of internal and external fire breaks, and controlling weeds in and around the plantation.

Figure 193: A- Very high fuel loads within the plantation pose extreme risk, particularly as fire danger rating increase from start of the dry season. B – Inside the plantation, gaps caused by host mortality will grow weeds and should be controlled to reduce fire risk. The weed level in these pictures would fuel a fire that would cause catastrophic damage to this nine year old plantation and is an unacceptable risk, (Cadell, Robson, 2024)

c) Cyclone and wind

Crown damage from wind is a common issue across the entire Sandalwood estate, though it typically does not lead to mortality. The primary forms of damage are stem breakage and tree leaning. High winds from tropical storms and cyclones can cause Sandalwood trees to lean significantly or result in snapped stems and branches. Such breakage can make trees more vulnerable to pests and serve as entry points for pathogen spores that may lead to cankers and stem dieback.

Wind damage is usually influenced by three factors: plantation age, wind speed, and wind intensity (gusts). In the first four to five years of rotation, while still relying on short-term hosts or transitioning to long-term hosts, Sandalwood is particularly susceptible to wind-induced leaning. During this early growth stage, Sandalwood often has an unbalanced shoot-to-root ratio, with exceptionally large crowns and small root systems. Although the nutrient and water supply from the short-term host and irrigation support rapid growth, but the Sandalwood does not develop an extensive root system. The roots primarily run along the row with minimal expansion outward, making the trees highly vulnerable to crosswinds and intense gusts. Additionally, the large crowns act like sails, catching wind gusts, although the overtopping sesbania does provide some degree of shelter. Winds exceeding 60 km/h can push over young Sandalwood trees.

Once Sandalwood transitions to long-term hosts, their root systems, though still shallow, become quite expansive, and their crowns thin out. During this phase of the rotation, Sandalwood becomes reasonably resilient to being blown over in storms or low-category cyclones.

Due to the potential for cyclone damage, most plantation companies adhere to a general rule that plantations should not be developed closer than 70 km to the coastline. Since cyclones typically weaken rapidly upon making landfall, bringing heavy rain but reduced winds, this policy has generally been effective. Since 2012 when the majority of the plantations were established, there has been eight cyclones that have tracked near the plantations. (Figure 194). Apart from TC Debbie, these were mainly category 1 (the weakest level) by the time they approached the plantations.

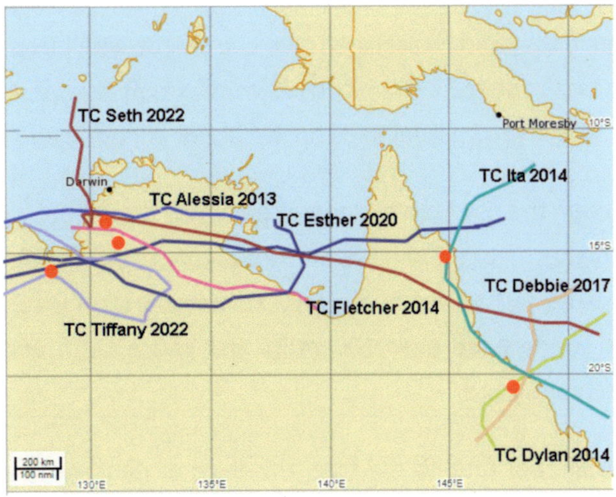

Figure 194. Cyclone track map for cyclones that tracked near plantations between 2012 and 2024, (modified from BOM). Red dots represent locations of main Sandalwood plantations.

In March 2017, Tropical cyclone Debbie passed over the Whitsunday Islands as a Category 4, with winds of 165 km/h. The cyclone weakened to Category 3 before making landfall in Airlie Beach with winds of 150 km/h. The cyclone tracked 50 km south of the Sandalwood plantations as a Category 2, with the plantation recording wind speeds of over 100 km/h. In the stands that had transitioned to the long-term hosts, there was minimal damage from the high winds, other than the occasional broken branches. However, in the youngest plantings, established the previous year, a high proportion of the Sandalwood trees were leaning at a 45 degrees or greater angle (Figure 195A).

Figure 195: A-18 month of Sandalwood tees leaning due to TC Debbie. These were all cut off 50cm above the ground (Robson, 2017). B- recovered tree seven years after wind event. Note that the old stem can still be seen. (Robson, 2024).

When young plantations are subjected to this level of damage they can be salvaged. For the stands with blown over trees due to TC Debbie, trees were cut off at approx. 5cm above grown level and coppice shoot encouraged.

Although the risk from cyclones is generally thought to be the highest from a wind event, frequent tropical storm can cause just as much damage. Commonly storms can have wind speed over 100km/hr and cause stem and branch breakage.

Although not fatal to the tree, this level of damage to a high number of trees in a stand will cause a range of problems. Broken crowns as shown on figure 196 will disrupt machinery and staff working though the plantation. With snapped stems like this, the tree will stop growing and will be susceptible to heartwood rot. This should be remedied by removing the broken stem, cutting it off just below the break at an angle and then painting with a wound sealant. Although a healthy tree will eventually regrow a new crown, it will take many years but will not fully recover.

Figure 196: Stem breakage in 13-year-old Sandalwood tree caused by tropical storm, (ORIA, Sawyer, 2024).

d) Heat

Elevated temperatures are increasingly becoming a significant abiotic stressor that limits plant growth and productivity, particularly as global temperatures are expected to rise by 2050 (Houghton, Ding, & Griggs, 2001). As noted in the climate section, there has been a noticeable trend towards more extreme temperatures within plantation estates, with instances of daily maximum temperatures exceeding 40°C becoming more frequent.

Heat stress affects plants at various developmental stages, potentially causing leaf wilt or drop, reduced stem growth, stem sunburn, and even plant mortality. Key responses of plants to heat include alterations in phytohormone levels, primary and secondary metabolites, the synthesis of heat shock proteins (HSP), and the production of reactive oxygen species (ROS) (Iba, 2002). ROS are highly reactive molecules produced from the reduction of molecular oxygen under high-energy exposure or through electron transfer reactions. Environmental stress exacerbates ROS generation, disrupting cellular homeostasis and posing a risk to cell structures by promoting lipid peroxidation, protein oxidation, damage to nucleic acids, enzyme inhibition, activation of cell death pathways, and ultimately leading to cell death.

One of the earliest responses in leaves to mild water/heat stress is stomate closure. Stomates are small valve-like openings on undersides of leaves which allow for gas exchange and water loss. Stomates often close during initial stages of heat stress, long before leaf wilt. Like many trees, Sandalwood will close stomates temporarily at a point during the day in response to rapid water loss due to extreme heat. Midday stomatal closure is generally followed by reopening and increased transpiration in late afternoon (Coder, 2018).

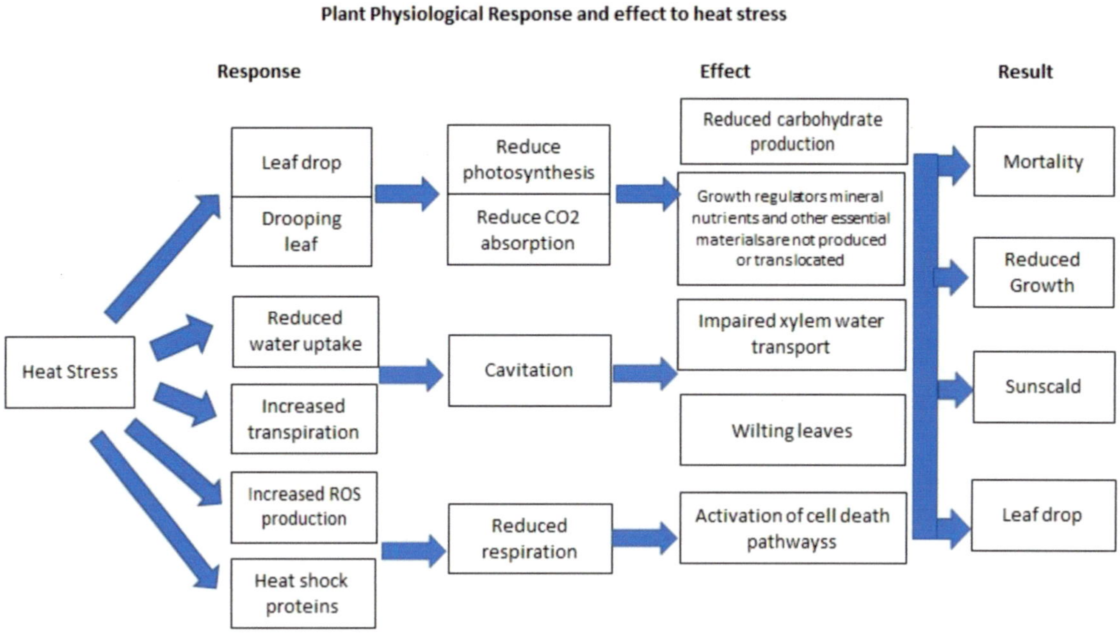

Figure 197: The response to a heat stress in plants can vary the effect and ultimately the result , which can range from a slowdown in growth to tree death, (Robson K. , 2021).

A study conducted in ORIA plantations in 2005 investigated plantation water use (Lewis, Wildy, & Higgins, 2006), it found that when leaf water tension increased as the temperature increased during the day, Sandalwood would close stomata to conserve water loss. During this period, the trees had reduced photosynthesis, respiration and potentially reduce growth.

Quintis sapflow meter data supported this by showing a slowdown in transpiration (sapflow) by Sandalwood in the hotter months. In July, the example tree in Figure 198 had a sapflow of 1.4 litres per hour and this had reduced to 1.04 litres per hour in September. Although this tree has not completely stopped transpiring, it seems it has reduce its water use during the time of year where evapotranspiration and temperatures have increased.

Continuous extreme heat can cause injury in trees, such as scorching of leaves and twigs, sunburn on branches and stems, leaf senescence and abscission, acute leaf death, and shoot and root growth inhibition. Wilting is one of the first major symptom of water loss excesses and heat loading. Leaves under heavy heat loads may progress through senescence, brown-out and finally abscise. Leaves quickly killed by heat are usually held on a tree by tough xylem tissues and lack of an effective abscission zone. The sunburn / sunscald of Sandalwood stems can be a significant problem.

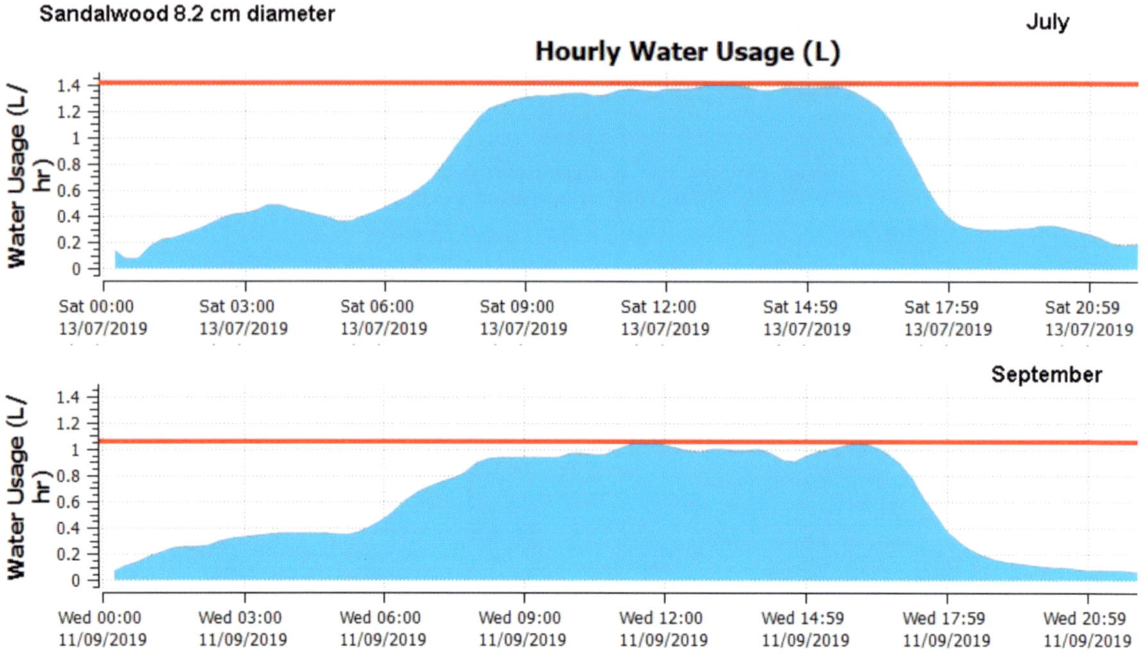

Figure 198: The variation in sapflow for an eight year old 8.2 cm Sandalwood tree growing at Kingston Rest in July (top) and September (bottom). Charts copied from ICT instrument software.

Stem sunscald

In the context of Sandalwood plantations, due to extreme heat from the sun on exposed Sandalwood stems, sunscald can occur. This is when stems are consistently exposed to intense afternoon sunlight, particularly in the Australian tropics where the northwestern side of trees are exposed to the afternoon sun and are most vulnerable.

Sunscald can typically emerge when there has been significant host mortality, which reduces the protective canopy over the lower portion of the Sandalwood. This condition is rarely found in young trees and is more typical during mid-rotation, from ages six to ten years. As Sandalwood transitions from short-term to long-term hosts, above-ground growth slows and reduced sap flow under intense sunlight can cause the bark to thicken and crack.

The severity of sunscald can range from very mild to extremely damaging. Trees with low (Figure 199 A) to mild (Figure 199 B) levels of sunscald have thicken 'crocodile' like bark but the bark is still intact and covers the cambium layer. This level of damage is usually observed in trees with stems that have been exposed for only a brief period. The time period of exposure for this level of damage to occur will vary dependant on the individual trees. Although tree size does not seem to be a factor, as trees of all sizes can be found with varying levels of sunscald.

Figure 199: Sunscald at low (A) and mild (B) levels on eight year old Sandalwood trees. For both trees, the LTH had died and they were exposed to the afternoon sun. However, as the bark is still intact the sunscald has not been damaging to the underlying wood, (Florina Rd, Robson, 2024).

When exposed constantly for extended periods the sunscald will become more damaging. The bark will thicken and then blister and crack, exposing the wood underneath (Figure 200 C). This effect can be along the entire side of the tree exposed to the afternoon sun (Figure 200 D). Sunscald to the level shown in example D will more than probable cause tree mortality. Although it is usually the smaller trees that are afflicted to this degree, there are cases of the same happening to larger trees. Scalding at the level in example C will not cause mortality, as in many cases the tree will completely recover and will grow new bark to eventually cover the wound. However, during the period that the wood was exposed in all cases the stem has been affected by wood rot pathogens, making that section of the stem uncommercial for heartwood production, this can be clearly shown in example E.

Several strategies can be used to minimise stem sunscald in sandalwood. Retaining lower branches helps shade the stem and protects it from direct sunlight. However, the most effective approach is to ensure that the stem, particularly its western side, is not exposed to intense afternoon sun. Continuous monitoring of long-term host health is therefore essential, allowing declining hosts to be replaced before they die, thereby maintaining adequate canopy cover and ongoing protection of sandalwood stems.

However, the loss of hosts on the northern side can be unpredictable and or unpreventable. If the loss of host happens, one solution implemented by TFS was to paint the exposed side of the tree with white paint. An assessment 12 months after treatment application did indicate that the level of sun scald on tree stems had not increased.

Figure 200: Sever sunscald (C) at the base of an otherwise health tree showing exposure of the underlying wood to the elements. A severely afflicted tree (D)with sunscald along the entire stem exposed to the afternoon sun. This tree will more than likely succumb and eventually die (Florina Rd, Robson, 2024). When cut, the damage from the severe sunscald is very obvious (E), decay has penetrated into the heartwood and rendered a large section of the stem worthless, (Florina Rd, Haverkorn, 2024.)

4. THE ROLE OF STRESS IN HEARTWOOD DEVELOPMENT

Physiologically, heartwood develops within trees to shield their inner core from pests and diseases by depositing protective chemicals, such as Sandalwood oil in the case of Sandalwood. As the tree matures and the vascular cambium layer produces new sapwood, the inner xylem cells become compact and hardened, no longer actively transporting water. This transformation forms the central support structure known as heartwood, primarily composed of cellulose fibres bound by lignin.

The deposition of chemical compounds in heartwood is a common phenomenon across tree species (Taylor, Gartner , & Morrell , 2022). Ecologically, these deposits offer long-term protection against pathogens and microbial decay (Celedon, et al., 2016). In Sandalwood, these deposits consist of sesquiterpenes, which constitute the prized Sandalwood oil.

As Sandalwood ages, it accumulates oil in the heartwood, starting from the base of the tree and gradually moving upward along the stem. Historical reports from India suggest that heartwood initiation in Sandalwood typically begins around age ten, though this can vary significantly, with initiation occurring as late as age 46 (Weiss, 1997) (Rai S. N., 1990). The process of forming heartwood necessitates diverting resources from stem growth to the deposition of oil (Barbour, Plummer, & Norris, 2011).

a) Heartwood initiation and development

It is widely believed in the industry that stress plays a crucial role in initiating heartwood development. Early theories suggested that maximising tree growth during the first half of the rotation, by providing ample water, fertiliser, and space, would result in large stem volumes. Subsequently, during the latter part of the rotation, gradually reducing these luxuries would stress the trees, prompting them to accumulate heartwood.

This theory is supported by studies indicating that tree size and growth are correlated with heartwood development, a relationship corroborated by Quintis harvest data.

Harris (1954) proposed that growth rate is likely the primary factor influencing the onset of heartwood formation. Rapid growth increases water demand. When water availability cannot match this demand, the moisture content in the inner sapwood decreases, initiating the formation of heartwood. Thus, hydraulic requirements of the tree are critical drivers of heartwood dynamics.

This theory aligns with the observations from early Sandalwood plantations, where heartwood initiation often coincides with the transition from Short-term Host (STH) to Long-term Host (LTH). During this transition, which can take up to three years, the tree's crown diminishes in size, and sap flow decreases as the need for extensive xylem volume diminishes, triggering the conversion to heartwood.

During the transition period, the tree crown diminishes in size and sap flow is reduced. The tree no longer requires a large volume of xylem and initiates conversion into heartwood. Recent observations indicate that nearly all trees within Australian plantations by age eight, possess significant levels of heartwood, albeit not necessarily oil-bearing heartwood.

For instance, Figure 201 illustrates the presence of heartwood (HW) in two trees grown under different plantation configurations. Tree A, grown in a Standard configuration, experienced an extended transition period, while Tree B, grown in a Mix configuration, underwent a milder transition from STH to LTH. Both trees exhibit transitional heartwood (tHW) covering 90% of their volume, with Tree A additionally developing oil bearing heartwood (OBHW) containing Sandalwood oil. Over time, transitional heartwood changes into oil bearing heartwood. This comparison underscores that the Standard configuration (STD) grown tree initiated heartwood at an earlier period than the Mixed configuration(MIX) grown tree.

Heartwood development provides several advantages to trees, including structural support as they mature, increased resistance to pests and diseases due to nutrient removal within the heartwood, and reduced demands for nutrients and water as sapwood converts to heartwood.

While the timing of heartwood initiation varies among tree species, growth rate appears to be a primary determinant. Other factors, such as environmental conditions like stress, tree stature, and forest management practices, may also influence the speed of oil-bearing heartwood development (Chadrack , et al., 2024).

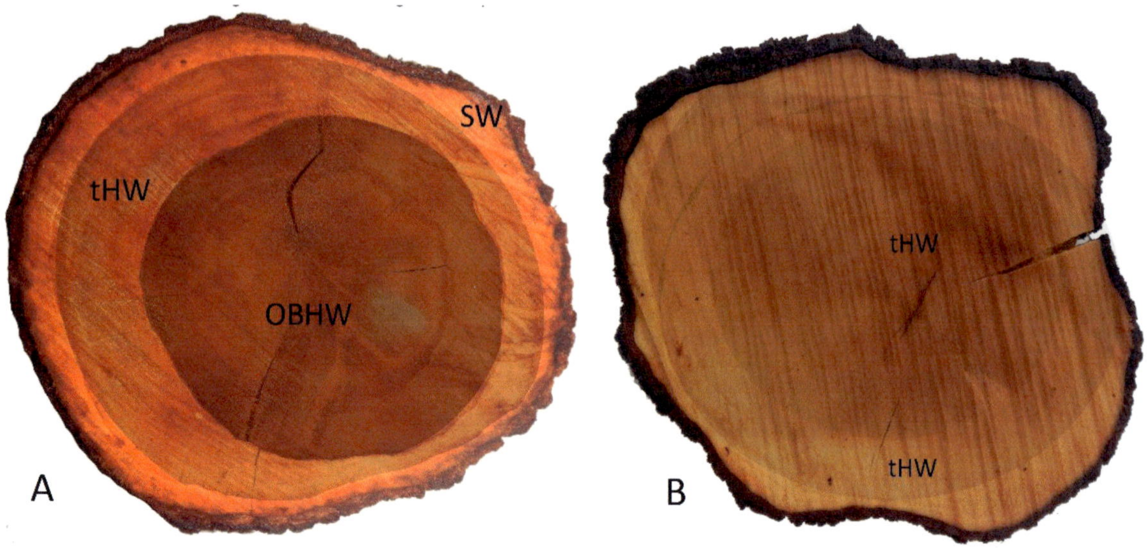

Figure 201: A – A stem biscuit from a nine year old tree grown in a standard (STD) configuration that underwent significant transition stress. B- A stem biscuit from an eight year old tree grown in a mixed (MIX) configuration that had a short, less stressful transition event. Note that the images have been slightly manipulated to enhance the heartwood view, (Robson, 2024).

In summary, the development of heartwood in Sandalwood involves a complex interplay of growth conditions, stress responses, and resource allocation strategies, all contributing to the formation of valuable heartwood containing Sandalwood oil.

b) Abiotic stresses to increase heartwood development.

In Kununurra plantations, harvested 15-year-old Sandalwoods typically contain approximately 30% of commercial oil-bearing heartwood by volume, though this can vary significantly. Some trees yield as high as 60%, while others produce nearly 0%.

Several research projects have explored methods to artificially enhance heartwood conversion. Key treatments include nutrient and water manipulation, along with the injection of "irritants" to stimulate heartwood formation.

Water and nutrient stress

An operational trial in a Packsaddle stand involved withholding irrigation during the final two years of a 15-year rotation. This resulted in significantly lower heartwood yields, approximately 30% less than in irrigated stands, and substantial mortality during the non-irrigation period.

NDVI images indicated reduced photosynthetic activity over the site without irrigation compared to neighbouring irrigated stands. This reduction in photosynthesis is crucial since the production of Sandalwood oil, a secondary metabolite, is an energy-intensive process reliant on photosynthesis and sap flow. Drought conditions diminish the tree's ability to convert transitional heartwood.

Physical damage to tree crown

Observations during Quintis harvests revealed that trees with damaged crowns, caused by wind events, developed heartwood higher up the bole or even in the crown itself. An experiment intentionally damaged tree crowns in a 17-year-old stand, resulting in an average of 8% more heartwood compared to undamaged controls (Brown, 2020).

Artificial stressing the tree

Experiments injecting various irritants into Sandalwood have sought to induce heartwood formation. Wounding a Sandalwood tree triggers a stress response in the cambium tissue, prompting deposits around the wound as a protective measure against infection, similar to responses during damaging wind events.

The use of chemicals induces a stress response that shares some chemical characteristics with naturally derived heartwood formation chemicals, the outcome via 'stress response' is not necessarily oil-bearing heartwood (Shain, Hillis, & Hillis, 1973).

Products such as carbon dioxide, ethylene, nitrogen (Liu etal 2022), indole-3-butyric acid, methyl jasmonate, methyl salicylate, paclobutrazol, ethephon, 6-benzylaminopurine (Barbour E. L., et al., 2020), Paraquat, ethrel (Radomiljac A. M., 1998), methyl jasmonate (Manson, 2016) have been injected into Sandalwood branches to induce heartwood production.

Liu (2022) found that all their treatments induced fragrant heartwood formation around the input site. Barbour (2020)found that the whole tree treatment with ethrel and paclobutrazol stimulated heartwood production. In addition, Mason (2016) found that methyl jasmonate induced fragrant heartwood formation around the input site.

These experiments have been purely academic, in that any artificial inducement will most likely not be commercially suited to induce heartwood in plantation trees. The use of these experiments is to gain a better understanding of heartwood formation.

Chapter 6. BREEDING AND SEED PRODUCTION

1. SANDALWOOD BREEDING PROGRAM AND SEED PRODUCTION

Santalum album belongs to the family Santalaceae of which there are 16 species (Harbough & Baldwin, 2007). *S. album* has a native geographic range across the tropical belt in India, in Eastern Indonesia province from Java, across to east Nusa Tenggara into Timor Leste. *S. album* also occurs naturally in Australia and can be found in monsoon vine thickets along the coastal sand dunes and laterite in dry rainforest around the Northern Territory coastline (Plonk, 2022) (McKinnell F. H., 1990).

Although the precise origins of seed used in the early Kununurra plantations have largely been lost over time, Peter Kimber and Chris Done, both integrally involved in establishing the Western Australian sandalwood program, recall that the initial Quintis plantings were sourced from a mixture of provenances. These included seed supplied by the Institute of Wood Science and Technology, as well as seed acquired from the former Department of Conservation and Land Management (now the Forest Products Commission). Much of this material originated from extensive trial plantings at the Frank Wise Research Station, Kununurra, which themselves represented a range of seed sources and geographic origins. There has also been speculation that Quintis imported seed directly from India, although this has never been conclusively verified. Consequently, despite the apparently broad genetic base represented at Frank Wise Research Station, the provenance and genetic background of many of these early accessions remain uncertain.

2. GENETIC DIVERSITY

The *Santalum album* population genome is not diverse. A key determinate for this is to investigate the number of alleles per loci and heterozygosity. The terms important here are <u>Alleles (*A*):</u> One of the several alternative forms of a specific gene, each possessing a unique nucleotide sequence, <u>Locus</u>: The specific position or location of a gene on the chromosome and <u>Observed heterozygosity (Ho)</u>: having two different alleles of the same gene.

In their paper on Indian genetic variation, Angadi, Jain & Shankaranarayana (2003) determined that there was variation between provenances within India and that Ho ranged between 0.50 and 0.18, with greatest diversity at the Bangalore provenance. The mean number of alleles ranged from 3.6 to 1.6. Brand (1994) study focused on Timor provenances and found that the Ho ranged from 0.11 to 0.17. His results for a Bangalore collection only showed 0.18, a fair bit less diverse than Angadi' s analysis (and recent data shows that this is more likely). The mean number of alleles ranged from 1.7 to 2.0. The work undertaken by Quintis demonstrated that with individuals from the Quintis breeding population (n=108) the Ho was 0.14 and the number of alleles was 1.5. This will be discussed in more detail later in the chapter.

When comparing the Ho and *A* results with other Australian forestry species, there is a lack of genetic diversity in *S. album*. For example, *Eucalyptus camaldulensis* (Arumugasundaram, Ghosh, Veerasamy, & Ramasamy, 2011) has a mean Ho 0.48 and alleles 3.91, *Tectona grandis* mean Ho 0.70 and alleles

16.7 (Verhaegen, Ofri, Fofana, Poitel, & Vaillant, 2005) and *Eucalyptus globulus* mean Ho 0.84 and alleles 17.6 (Jones T. H., 2005).

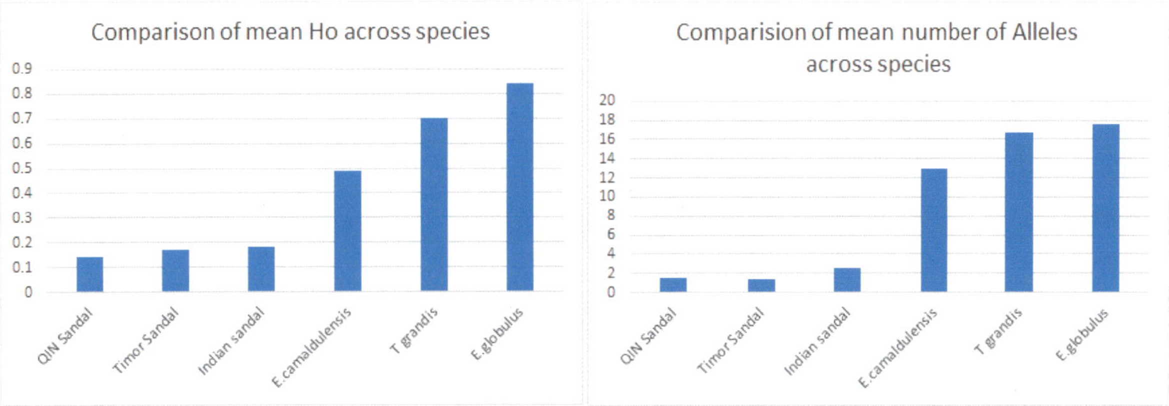

Figure 202: Graphic illustration of Ho and A between S. album and three other much more genetically diverse tree species, (data from various sources).

Although the small genetic diversity of *S. album* is well described, and the diversity of this species in Australia is even smaller (Jones , Plummer , Barbour, & Byrne, 2008). It is widely accepted and generally reported that there is a wide morphological variation in the physical appearances within Sandalwood species (Angadi, Jain , & Shankaranarayana, 2003) (Doran, Thompson, & Brophy, 2002). This seems to be somewhat counter intuitive and points to the significant role environment, hosting and physiology play in a Sandalwood tree's growth and development.

3. SANDALWOOD BREEDING PROGRAM

The key to increasing desirable plantation traits (stem volume, form, heartwood ratio, oil characteristics) is selecting for desirable phenotypic traits and potentially increasing the diversity of the breeding population. With the known low diversity, this last point is somewhat important for continued sustainable improvement.

The improvement of the Sandalwood breeding population is a critical component to seeing advances in stem volume and oil production within the operational population.

Although components of the breeding program started in 2005, the actual Quintis breeding strategy was not developed until 2012. The first generation breeding population was selected within five year old plantations, based on desirable phenotypic traits. These traits primarily being stem form, stem volume and an indication of heartwood aromatics.

With the development of the breeding strategy, it became necessary to determine the genetic diversity within that population it would look at techniques to increase the diversity of genes (increase Ho) in both the current breeding population and the propagation population. In 2022, DNA analysis was undertaken of the complete first generation breeding population. The results of this analysis are discussed later in this chapter.

Breeding concepts for tree species have been well studied and developed. An effective breeding strategy involves maintaining a hierarchy of three population types, base, breeding and propagation populations (Figure 203). The improvement plan encompasses traditional breeding strategies. Improvement is accomplished through selection of elite individuals and the establishment of grafted clonal seed orchards and seedling progeny trials/ seed production areas.

a) 1st Generation

The base populations for selection came from within the early Quintis plantation, from the oldest plantings at the time. These base populations were searched, and selected trees were chosen based on phenotype (form and vigour). These were relatively tall, single-stemmed, straight trees. Although the plantation had been form-pruned, a selection aimed to choose trees exhibiting minimal evidence of branches prior to pruning. Form and vigour are selection traits composed of several traits that could have been recorded separately - height, diameter, straightness, presence or absence of forks, branch size and angle.

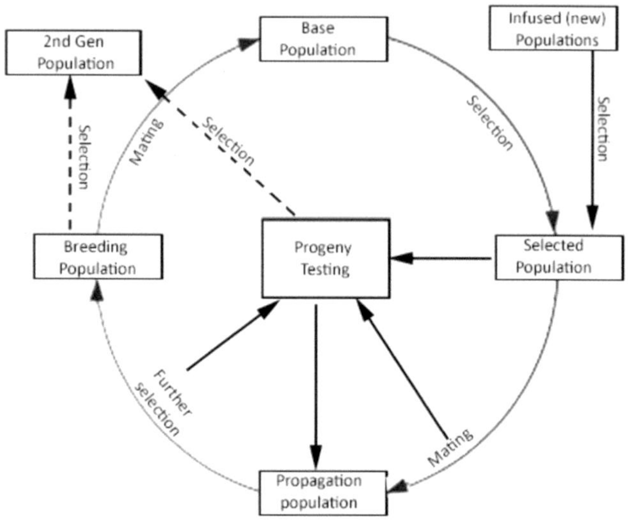

Figure 203: The major components and activities of a breeding cycle as represented by (White 1987). The Quintis Sandalwood breeding plan is based on this concept. Diagram modified out of (Eldridge, et al. 1993).

A third trait was the early indication of an oil aroma. This was accomplished by using an electric drill with a small diameter bit to sample the inner wood for an indication of the presence of oil. This test was only to give an indication for the presence of oil, it did not necessarily mean the tree did or did not have oil. However, if there was no aroma in the 'sawdust' the tree was rejected. This test of young trees aimed to try and predict the early production of oil, which could indicate oil at harvest age.

The select trees (the ortets) were systematically numbered and marked and mapped in the plantation. The selection intensity for the traits, form and vigour was 1 in 150. The selection intensity for the early indication of aroma was 29 in 100, or about 1 in 3. This two-stage selection procedure is described as independent culling (Cotterill & Dean, 1990).

In 2005, an initial search of an early plantation established in 1999, of about 15 000 trees, 84 ha, at Stockroute plantation compartments 3, 4 and 5 were undertaken. The following year, a further search of the 2001, 2002 and 2003 plantation of 281 ha, at Packsaddle plantation, compartments 1, 2, 3, 4, 5 and 7 (66511 trees) was undertaken. There were 108 selections made that make up the 1st generation breeding population.

Seed Orchards

The plan includes the establishment of both seedling and clonal seed orchards. The CSO's are grafted plantings of the selected trees. The location of clonal ramets with each CSO is specifically designed to ensure maximum outcrossing of clones. Scion material was gathered from the plus trees and grafted. This methodology is described later in this chapter.

The original CSO was established in the Kununurra Packsaddle Plantation in 2007. This consisted of 24 families, all selected from elite trees in the Stockroute plantation. There were only small numbers of ramets per clone (one to seven) and the design was the randomised location of families. The layout was as per standard plantation with alternating Sandalwood and host rows. This CSO was terminated in 2015.

A second CSO was established on a leased site called Dessert in Kununurra. There were two blocks established 2010 and 2011. These sites included 67 families. The design was a randomised location of families. The layout was as per standard plantation with alternating Sandalwood and host rows. This CSO was terminated in 2017 due to poor seed production and high cost of land lease.

With the genetic analysis completed in 2012, there was a better understanding of the relativeness of all the families within the first generation. This provided an opportunity to better develop the layout and design of orchards established after 2011. The CSO's that were current in 2024 and their design are described in the table below.

Quintis 1st Gen clonal seed orchards

Table 36: There are currently 3 x 1st generation Clonal seed orchards.

Dalbeg CSO	Taylor Park CSO	Woods CSO
• QLD Dalbeg Stand 42 • established in 2012 • 106 clones x 15 (av) ramets = 1630 ramets. • Area: 4.8 ha	• NT Taylors Park Stand 89 • established in 2013 & 2014 • 107 clones x 15 ramets = 1605 ramets. • Area: 5.2 ha	• QLD Dalbeg Stand 59 • established in 2015 • 59 clones x 12 (av) ramets = 710 ramets. • Area: 2.2 ha

b) 1.5 Generation

In 2016 all progeny trials were assessed for family volume performance. The data from two sites was analysed and the best 20 families were selected. Selection was based on progeny volume performance across the sites, and plus tree (ortet) heartwood findings (actual oil yield over predicted, % of heartwood diameter) (table 39). The family genetic data was also used to determine similar genotypes/relationships to cull out half or full sibs. Although the number of families has been reduced in the orchard, the analysis indicated an increased observed heterozygosity (Table 38).

Table 37: Details for the 1.5 CSO

Windsor CSO
• QLD Millaroo Stand 119 • Established 2017 • 20 clones x 130 (av) ramets = 2600 • Area: 7.0 ha

Table 38: The genome analysis has been conducted for all of the Quintis breeding population. Results and comparisons shown below. Data analysed by M. Byrne UWA.

Collection	N	A	P	Ho
Quintis (1.5 gen)	20	1.6	63%	0.21
Quintis Breed pop (1 gen)	108	2.0	44%	0.14

Table 39: Selected families included in the 1.5 Orchard, shows progeny improvement over unimproved, heartwood information form Ortet and related families in the breeding population.

Family	2012 Gene Group	Mean % Progeny Vol > than Unimproved	% of diam HW	Related families
P3.02	B	56%	57%	P2.20
O	A	49%		*
P2.15	A	43%	73%	P2.13, P4.16
P4.20	B	40%	66%	P1.06
P4.27	B	38%		*
D	A	36%	32%	S
P4.32	B	39%		M, P4.25
P3.07	B	32%		P3.08, P3.13
P3.14	B	30%		*
P2.06	B	30%	49%	P1.03
P2.07	B	28%	50%	*
P3.01	B	26%	50%	P2.14
P3.10	B	25%		*
P4.26	B	25%		*
P3.09	B	25%		*
P3.15	A	25%		*
P4.24	B	37%		P3.06, P4.23
P1.01	B	24%	58%	*
P2.16	B	24%	60%	P3.12
P2.05	A	22%	76%	*

c) 2nd Generation

The second generation breeding selection program commenced in 2022. The selection of this consists of selections of elite individuals within the progeny trials and improved operational stands. Selections have been made based on diameter over progeny trial average, bole length, overall bole form, presence of any defect and family (if known), heartwood volume and oil quality. The selection rate in the progeny trials is about 1 in 150. This has resulted in 38 individuals selected in the progeny trials to go into the 2nd gen breeding population. A further search could be undertaken in improved operational stands for outstanding individuals to be included. The original plan was to establish the selected individuals into a new clone bank in 2024. Due to the insolvency in 2024 this never occurred.

d) Progeny trials

The seed collected from individuals within the CSO's has been tested in progeny trials. The progeny trials look at the growth performance of each clonal family. Progeny from each family within the existing trials was raised and planted out in progeny trials in WA, NT and QLD. Progeny trials are planted in Stand 42 in QLD and in Taylors Park 9 stand 88 in NT, HCJB and Chapmans in WA.

Early Progeny trials (2008 to 2010)

There were three progeny trials established from the breeding that occurred in the Packsaddle CSO. These were all planted in the Kununurra plantation region in 2008, 2009 and 2010. Although these trials did provide some insight into the improved performance of improved progeny, there were flaws in the trial design and management. There were only a small number of replicates, trial pegs were lost, and irregular assessment was undertaken.

The first progeny trial was planted into Voyager 9 in 2008. This experiment had progeny from 20 clones with seed collected from the packsaddle CSO. The design was 22 un-randomised line plots across the stand. Unfortunately, no unimproved controls were included in the trial, so it was impossible to determine any growth benefits from the progeny. Assessment at 18 months did show that families, 'I' and 'AD' were significantly taller than all others. An attempt to undertake an assessment in 2020 could not be completed because all the identification pegs were missing, and it was not possible to determine where the progeny line started.

In 2009, a second progeny trial was planted in Chapman's plantation. This consisted of 29 families from the Packsaddle CSO and an unimproved seedlot. The design was row plantings of progeny randomly allocated down each row. The planting configuration is Mixed (hosts and Sandalwood in the same row). The results from the 2015 assessment clearly showed that for 26 of the 29 improved progeny, had larger diameter than the unimproved. The best families, L, D, V and F were more than 15% bigger.

In 2010, a third progeny trial was planted in HCJB plantation. This consisted of 26 families from the Packsaddle CSO and an unimproved seedlot. The design was line plantings of progeny randomly allocated down each row. The planting configuration is Standard. The results from the 2015 assessment showed that only 15 families were on average bigger than the unimproved. The best families were M, Q, O and I.

Progeny trials (post 2012)

In 2012 and 2013, there were two progeny trials established. A trial in Dalbeg 42 consisted of 41 CSO progeny families collected from Dessert CSO, one seedling improved bulk collected out of the Chapmans progeny trial, and one unimproved seedlot that was collected from Frank Wise Research Station.

The 2nd progeny trial was established at Taylors Park 9 Stand 88 consisted of 83 CSO progeny families, 61 collected from Dessert CSO and 21 collected from Packsaddle CSO, and one seedling improved bulk collected out of Quintis plantation. The trial also included 30 seedlots of improved Timor families.

Dalbeg 42 Progeny trial

This trial was a randomised incomplete single tree plot design. There were 18 replicates and 54 incomplete blocks, not all treatments are represented in each replicate. The configuration was standard 1f with 505 Sandalwood per hectare. An assessment at age five was undertaken, when DOB at 20cm was considered, the growth for the 41 improved individual families. There were 24 families with diameter greater than the unimproved, four with DOB at 20cm the same and 15 with DOB at 20cm smaller than unimproved. Six of the families had up to 20% greater diameter than the unimproved. There were 15 families that grew less than the unimproved. The best family average is P2-12, being an average 20% bigger than unimproved. In 2022 a search was made of the trial for elite trees to be included in the 2nd generation breeding program and eight trees were selected. These trees are on average 60% bigger diameter than trial average.

TP9 Progeny trial

This trial was a randomised incomplete single tree plot design. There were 20 replicates with 80 incomplete blocks, not all treatments are represented in each replicate. The configuration was standard 3d with 505 Sandalwood per hectare.

A 2021 assessment (age eight) of the 83 families in the breeding population showed that only one family had grown less than the unimproved. Family P3-2 the best family, was on average 17.2cm DOB at 20cm (90% bigger than unimproved). However, when the diameter increment growth between 2016 and 2021 is reviewed, 13 families had increased less than the unimproved. One of the Timor families had also grown less in that period. The incremental increase for the unimproved over the five year period was 3.5cm, and average of 0.7cm/year. Thirty of the improved families increased DOB at 20cm by more than 1cm/year (50% greater than unimproved), with family P2-8 growing an average of 1.5cm/year DOB at 20cm (107% more than the unimproved).

In 2021, the breeding population exhibited a mean diameter over bark (DOB) that was 25% greater than the unimproved population at 20 cm above ground, while the Timor provenance averaged 22% greater. When the top 20 elite families were analysed separately, the remaining breeding population averaged 21% greater than the unimproved population, whereas the elite families were, on average, 35% larger.

Heartwood development in Progeny

As part of the second generation breeding program, heartwood sampling was undertaken in 2022, for trees selected to be included in the second-generation breeding population. These initial selects were all within established progeny trials. These trials are in QLD (Dalbeg 42), NT (Taylors Park 88), and WA (HCJB and Chapmans). As the progeny trials were established between 2009 and 2013, the trees sampled varied in age.

All the trees sampled had good heartwood development, with only three trees with less than 20% of the core heartwood. When the percentage of heartwood is ranked across all the sites, of the top ten trees with greater than 40%, seven are in Dalbeg, two in Chapmans and one at HCJB. The next three best trees are from TP9 with 39%. A comparison with the 2023 ORIA commercial harvest of 29% heartwood, has shown that the breeding population is significantly above the mean unimproved.

When transitional heartwood is combined with oil-bearing heartwood, Chapmans has the greatest representation in the top ten, with four trees, followed by TP9 with three. Chapmans is the oldest site, and we would expect that transitional heartwood would be the greatest in these trees. If age were the sole determining factor, HCJB should exhibit similar levels. However, this is not the case; instead, TP9, despite being three years younger than HCJB, shows a greater proportion of transitional heartwood.

The percentage levels of volatile oil in the trees were lower than expected, considering when harvested the expected oil levels are around 3%. Although some sampled individuals had over 4%, 21 of the 34 trees had less than 2% and only three trees above 3%. This lower figure could be due to analysing a small core sample and not a complete biscuit, which could provide more material from which to extract oil.

The concentrations of α-santalol and β-santalol were generally satisfactory across samples, though several trees exhibited levels exceeding those specified by ISO standards. Genetic analysis confirmed that silverbacks tend to possess a higher percentage of heartwood. However, these specimens typically display lower volatile oil content within the heartwood.

The selection of the same families on different sites seems to indicate that sites have the largest influence on diameter. Trees in the same family had similar diameter when grown on the same site, but when planted on different sites, they were more different.

However, for heartwood development there was a similar variation within families irrespective of site. There were some big differences within families, for example P3-09 grown in TP9 and Dalbeg had similar diameter, but the Dalbeg site had 66% more heartwood.

Table 40: Heartwood core results for West Australian Progeny trials.

Tree ID	Ortet location	Age	Family	DOB at 20 (mm)	% of HW	HW + TW (%)	a-santalol	b-santalol	Volatile Oil %
S2-01	HCJB	12	P	225	38%	74%	53.73%	21.30%	0.85%
S2-02	HCJB	12	B	240	17%	42%	50.90%	23.81%	4.31%
S2-03	HCJB	12	M	263	31%	61%	54.51%	22.66%	1.83%
S2-04	HCJB	12	S	253	52%	65%	51.73%	18.19%	0.71%
S2-08	HCJB	12	L	251	32%	65%	52.50%	23.76%	2.00%
S2-09	HCJB	12	C	201	36%	84%	51.41%	22.55%	2.52%
S3-02	Chapmans	13	L	262	62%	100%	52.99%	22.37%	1.05%
S3-03	Chapmans	13	W	235	21%	82%	51.57%	20.57%	1.25%
S3-04	Chapmans	13	M	227	28%	81%	53.55%	24.14%	2.03%
S3-05	Chapmans	13	AD	214	40%	80%	51.78%	22.69%	2.17%
S3-06	Chapmans	13	AD	213	20%	67%	52.45%	24.05%	2.15%
S3-07	Chapmans	13	H	222	14%	69%	41.76%	19.30%	2.17%

However, with family AD, the two trees both grown at Chapmans, had almost the same diameter but one tree had twice the heartwood. This seems to indicate that genes possibly do not play a leading role in heartwood development, but it is more dependent on site.

When the genetics is compared, the trend for silverbacks to have more HW% is confirmed, however they do have lower levels of volatile oil content in heartwood. The term 'Silverback' will be explained later in the chapter.

Table 41: Heartwood core results for Northern territory and Queensland Progeny trials.

Tree ID	Ortet location	Age	Family	DOB at 20 (mm)	% of HW	HW + TW (%)	a-santalol	b-santalol	Volatile Oil %
S5-01	TP9 R22	9	P3-05	209	39%	60%	47.75%	22.20%	4.47%
S5-02	TP9 R22	9	P1-11	218	21%	81%	53.19%	22.69%	2.11%
S5-04	TP9 R22	9	P4-24	215	35%	60%	50.18%	23.72%	4.84%
S5-05	TP9 R22	9	T11	209	27%	61%	52.14%	23.43%	1.70%
S5-06	TP9 R22	9	P3-11	215	29%	62%	51.52%	23.18%	1.18%
S5-07	TP9 R22	9	P3-15	206	20%	63%	51.17%	22.29%	1.44%
S5-08	TP9 R22	9	P4-19	201	29%	67%	51.13%	22.44%	1.73%
S5-09	TP9 R22	9	P2-06	214	32%	68%	53.14%	22.53%	2.01%
S5-10	TP9 R22	9	P3-05	212	20%	55%	52.67%	21.50%	0.41%
S5-11	TP9 R22	9	P2-20	188	39%	53%	50.28%	23.83%	1.71%
S5-12	TP9 R22	9	P1-02	270	39%	80%	54.04%	20.98%	0.57%
S5-13	TP9 R22	9	P3-09	209	35%	63%	52.34%	23.25%	1.20%
S4-01	D42-R20	10	P1-03	225	47%	74%	52.94%	25.12%	1.22%
S4-02	D42-R20	10	P2-01	246	65%	73%	52.07%	23.09%	1.13%
S4-03	D42-R20	10	P1-10	198	34%	74%	53.28%	24.62%	1.49%
S4-04	D42-R20	10	P3-09	227	53%	71%	51.50%	27.50%	1.82%
S4-05	D42-R20	10	P1-09	210	46%	74%	51.84%	24.41%	2.99%
S4-06	D42-R20	10	P2-06	242	56%	67%	54.99%	22.67%	0.41%
S4-08	D42-R20	10	P4-23	227	47%	53%	51.58%	27.36%	2.22%

4. CLONAL SEED ORCHARD DESIGN AND LAYOUT

There has been little published work investigating the optimal spacing for pollination of *S. album*. However, it is well known that Sandalwood is primarily an outcrossing species (Badami & Venkata Rao, 1930) and that pollen is transferred via insects (personal observations). The main pollination vectors for Sandalwood in the orchards are insects such as bees, wasps, butterflies, moths and ants. Flying insects travel from tree to tree carrying pollen and effecting the out cross required for good quality genetically diverse seed. The configuration of individuals within the orchard will have an impact on how well that out crossing occurs.

Although there are no published papers for Sandalwood seed orchard design to encourage out crossing there are number of assumptions that can be made. To actively manage pollination, it is necessary to understand the behavioural characteristics of major pollinators of specific species. The main assumption is that the pollinators on Sandalwood are all operating by sight in the daytime. The reason for this is that Sandalwood flowers are weakly scented and rarely visited at night. Trees that are closer together and within line of sight are more likely to experience cross-pollination.

Therefore, with an assumption that as distance doubles, the chance of outcrossing halves. Figure 204 shows the chance of out crossing between individuals, starting with 3 metres being 100% chance and then halving each time that distance doubles.

Using Figure 204 it is possible to calculate optimal out crossing score in a 10 metre radius from each individual ramet. Although bees range everywhere, other pollen vectors not so. For this reason, a 10 metre radius for each individual ramet within Line of Sight (LOS) is used as a limit for best chance out crossing.

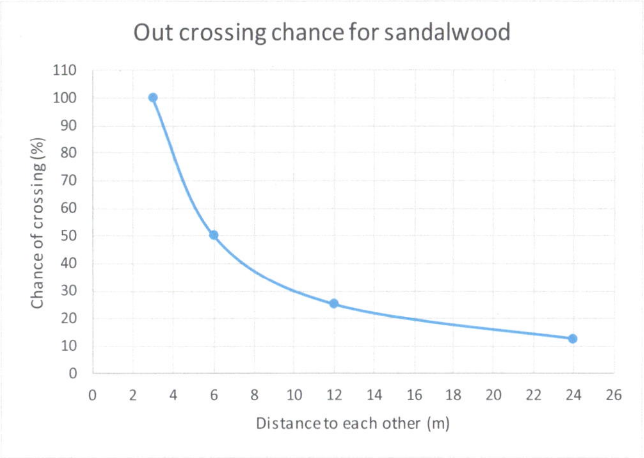

Figure 204: Chance of outcross decreases as distance increases.

If this is the case, all seed production areas should be planted at optimal spacing of 3m between individuals, and this would be the case, except for the need for long-term hosts. It is anticipated the life span of a Seed Orchard is up to ten years. During this time, for the clones to remain productive good hosting is essential. This make's continuous 3 metres along row stocking not possible. Using the assumptions mentioned, we will look at best options for SO establishment.

Table 42: Out crossing score for distance from pollen source.

Distance (m)	Crossing Probability
3	100
5	60
6	50
7	43
8	38
9	35
10	30

Figure 205: Large black wasp pollenating flower, (Dalbeg Robson, 2014).

a) Orchard Configuration models

To determine the best layout for the clonal seed orchard, six models are examined. Pictorial diagrams of each model are shown.

Model 0

This model shows the optimal configuration for an orchard. Horticultural fruit orchards would be planted in this configuration. Due to the need for hosts this model is not suited to Sandalwood orchards. It does however show that with optimal outcross, an outcrossing score (OS) of 928 with 16 potential partners is possible.

Model 1

This is the layout for standard Sandalwood plantation, with alternate rows of LTH. Dessert CSO was planted in this configuration. The main constraint with this design is reduced line of sight (LOS) and only single rows of available outcrossing partners. For this system to be viable, so that there is LOS, the LTH row would need to be continually reduced in height. The OS is 200 with 2 potential LOS partners.

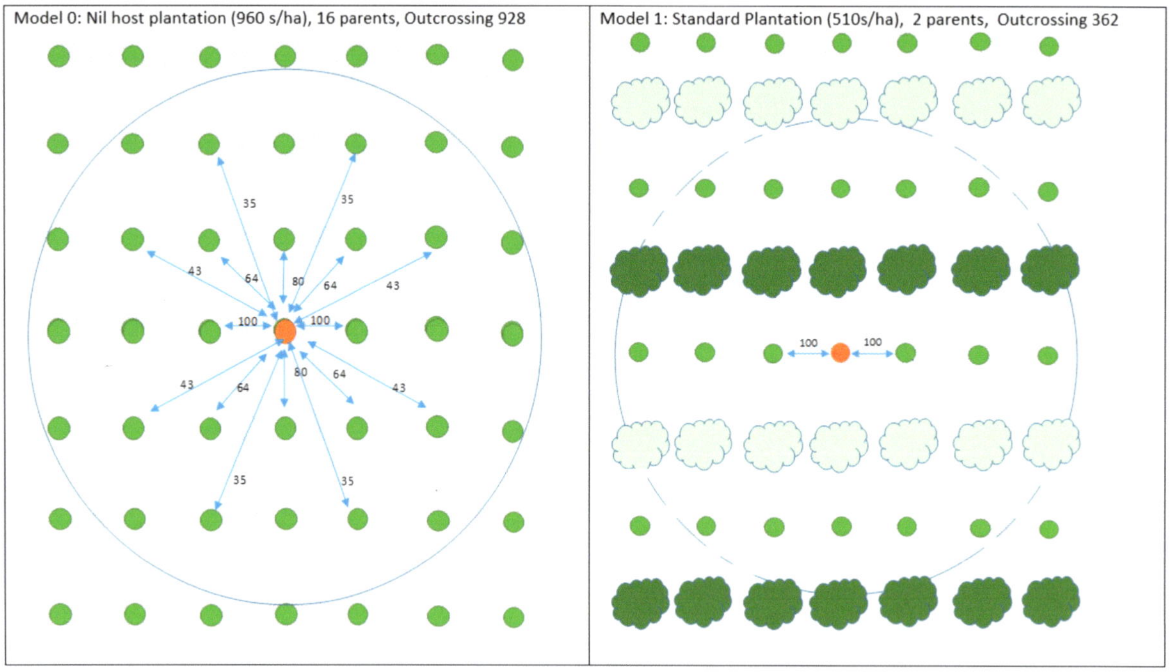

Figure 206: Concept models 0 (no hosts) and model 1 (STD configuration).

Model 2

This is a close approximation of the standard "MIX" configuration used operationally. For seed production this layout is particularly good for LOS. During the first two years the LOS is 10 potential partners, then as the LTH grow this may be reduced to 7. Due the higher stocking the OS is also particularly good, 582 and then 448. When crown sizes have doubled, LOS partners are again reduced to 420 and partners six.

Model 3

This layout has been used for seed Orchard design from 2012 and was utilised in QCSO and TPCSO. LTH are planted in the same row as clones and clones are planted in groups of four equal distance apart. The OS for this design is 389 with six LOS potential partners.

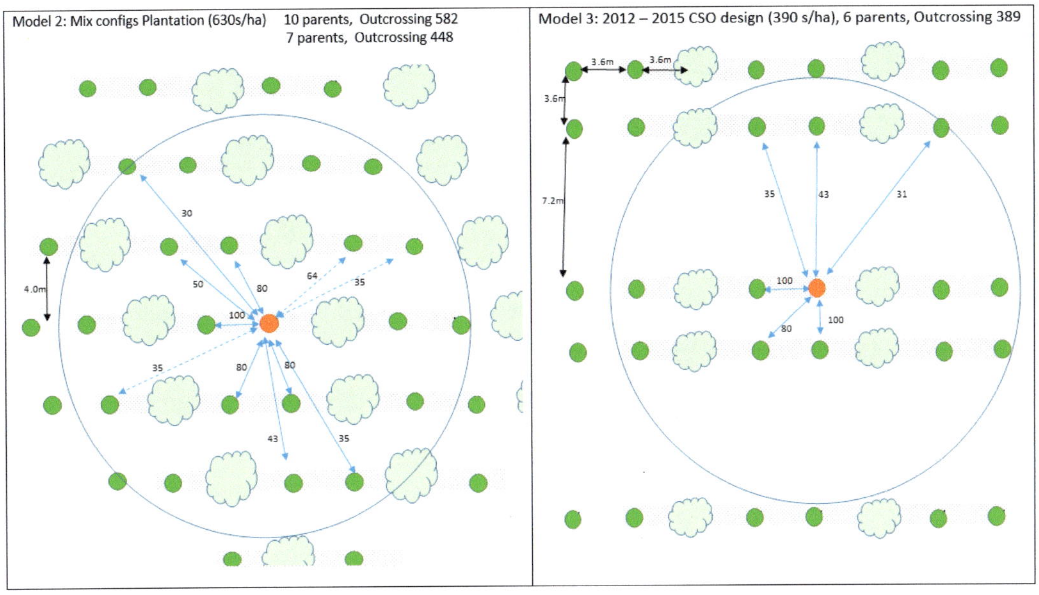

Figure 207: Concept models 2 (MIX configuration) and Model 3 (double row).

Model 4

With this design, each row has a spacing of 3m between pairs of Sandalwood clones with 8m between pairs along the row. Sandalwood is at right angles to each other across the row. This layout has 7 LOS Sandalwoods with 10m and calculates a total of 388 points for out crossing.

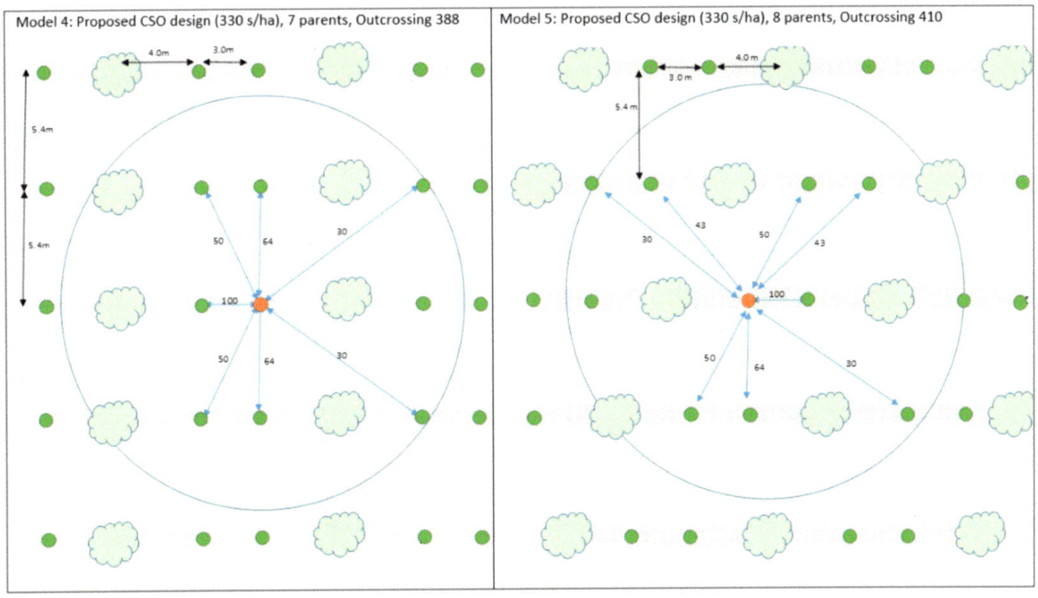

Figure 208: Model 4 and Model 5 (proposed CSO design).

Model 5

This is the same spacing as model 4 but each row is offset by 3m. This increases potential LOS partners to 8 Sandalwoods, which calculates an OS of 410.

Table 43: Summary of LOS Partners and OS score for various Models.

Model No.	LOS Partners	outcrossing score	SPH	Comments
Model 0	16	928	960	no hosts
Model 1	2	200	510	STD QIN config
Model 2	10	582	630	Chapmans
Model 3	6	389	390	QCSO &TPCSO
Model 4	7	388	330	
Model 5	8	410	330	WCSO (2017)

Although there have been many assumptions on insect vector behaviour in regard to their pollination habits, these assumptions are based on observed generalities, such as line-of-sight travel. In using these assumptions, with the configuration models discussed, it is clear layout and design does have an influence in the availability of individuals to maximise their out-crossing potential. That said, the best design, that includes required hosts, is the operational mixed species configuration, Model 2. Followed by the proposed Model 5, which is a wide spaced Mix configuration.

A key subject not discussed so far which can play a role on seed orchard design is stocking, the main difference between Model 2 and 5. A wider spacing model, although not as efficient in the early years, due to available room for crown development will not eventually have a reduced number of crossing partners due to crown development. Figure 209 shows how model 2 has reduce available crossing partners compared to Model 5 which has not changed.

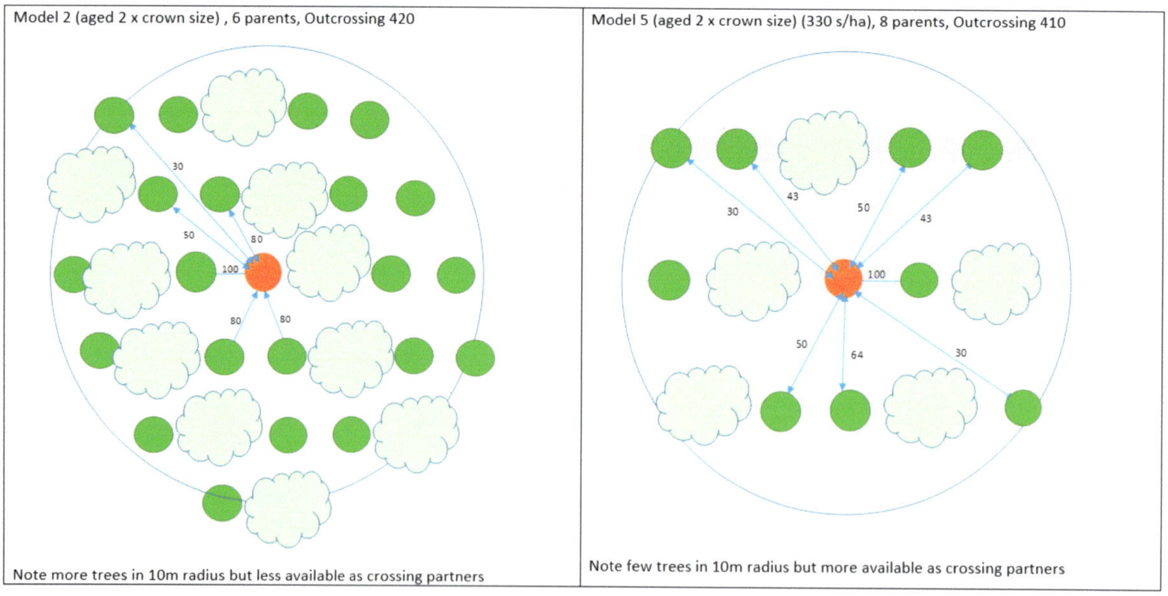

Figure 209: Stocking and LOS comparison between Model 2 an operational MIX configuration and Model 5 a specific SO design.

Table 44: Summary of LOS Partners and OS score for 2 "aged" models (2x crown width).

Model No.	Partners	outcrossing score	SPH	Comments
Model 2	6	420	630	Mix config
Model 5	8	410	330	WCSO (2017)

Over the years, Quintis progressively reduced orchard stocking from 510 to 390 and ultimately to 330 sandalwood trees per hectare, with the aim of increasing seed yield and enhancing genetic diversity. The proposed Model 5 is expected to further improve outcrossing by increasing the number of potential mating partners. Wider spacing should also promote better airflow, improve operational access, and extend the productive lifespan of the orchard.

b) Flower and fruit development

Like Sandalwood in other tropical environments such as India (Radomiljac & Vernes, 2019), Sandalwood in Australia tend to flower twice a year but this is dependent on locality. In the Queensland seed orchards and plantations, the main flowering occurrence is from early February through to late March, with fruit maturity about 90 days after. A second flowering, albeit generally a smaller event is from late September to early November. Flower, buds, young and mature fruit can be found on a single tree at the same time.

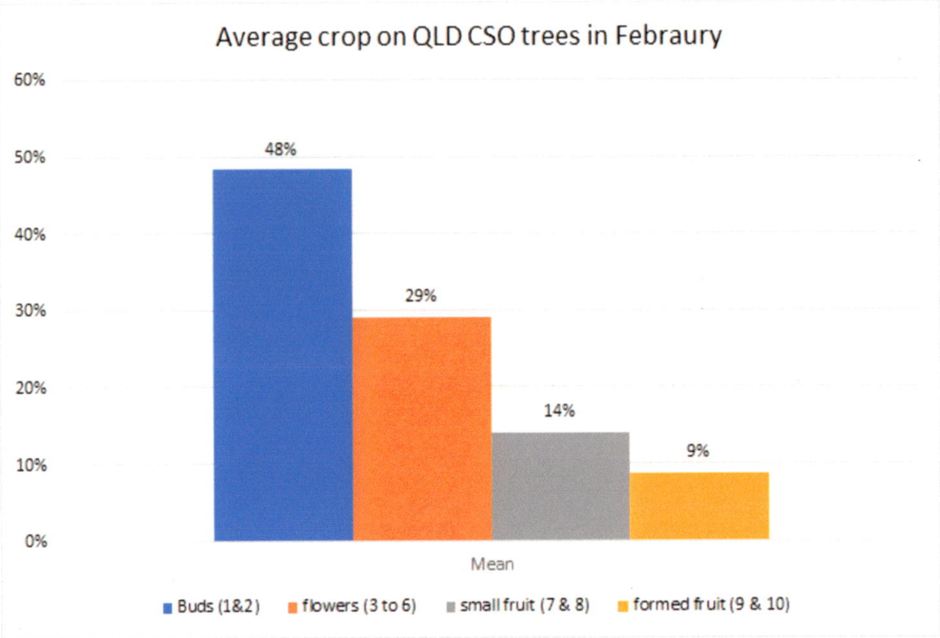

Figure 210: The mean percentage of flower/fruit stages found on individual QLD seed orchards trees in February.

The flower to fruiting can be broken down into ten stages. The process from initial budding to ripe fruit takes about 90 days. Figure 211 details the various stages with pictures and descriptions.

Newly formed buds (stage 1)
Mid-term buds (stage 2)
Mature buds (3) and newly emerged flowers. Start white (4- viable) and gradually turn pink (5-viable)
9 & 10. Fully formed but green fruit (9), with ripe (10) deep purple fruit
9
10
8.
7.
Small fully formed but immature fruit
New small fruit with old flower excised.
Total days 80 to 90
25 to 35 days
25 to 35 days

5. VEGETATIVE PROPAGATION

a) Sandalwood

Although the preferred method of propagation is via seed through the described breeding program, the use of clonal propagation can provide some 'leaps' in volume and can be used to better understand the parasitic nature of Sandalwood. Vegetative propagation is any form of asexual reproduction occurring in plants in which a new plant grows from a fragment or cutting of the parent plant or specialised reproductive structures. This includes macro cuttings, tissue culture and grafting.

As discussed, grafting is utilised to capture genetic material and progress the breeding with the development of grafted (clonal) seed orchards. However, the form of Sandalwood grafts, or at least the way they are managed, makes this method unsuitable for operational clonal forestry. They tend to not be apical and do not develop a significant bole. There is some evidence that the graft union disrupts the heartwood development. This was evident when Packsaddle grafted orchard was terminated. In an investigation into heartwood developed in the grafted trees, it was found that there was no heartwood above the graft union in these eight year old trees.

Tissue culture facilitates the exponential multiplication of plants in a sterile laboratory environment and allows for clonal replication on a large scale at any time of year and is not restricted by seasonal conditions. It has been shown to produce high health, high vigour, superior quality (Baltunis & Brawner, 2009) as all the plant's nutrition and environmental requirements have been provided for during the culture process. Cloning is the exact duplication of the same genetic material and physical characteristics resulting in more uniform results when planted out.

Other forest industries have made use of tissue culture to huge economic gain. The hybrid eucalypt industry in Brazil has planted over 600 000 ha of tissue cultured eucalyptus seedling in plantation (Beltrame, et al., 2023).

There are immense potential benefits for plantations to producing seedling by tissue culture. It has been observed that environment plays a role in heartwood development. Observations of trees that have had little, or no stress, grow to a large diameter with small heartwood, and the opposite is true for smaller trees that have had significant stress with substantial amounts of heartwood. The development of plantation establishment methodologies and use of improved seed sources has meant reasonably uniform development of new plantations. The variability of establishment is limited to the site inconsistencies. This potentially means that in many cases the effect of environment on heartwood development between individuals is limited.

The other key factor determining heartwood development is genetics. The breeding program selected individuals that had larger than average heartwood and volume production. Moreover, progeny trials have indicated that some families are up to 60% better than the unimproved. When the mean benefit is calculated, due to families with lower potential, the overall gain is closer to 30%. If only families or individuals with greater than 60% increase over unimproved were selected and propagated, there is huge increase in potential volume. With developed tissue culture, this option becomes commercially available.

In their review on Sandalwood vegetative propagation Teixeira et al. (2015), report mixed results when attempting propagation by macro cuttings. In an extensive study by Collins and Haines (2000) where they looked at the amenability of four Santalum species to macro cuttings, they demonstrated the *S. album* was the least likely to root using this methodology with less than 10% success.

Tissue culture remains the most promising system to clonal reproduce seedling of *S. album*. Tissue culture is the *in vitro* regeneration of axillary buds, adventitious shoots or somatic embryogenesis by using MS medium with a varying array and concentrations of auxin, cytokinin and organic acids. Literature shows work started on experimenting with *S. album* in the late 1970's (Rathore, Rangaswamy, & Goyal, 2022). However, *S. album* has been recalcitrant to *in vitro* propagation, particularly the development of roots (Sanjaya, Anathapadmanabha , & Rai, 2003). There are many examples of work undertaken that have successfully produced callus and shoots using somatic embryogenesis, and a few with development of poor root systems.

Although it has proven to be difficult, it has not been impossible. JA McComb from Murdock University successfully produced tissue cultured seedling for field trials in 1990 (McComb, 2009). The clonal selections were unimproved with untested ortets (parent trees), the clones did not express any superiority over seedlings across the sites.

Observation of a small *S. album* trial established at Frank Wise research Station in Kununurra, with tissue cultured plants in 2008 demonstrated there was great variation in growth and development within the plot (without knowing any clonal makeup of the plot). There was a below ground root biomass taken and the observation was that root development of the clones was poor, compared to the identified seedling. This would have contributed to poor above ground development.

In 2020, Quintis entered into an agreement with RAMM botanicals, a tissue culture lab on the central coast of NSW. This lab specialises in the production of Blueberries. Blueberries have been bred and many high yield hybrids exist. RAMM have developed tissue culture protocols for the multi propagation of this species. Their experience in developing new protocols, such as root media, varying environments, combinations of auxins with cytokinins, pH and carbon sources for tissue culture will be investigated to find a successful protocol for *S. album.*

Figure 212: RAMM experimental work with Quintis supplied S. album has stated to show some results after 6 months of initial trials, (RAMM report, 2023).

A cautious approach will be essential if tissue-cultured plantlets become commercially viable. Any cloned family should undergo comprehensive field testing against improved seedling stock, with particular attention given to phenotypic performance and heartwood development. Such evaluation would require at least six years, sufficient time to determine whether clones initiate and develop heartwood at a rate comparable to conventionally propagated seedlings.

b) Hosts

Quintis research conducted some experimental work to look at the amenability of *Dalbergia lanceolaria* and *Cathormion umbellatum* to vegetative propagation, specifically macro cutting. A trial investigating the use of two rooting auxins, IBA (Indole-3-butyric acid) and NAA (Naphthalene acetic acid) with nodal cutting from these species resulted in 30% rooting for Dalbergia and between 78% and 28% root success for Cathormion.

Figure 213: Good rooting success for nodal cutting from Cathormion and D. lanceolaria show that these species could be propagated in a clonal forestry program, (Robson, 2016).

The rooting cuttings from this trial work were planted out at Early Storms in 2017. After six months it was impossible to tell them apart from normal seedlings.

6. GENETIC ANALYSIS OF AUSTRALIAN PLANTATION POPULATIONS

To determine the genetic relationship between each of the trees within the 1st generation breeding population, an analysis of the DNA using micro-satellite technology was undertaken.

DNA based molecular markers are an effective means of determining genetic relatedness and provide estimates of genetic diversity (Parker, Snow, Schug, & Booton, 1998). Nuclear DNA markers, particularly RFLPs, microsatellites or short sequence repeats (SSRs) have been utilised to determine both observed and expected heterozygosity (Ho and He). Heterozygosity denotes the possession of two different alleles of a SNP by an individual. The lower the heterozygosity, the less diverse the gene pool.

This work was undertaken by Professor Margaret Byrne and Dr Rachel Binks from the WA Department of Biodiversity, Conservation and Attractions. Four separate analyses were completed, in 2012, 2014, 2017, and 2022, each discussed later in the chapter.

A study conducted in the Western Australian government owned Frank Wise Research Station gene pool with nuclear RFLPs by Jones, et al (2008), revealed low levels of genetic diversity for a tree species, with an observed and expected heterozygosity (Ho and He) of 0.047. However, Byrnes quoted a Ho figure of 0.169 for the FPC collection in her 2012 paper for Quintis and 0.14 for the Quintis 108 families in the base breeding population. Tested samples from Timor Leste displayed similar levels of Ho averaging 0.15 and 0.13 for a limited Indian collection (Brand, 1994).

a) The 2010 analysis

This was an analysis of the Elders/Santanol breeding population undertaken by M Byrne WA Department of Environment and Conservation. The DNA was extracted from the leaves using a Qiagen DNAeasy extraction kit. Each DNA sample was assayed with nine microsatellite loci that amplify clear single loci in *S. album*. These loci were selected from a microsatellite library developed for *S. spicatum*. Genetic relationships among the individuals were determined through genetic analysis using GenAlEx v6.2 based on genetic distance and genetic relatedness (Byrne, Genetic diversity: a basis for forestry breeding programs., 2009).

Leaf samples were obtained from 32 trees of *S. album* from the Elders Forestry collection. There were 30 unique genotypes within the 32 samples. Two of the pairs of samples had the same genotypes. Comparison of the level and pattern of genetic diversity in the Elders Forestry collection with that in the Forest Products Commission (FPC) collection shows that the level of heterozygosity is similar, but the FPC collection has higher levels of allelic diversity as would be expected from the larger size of the collection.

The diversity captured in the Elders Forestry trees is less than that in the FPC collection. It is suggested that the genetic diversity of the collection could be improved by the incorporation of some germplasm from the FPC collection. In particular trees from the FPC collection occupying the "middle" of the ordination space in the PCA (Figure 214).

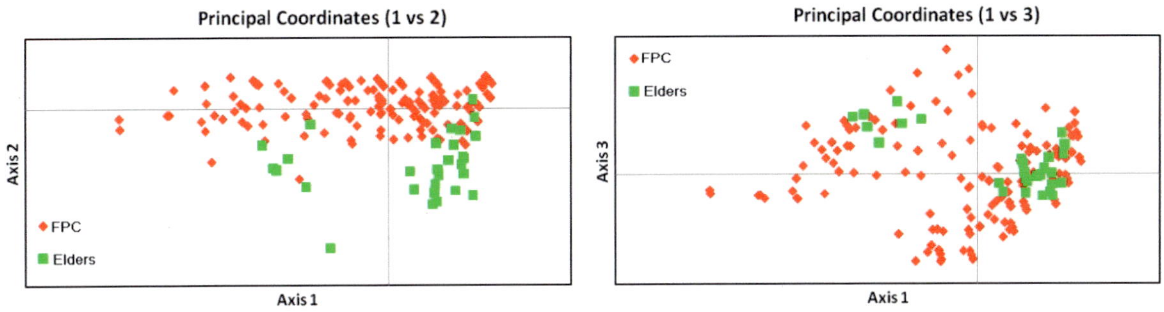

Figure 214: Principal coordinate analyses of the genetic relationships among individuals of Santalum album in the Elders and FPC population.

b) The 2012 analysis

This was the first analysis for the TFS/Quintis breeding population, also undertaken by M. Byrne. WA Department of Environment and Conservation. Like the previous analysis the DNA was extracted from the leaves using a Qiagen DNAeasy extraction kit. The same process as the 2010 analysis was used with nine microsatellite loci to amplify the loci in *S. album.*

The genetic relationships between 108 trees of *S. album* were determined. There were 62 unique multilocus genotypes in the 108 samples. The genetic diversity was found to be low with a large number of trees having the same multilocus genotypes at the loci assayed. As mentioned in the introduction, genetic diversity in Australian collections of *S. album* in known to be relatively low compared to other tree species. The relatively low diversity has emphasised the two genetically distinct groupings found in the Principal Coordinates Analysis (Table 45).

Table 45: Genetic diversity parameters for trees of the TFS collection and the FPC collection. A, average number of alleles per locus; P, percent polymorphic loci; Ho, observed heterozygosity.

Collection	No. Trees	A	P	Ho
TFS	108	2	44%	0.14
FPC	241	6	89%	0.169

Separation of individuals into two groups based on homozygosity at this locus is also evident in the collection when considering the genotypes at this locus. However, the greater diversity at the other loci reduces the differentiation of the two groups in the PCA (Figure 215).

Byrne determined that the fixed difference at this locus is somewhat odd for a species with a mixed mating system. It suggests there maybe two discrete origins of the trees in the Australian collections. A small number of trees in the Quintis collection are heterozygotes for the two alleles. This suggests they may have arisen from crossing between individuals from the two groups at some stage. Crossing between trees in these groups within the Quintis collection would allow recombination of the diversity in the collection

The observed and expected heterozygosity (Ho and He) for these samples was 0.19 and 0.14 respectively.

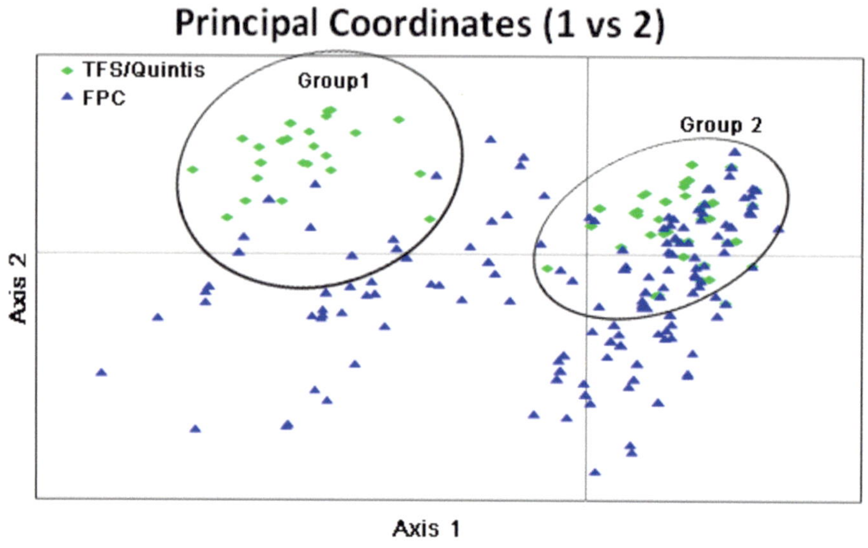

Figure 215: Principal coordinate analyses of the genetic relationships among individuals of Santalum album in the TFS and the FPC population.

Table 46: Groups of related families and those with unique genotypes. The families shaded yellow are the ones with the top performing progeny and included in the 1.5 CSO.

Shared multilocus families

1	1.05	1.09	5.01	E		
2	3.02	2.01	2.12	2.20	1.04	
3	2.06	1.03				
4	1.12	2.08	5.03	AD	F	
5	1.10	2.11	7.03	7.04		
6	3.01	2.14	2.19	5.04	1.02	
7	2.15	2.13	4.16	4.28	Y	
8	3.07	3.08	3.13			
9	4.20	3.11	1.06	4.21		
10	2.16	3.12				
11	4.32	4.18	4.25	3.04	M	W
12	4.24	4.23	3.06			
13	4.30	4.31				
14	5.05	7.01				
15	7.05	C	X			
16	H	K				
17	AC	L	Z			
18	2.1	Q				
19	N	R				
20	D	S				

Unique multilocus families

21	1.01	42	4.27
22	1.07	43	4.29
23	1.08	44	5.02
24	1.11	45	5.06
25	2.02	46	7.02
26	2.03	47	7.06
27	2.04	48	7.07
28	2.05	49	7.08
29	2.07	50	7.09
30	2.09	51	A
31	2.17	52	AB
32	2.18	53	B
33	3.03	54	G
34	3.09	55	I
35	3.10	56	J
36	3.14	57	KS100
37	3.15	58	O
38	4.17	59	P
39	4.19	60	T
40	4.22	61	U
41	4.26	62	V

c) The 2014 analysis

The 2012 extraction methodology was Qiagen DNAeasy. This analysis used the Doyle and Doyle extraction method with the addition of sodium sulphite and examined seven new samples from the Timor provenance seedlings and compared them with the existing first generation population (Byrne, 2014). The DNA samples were assayed with the same nine microsatellite loci use in 2012, that amplify clear single loci in *S. album* but based on *S. spicatum*.

Each of the seven additional samples had a unique multilocus genotype. Four loci (A003, A106, B112 and B128) remained monomorphic despite the inclusion of the seven additional samples to the Quintis dataset. There were four alleles unique to the seven new samples. Allelic diversity was higher in the new samples than in the larger initial 1st generation population. The observed and expected heterozygosity (Ho and He) for these samples was 0.25 and 0.15 respectively.

Summary

The 2012 and 2014 analysis were limited, in that only nine SNP loci were used. However, this was enough to determine relatedness between selected clones within the breeding population. This indicated how diverse the breeding population is compared to other seed sources. Of the 108 selections within the breeding population, there are only 62 unique genotypes. This means that for 20 groups, that include 66 clones, these are either full or half sibs (Table 46).

The grouping based on homozygosity was evident not only in the Quintis collection, but also in that of the Forest Products Commission. Given that much of the Quintis germplasm originated from the FPC, this result is unsurprising. The analysis also confirmed that the Timor seed source is genetically distinct from the remainder of the Quintis population, potentially offering valuable opportunities for further genetic gain within the breeding program (Figure 216).

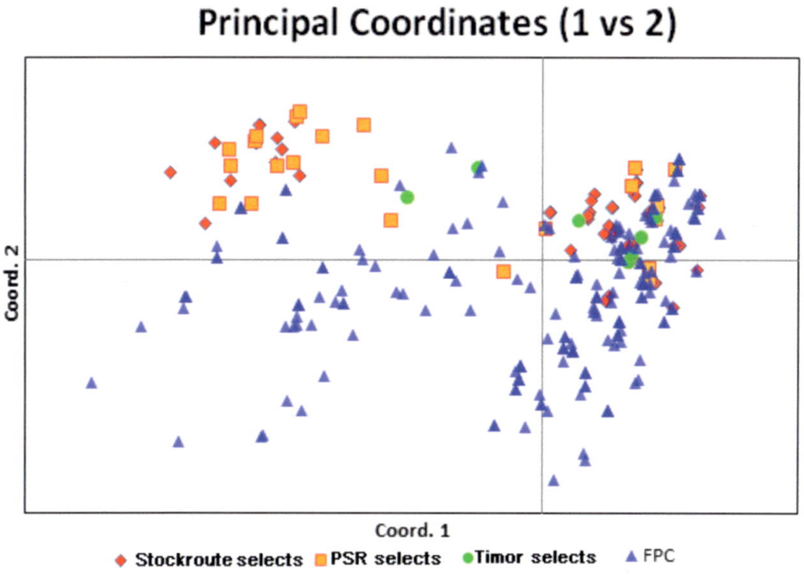

Figure 216: Principal coordinate analyses of the genetic relationships among individuals of Santalum album in the TFS selects, Timor selects and the FPC population.

d) The 2017 analysis

The objective of this analysis was to determine provenance identity of the Quintis breeding population (Binks & Byrne, Provenance assessment of Santalum album, 2017). Collections were made from external sources of *Santalum album* from known provenances stands, Northen Indian (Seoni, identified as Davis in analysis), Central Indian (Kolar, Bangalore, Mulbagal), Sri Lankan (Passara), and of Timor origin. To compare these provenance samples, 20 samples from the Quintis 1st generation breeding population was included. A total of 48 samples were tested.

The analysis methodology has improved significantly since 2014. The DNA was extracted from the leaf samples using the Doyle and Doyle extraction method with the addition of sodium sulphite. All 48 DNA samples were sent to Diversity Arrays Technology (Canberra, ACT) for DArT-seq analysis. This resulted in the generation of a large, codominant dataset consisting of 7008 SNP loci across the genome of *Santalum album*.

The results showed that tight clustering patterns found for the collections of known provenances demonstrate that these collections were sufficiently differentiated for assigning the unknown collections into Timorese, Indian or Sri Lankan provenance without ambiguity (Figure 217).

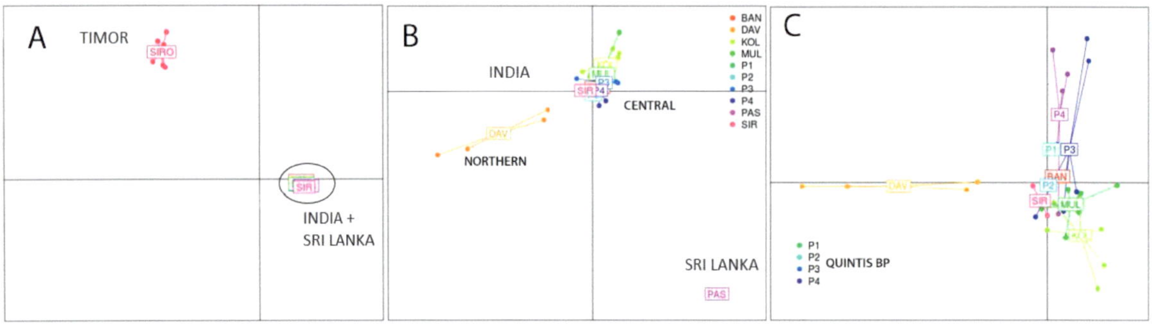

Figure 217: Principal coordinate analyses of the genetic relationships among individuals of Santalum album. Each graph represents a different subset of the data: A) the complete dataset of 48 samples from all known and unknown provenances, B) a subset of the data to focus on the samples closely associated with Indian and Sri Lankan provenances, and C) a subset of the data to focus on the samples closely associated with only the Indian provenances.

13 of the 20 Quintis samples clustered tightly with the Indian collections, strongly suggesting that they originated from there. However, the remaining seven samples (seen as blue in figure 218) did not cluster strongly with any of the collections of known provenances. Instead, these samples showed mixed affinity to the Indian provenance and an unknown source.

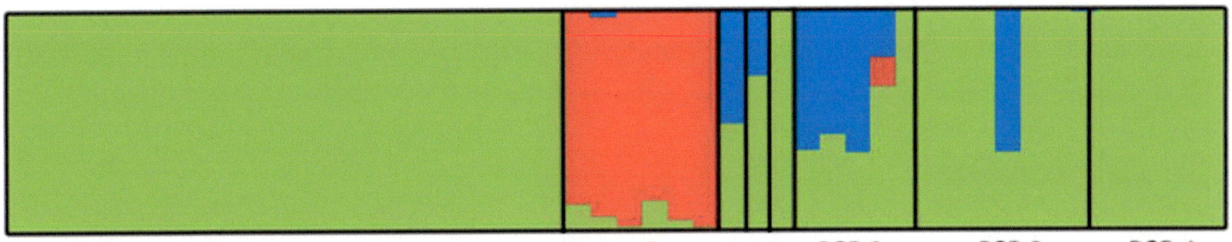

Figure 218: STRUCTURE analysis showing assignment of all 48 Santalum album samples to three genetic clusters. Each individual is represented by a vertical bar that is partitioned into the proportions of its affinity to each genetic cluster, as indicated by the three colours. Individuals are arranged in order of known provenance, as indicated.

The patterns shown in the STRUCTURE analysis were consistent and indicated three genetic clusters in the dataset (Figure 218). All samples of known Indian or Sri Lankan provenance were assigned 100% to Cluster 1 (green), while the Timorese samples were assigned to Cluster 2 (red). Cluster 1 included the same samples of unknown provenance that displayed strong Indian affinity in the PCoA. While the remaining samples that clustered separately in the PCoA, showed mixed affinity between Cluster 1 and Cluster 3.

Consistent with the above clustering patterns at the individual level, the Timor collection was highly differentiated from all remaining collections. The Indian collections exhibited low differentiation with each other and with unknown collections PSR3 and PSR4. The PSR2 collection demonstrated elevated levels of differentiation from the PSR3, PSR4 and Indian collections, reflecting the separation shown in most of the PSR2 samples in the clustering analyses.

Genetic diversity was similar across the PSR2, PSR3 and Timorese collections and significantly higher than the levels of genetic diversity detected in the PSR4 and Indian collections (Table 47). There was little indication of inbreeding in any of the collections.

Table 47: Genetic diversity parameters for collections of Santalum album consisting of more than four samples.

	He	Ho
Seoni (Ind)	0.158	0.158
Kolar (Ind)	0.17	0.149
Mulbagal (Ind)	0.173	0.162
Timor	0.258	0.191
PSR2	0.334	0.283
PSR3	0.229	0.192
PSR 4	0.172	0.169

Summary

This analysis shows how far DNA analysis has come since 2012. This analysis utilised 7008 SNP loci used to determine differences in genome, in comparison the 2012 analysis were nine loci were used. The analysis clearly showed that the original seed source for the Quintis base populations was from central India. Additionally, it also indicated that although not very diverse when compared to many other tree species, the Quintis base population is actually more diverse than some of the native collections (comparison of Ho). This analysis also indicates that there is an unknown gene source in some of the Quintis selects. At the time there was speculation that this could be inter species hybridisation.

e) The 2022 analysis

The objective of this analysis was to determine the unknown source discovered in the 2017 analysis. It was suspect that these anomalies found in the 2017 analysis are the known 'silverback' observed in the plantation.

Silverbacks are individuals within the general plantation population that exhibit slightly different morphological characteristics to the general population. They tend to be bigger, have leaves with a wavy margin and show a shine to the bark, giving them a silvery tinge.

Figure 219: Silverbacks trees tend to be larger than cohorts and have wavy leaf margin, (Robson, 2022).

For this analysis, there was a total of 106 samples tested (Binks & Byrne, 2022) . This was made up of; the various provenances supplied in the 2017 analysis, 21 samples from the Quintis breeding population. As well as new known sources of other Santalum species (*S. austrocaledonicum, S. lanceolatum* and *S. yasi*) and a local native provenance collection of *S. album* from Darwin. A further 48 samples of trees selected in the plantation (Voyager, Packsaddle, Farm Hill and Drovers) that exhibited 'silverback' characteristics was supplied for the analysis.

The same analytical process was applied in this study as in the 2017 analysis. With DNA extracted using a CTAB method (Doyle & Doyle 1987), modified with the addition of sodium sulphite. The DNA was checked for quality and then sent to Diversity Arrays Technology (Canberra, ACT) for DArTseq analysis. This DArTseq process resulted in the generation of a large, codominant dataset consisting of 86,073 SNP loci across the full suite of samples.

The full dataset was investigated using the multivariate ordination method of principal coordinates analysis (PCoA) for a broad assessment of the relationships among the four Santalum species. In addition it would identify the next level of subsetting required to narrow down those relevant to the unspecified *S. album* samples. If any two species are hybridising, it was expected that there would be a clear genetic signal in the PCoA for a wide spread of samples in the intermediate ordination space between the divergent parent species clusters.

The broad genetic analysis across the full dataset clearly distinguished the four Santalum species and showed no indication of hybridisation between them in the silverback stock. The dataset clearly separated the samples of known species/provenance into four broadly distinct clusters representing each of the four species: *S. album, S. austrocaledonicum, S. lanceolatum* and *S. yasi*. Finer-scale structure was visible within the *S. album* cluster, with clear separation of three subclusters representing the Indian + Sri Lankan provenance, the Timorese provenance, and the Australian provenance. All the samples of silverback group fell within the bounds of the *S. album* cluster, confirming that they all belong to *S. album*. However, the spread of the samples overlapped these distinctive provenances to suggest admixture between provenances. For the silverback samples, some fell within the tight range of the Indian/Sri Lankan cluster and some within the Timorese cluster. However, the majority fell in the ordination space between the Australian and Indian/Sri Lankan clusters (Figure 220).

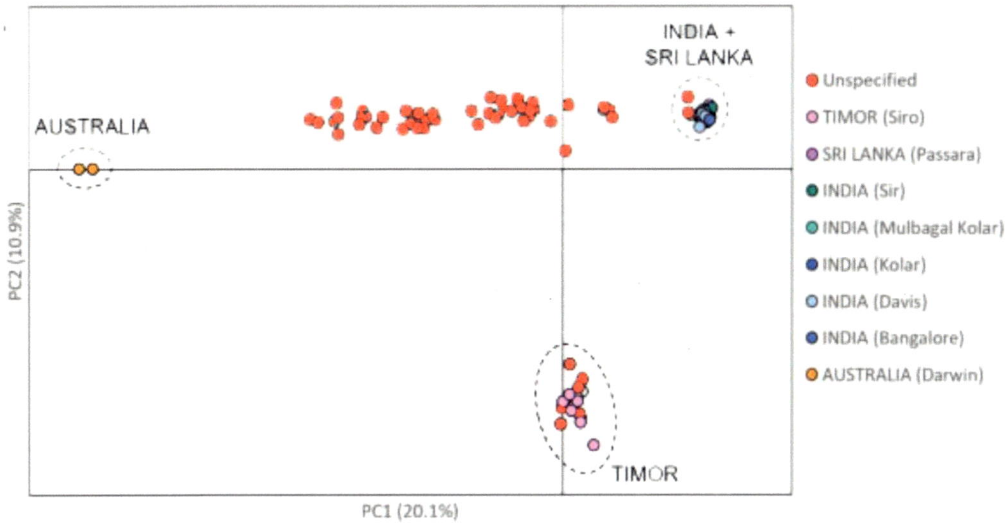

Figure 220: Principal coordinates analyses on subsets of Santalum album samples. A) Subset 1: all S. album samples (4,657 SNPs, 100 samples).

Heterosis

The first part of the analysis determined which individual samples were not pure. A further analysis (New Hybrids) was used to determine parental origin. All of the samples <u>not</u> determined to be pure were all classified with high probability (>99%) as being either second generation hybrids (F2) or hybrids backcrossed to the Indian/Sri Lankan cluster (Figure 221).

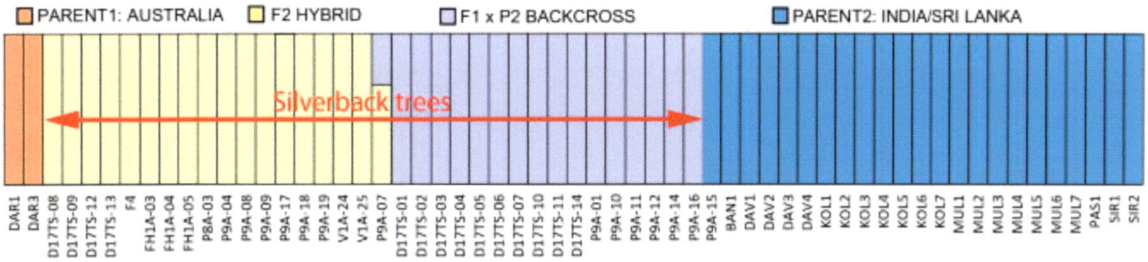

Figure 221: New Hybrids analysis of selected Silverback Santalum album samples, assuming the Australian and Indian/Sri Lankan provenances as the parents.

When we consider the hybridisation of trees within the breeding population, we only looked at 21 samples from the 62 genotypes within the breeding population. Of the 21 samples, seven are F2 hybrids and one was a F1 back cross (Figure 222)

When the relationships among the 108 families in the Quintis first-generation breeding population (as shown in Table 46) are considered, each family can be classified as either hybrid or pure. A sizeable portion (59%) of this population has unknown parentage. Of the families with known lineage, 10% are hybrids and 31% are of pure Indian parentage (Figure 223)..

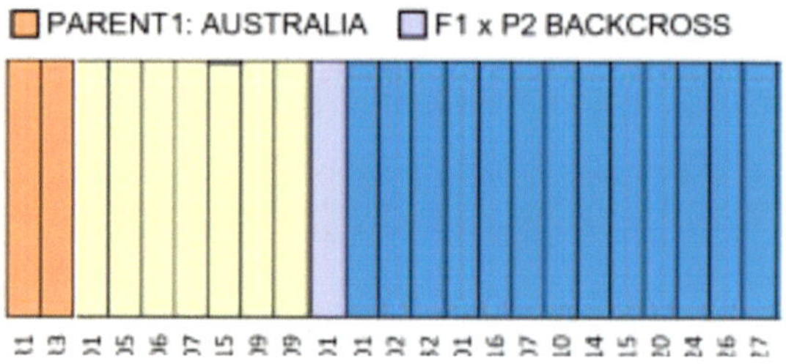

Figure 222: New Hybrids analysis of selected Santalum album Breeding population families.

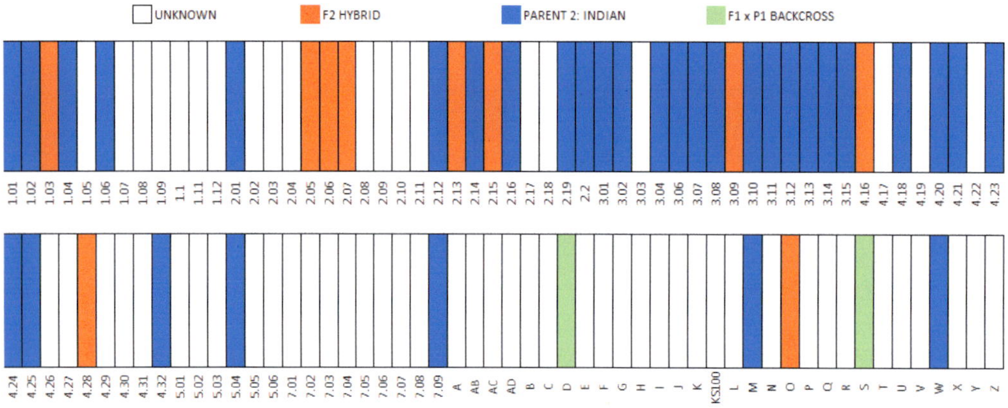

Figure 223: New Hybrids analysis of families within the Quintis 1st Generation breeding population. Each sample is represented by a vertical bar in which the colours indicate the assignment to each hybrid category: parental, second generation or backcross.

Summary

Based on the known provenances analysed, the signal present in the genetic data unambiguously pointed toward mixed ancestry between Australian and Indian *S. album*. Moreover, with three independently generated datasets (2017, 2021 and 2022) repeatedly showing the same signal. This means that if there is no logistical possibility for hybridisation between these two provenances, then the non-Indian parent remains an unsampled provenance that is closely related to Australian *S. album*. There was no indication of hybridisation between the Indian and Timorese provenances, nor between the Australian and Timorese provenances.

There is a considerable proportion of the first generation breeding population that has shown hybridisation. However, there are still nearly 60% with unknown parentages within the population.

7. GRAFTING TECHNIQUE

Collection of scion material

Grating success is dependent on collection of the correct material. This is small twigs that range between 3 mm to 6 mm thick. They should not be too green or too woody and just turned a light brown shade. After collection, remove all the leaves and trim to be 10 to 20 cm long. Collect at least two scions for every intended graft. Bundle and tie scions from each tree with flagging tape and record clone number on tape. These are then placed into a plastic bag, spray with fresh cool water, clone number recorded on bag and seal the bag. Place collected material into a cool bag or fridge until ready for grafting. The tools required include,

- Ladder and pole pruner,
- Secateurs,
- Pen, flagging tape and large plastic bags,
- Esky or cool bag with freezer blocks.

Shaping the scion

Hold the selected scion up to the stock plant to ensure a size match. Make sure the scion is the same size at the point of graft as the stock plant. The shaping of the scion is most important. The shaped point should be at least 20 to 25mm long and tapered evenly. Ensure that the surfaces of the taper are flat and not bowed.

Tools

- Very sharp knife or scalpel

Figure 224: Stock seedlings cut off and ready for grafting, (Robson, 2023).

Preparing the stock plant and taping

Use a nursery tag (or similar) to label the clone to be grafted. Ensure the pen is a paint pen (as it could fade over time). Match the scion and stock plant stem as closely to size as possible. At this point of same size, remove the top of the seedling stock plant and most of its leaves. Cut down the middle of the stem to the same length as the scion shape (Figure 224).

Completing the graft

Insert matching scion into stock plant and tape up firmly After the graft is taped, place zip-loc bag over new graft (Figure 225).

Figure 225: Stages of the grafting method, (Robson, 2023).

Removing the bags and survival count

After six weeks, the plastic over topping bags can be removed. Newly exposed grafts must be kept in heavy shade for at least a further two weeks. When bags are removed, it will be obvious which grafts have been successful and which have failed.

Removing the tape and survival count

After about three months, it is a suitable time to remove the grafting tape from the graft (Figure 226). Most surviving grafts should be strong enough to support themselves without the tape. However, this is dependent on the time of year and growth and healing of each individual graft. It is important not to leave grafting tape on for too long as it can strangle the stem and cause mortality.

Figure 226: Six weeks after grafting the tape is ready to remove (photo 1). Photo 2 shows the completed graft, (Robson, 2023).

Timeline

Grafting of the scion material must be completed no more than three days after collection. Every day after collecting reduces chance of a successful graft.

Bags must be left on grafts for at least six weeks. This will ensure that the area around the graft remains moist and humid. Remove bags at six weeks and a survival or 'take' can be recorded.

Tape should be removed at about 12 weeks after grafting but this will depend on growing conditions. Some individuals may need a few more weeks of support. Another survival count can be taken at this point, as some may have died due to incompatibility.

Table 48: Grafting timeline.

	Month 1								Month 2				Month 3			
	Week 1					Week 2	Week 3	Week 4	Week 5	Week 6	Week 7	Week 8	Week 9	Week 10	Week 11	Week 12
	day 1	day 2	day 3	day 4	day 5											
Collect scion	collect scions															
Graft		Graft (70 to 100%)	Graft (50 to 70%)	Graft (10 to 50%)	Graft (0 to 5%)											
Remove bags		Grafts cover with plastic bags														
Remove tape		Grafts held by tape														

8. HOST IMPROVEMENT

There is opportunity to improve the quality of hosts utilised within the plantations. Due to the perceived non-commercial value of hosts there has been no improvement completed. However, there is potential for the following species to have an improvement program. These species are, *Dalbergia latifolia and D. lanceolaria* and *Cathormion umbellatum.*

a) Breeding Plan

The breeding plan can use a similar but simpler approach to the Sandalwood. Tree selection would be based on desirable phenotypic traits that would be required from a host. These traits are, but not limited to, single stem, drought tolerant, and insect resistance.

Dalbergia

For both Dalbergia species, the original seed source and genetic makeup is unknown. When developing an improvement plan if using existing resources, it is beneficial to know how diverse that resource is. Unfortunately, with the existing population there is no information for either of these species.

One of the first objectives in a breeding plan would be to purchase several seed sources from external international seed merchants to broaden the genetics. These new seed sources should be planted out into provenance resource stands (PRS) or provenance trials. These are plantings that keep the different provenances apart. This PRS could eventually be thinned and used as a seedling seed orchard. Individuals from within the existing plantation with the desirable plantation traits could also be selected as plus trees.

Cathormion

Cathormion is a native and there is information on the original seed source. This seed was collected from the Keep River area in Western Australia. Although it is not known the number of parent trees, as the initial collection was over several years ago it would be safe to assume it would be several.

Cathormion has quite a broad provenance range (see Figure 20 in Chapter 2) and further seed collection should be made from these other provenances. These collections should be established in a PRS or provenance trials that would eventually be used a seed orchard. Individuals from within the existing plantation with the desirable plantation traits could also be selected as plus trees.

Deployment system.

The PRS and provenance trials could have trees with undesirable traits removed and these converted into seedling seed orchards to produce seed for operational plantings.

As discussed, both these species could be reproduced with vegetative propagation quite easily with macro cuttings.

Figure 227: Left photo is a vigorous sing stem Cathormion, and photo right is a D. latifolia with a long single bole and good vigour, (Robson, 2016).

Chapter 7. FUTURE GOALS

This book chronicles a quarter-century (1999 to 2024) of research and silvicultural advancements, significantly enhancing the understanding of plantation development. This progress has resulted in improved seedling survival rates and increased Sandalwood volume over the rotation period. Despite these advancements, many aspects of the host-Sandalwood relationship remain unclear. Greater insight into these relationships will further boost plantation productivity.

This chapter will explore some of the knowledge gaps still present in 2025 and propose a comprehensive plantation establishment and management plan based on best practices.

1. KNOWLEDGE GAPS

a) Stand management

There are significant gaps in the knowledge of stand management, especially in mid and late rotation stages. Research on optimal Sandalwood volume growth is limited, particularly concerning the significant impact of stand competition, water use, and pest control.

Effective management of the long-term host (LTH) is essential for commercial Sandalwood growth. A well-managed LTH supports aggressive growth without overwhelming the site. Australian plantations start with over 1000 stems per hectare, optimising tree form and height through competition among trees, hosts, and Sandalwood. Sandalwood's low apical dominance requires controlled sidelight to avoid broad growth. This would ensure competition reduces branching and promotes straight, single-stem growth.

Planting at higher stockings can lead to limited resources like light and moisture competition, reducing canopy size and efficiency in smaller trees, which can result in less diameter growth. Sapwood area reflects canopy health and competition (Yaemphum, Unawong, & Tor-ngern, 2021). The basal area measures the forest's cross-sectional area, including heartwood and sapwood, and varies with plantation age.

To maximise diameter growth, start with high stocking for competition (West, 2013). Thin trees as they grow to reduce competition but keep mutual benefits. In monocultures, thin to one-third of the basal area of a fully stocked forest. For form control, thin to half the maximum basal area (Reid, 2010).

In mixed species Sandalwood plantations, varied species grow at varying rates and face varying competition levels. The challenge is to identify when competition hinders Sandalwood growth and act before it slows. Additionally, managing the parasitic relationship between host and Sandalwood is complex; removing the host can cut off water and nutrients, reducing growth and making the tree vulnerable to pests and diseases.

Growing plantations in tropical regions involves managing various pests, including weeds and insects. The application of pesticides in Sandalwood cultivation needs to be carefully managed to avoid residues in the heartwood oil, which would affect its marketability. The current range of pesticides has been tested and found effective without entering the heartwood. Nevertheless, a growing number of clients are requesting a decrease in the usage of specific chemicals. Employing current silvicultural practices without essential pesticides, especially in the initial stages of the rotation, may render the commercial cultivation in northern Australia unsustainable.

To develop optimal crown management regimes a number of key questions still need to be answered:

- *When is the optimal timing and how many thinning events are needed achieve optimal growth for the Sandalwood?*

  Existing experiments clearly have indicated that thinning can significantly increase Sandalwood size. However most of these have been established late in the mid rotation phase, leaving a brief period remaining in the rotation to see results that will cover the costs of the thin.

- *What is the competition effect between Sandalwood and what effect will reducing Sandalwood stocking or crown size have on diameter growth and importantly heartwood development?*

  In 2020, Quintis commenced an evaluation of the benefits associated with thinning Sandalwood. However, the data did not conclusively demonstrate the extent of these benefits. In all instances, the thinning process was conducted late in the mid-rotation phase, which was too delayed to observe significant potential gains in the 15-year rotation stands.

- *What are the detrimental consequences, to removing parasitised hosts and does the reduction in competition outweigh any negative effects?*

  Early results from thinning trials show that careful host thinning is crucial. Removing too few hosts does not boost Sandalwood growth. While removing too many risks, over-hosting on the remaining hosts, making them vulnerable to pathogens and mortality, which can leave too few for effective hosting.

- *Although there is some idea for the optimal host to Sandalwood stocking for mid rotation (3 Sandalwood to 1 host, max 7m spacing), is this correct, and if so, how soon can thinning to this level occur?*

  Research trials have started to answer this question but many of the trials have started too late. Thinning should be started before competition become detrimental, not well after, as most of the existing trials have.

- *Currently, competition is managed with crown reduction by hedging but there is not any empirical data that concludes this is adequate to reduce the competition effect.*

  There is a need to determine how effective hedging is compared to a thinning regime. If hosts are thinned early (prior to hosting), this will encourage hosts to grow larger and impose a greater competition for above ground resources on Sandalwood?

- *The physical removal of the thinned host tree may be, at least in the short term, detrimental to the parasitic relationship, are there other systems that can be employed to reduce this effect?*

  The use of girdling (ringbarking) is potentially as solution. The host dies slowly allowing the Sandalwood time to look for other alternates. This method and potential others need to be explored further,

- *Is sunburn and sunscald triggered or caused by our pruning and if so, how can this be avoided?*

  As described, the effect of sunburn/scald on Sandalwood trees is severely detrimental to heartwood. More research needs to be undertaken to better understand the causes of sunburn/scald and how management decisions can reduce this occurring.

- *Use of specific chemicals for control of key pests.*

  The Sandalwood plantation industry currently depends on Fipronil, Clothianidin, and Glyphosate for pest control. These chemicals are under scrutiny by environmental groups, who are pushing for their reduction and eventual ban. Northern Australian Sandalwood plantations would struggle without these chemicals. Urgent research is needed to find new alternatives before these options are no longer available.

- *Commercial use for hosts at the end of rotation (Post harvest Host Utilisation).*

  Despite the potential commercial use of some host species, the lack of pruning has rendered them valueless for timber after 15 years. Future plantations with rotations over 20 years should manage hosts for commercial use. Currently, many hectares remain without a use. Quintis began researching the conversion of host biomass into biochar and vinegar oil, with promising initial results, but further research, potentially with external investors, is needed.

- *Seed production in older age seed orchards.*

  The development of clonal seed orchards is an expensive and time consuming process. Experience has shown that the seed orchards are very productive for the first four to five years after grafting but then seed productivity rapidly drops off. Research into why this occurs need to be undertaken. Are the reasons for this nutritional or physiological?

b) Nutrition

The soil and plant nutritional requirements have been discussed in the Chapter 5 section on Nutrition.

Soil nutrition analysis has demonstrated that most planted soils are nutritionally low. However, foliar analysis indicates no deficiencies in species up to age five on most sites.

Tropical Australian soils are low in phosphorus (Wild, 1958) and nitrogen, but many native species have adapted to thrive in these conditions. Although most long-term host species used by growers are not native, except for *C. umbellatum*, all are nitrogen-fixing.

These are trees that can convert (fix) atmospheric nitrogen into a form that can be utilised for in association with symbiotic organisms such as Rhizobium bacteria. Foliar testing across a range of the Sandalwood plantation trees at various ages has shown that there is not generally any deficiency of N.

A range of field trials have shown the application of granular fertiliser applied broadacre is the best way to increase plant nutrition. The research trials indicated that on most sites an application 60kg applied elemental P /ha is necessary to see a response. This evidence, as well as results for other species on other sites also indicate in Keenan et al (1998) paper.

Foliar and fertigation application generally will give an instant result and could be used in cases of extreme nutritional stress but for long term nutritional supplements the use of granular would be more beneficial.

Although some research work has been undertaken to understand the nutritional requirements of a Sandalwood plantation, there are still many unanswered questions.

- *Fertiliser timing, when should any fertiliser be initially apply – should it be applied before, during or after transition phase*?

  Sandalwood plantations typically does not face nutrition deficiencies initially, thanks to existing soil nutrients and low nutrient-efficient species. However, by age eight, phosphorus deficiencies often become evident. Research is needed to determine optimal timing for nutrient application to maintain adequate nutrition.

- *How often should fertiliser be applies? Will this be an annual application or every two or three years?*

  How long does a fertiliser application last? Research indicates that granular fertilisers remain effective for at least three years on Kandosol soils. Research into the application of liquid fertigation has just begun, and it is currently unclear whether fertigation will require annual application or if its nutritional effects will persist longer.

- *Does configuration type effect any ongoing fertiliser regime?*

  Determining fertiliser effectiveness can be challenging with various species. The aim is to boost host nutrition for Sandalwood growth, not merely increase host growth. Excessive host growth from fertiliser may negate benefits due to added competition. In a MIX setup, this could become a major management issue.

- *How does irrigation system effect any ongoing fertiliser regime?*

  All of the mid rotation fertiliser research has been in dripper systems and not flood systems. Using a broad acre application method would not be suitable of a flood system as any granular fertiliser would be washed out of the stand during irrigation.

- *The role of fertiliser types? There is obviously a role for both liquid and granular fertiliser, which are the best products for use?*

  Further research needs to be undertaken to determine which fertiliser type works best.

- *Although it is recognised that soil biota plays a critical role in facilitating nutrient uptake from host plants, the extent of this influence remains unclear. Similarly, it is not well understood how changes in soil biota, together with increases in soil carbon, may affect the nutrient status of both soils and trees.*

  While the benefits of soil biota and increased soil carbon for plant growth and health have been studied, there has been limited research within mixed Sandalwood plantations. The role of rhizobium and their symbiotic relationship with host plants is well-documented, but it remains unclear if a similar relationship exists with Sandalwood. Additionally, it is important to investigate whether other organisms, such as mycorrhizae and microbes involved in litter decomposition, interact with the feeder roots of Sandalwood in the topsoil.

- *How will second rotation nutrition, vary from first rotation?*

  When developing a second rotation plantation crop, it is necessary to investigate the nutritional value of the soil. Generally, the levels of organic carbon and fixed nitrogen will be higher than in the first rotation. However, there typically will not be a soil bank of nutrition as there was in the first rotation. It should be assessed if the application of phosphorous will be required from establishment.

c) Physiology

The below-ground physiological characteristics of Sandalwood with various hosts are not fully understood. For example, the extent to which Sandalwood utilises and sources water, either independently or from its hosts, remains unclear. Numerous studies indicate that Sandalwood sources water from its hosts (Tennakoon & Cameron, 2006) (Doddabasawa, Lokesh , & Lokesh , 2020) but none specify the proportions involved. It is well established that hemi-parasitic plants acquire solutes through xylem transfer from their hosts, yet the composition of this transfer is less understood. Sandalwood acquires some water, organic solutes, and amino acids by tapping into the host's xylem. Several studies also suggest that since Sandalwood connects to xylem pits via haustoria and lacks direct lumen-to-lumen connections, the direct flow of water or nutrients through tracheary connections is unlikely (Radomiljac, McComb, Pate , & Tennakoon, 1998). This finding implies that although Sandalwood may absorb some water from the host, it is likely a minor contributor to its overall requirements.

The requirement for acquisition of nitrogen is critical for above ground crown development and photosynthesis. It has been suggested that a host association is essential for Sandalwood as a source of nitrogen and phosphorus (Iyengar A. V., 1960) and this is supported by Quintis research as well. Nitrogen availability determines host suitability, there is substantial benefit in terms of nitrogen gain accruing from a well nodulated legume host (Radomiljac, McComb, Pate , & Tennakoon, 1998). The benefits of increased foliar N concentration, leaf chlorophyll contents and leaf photosynthesis, lead to larger increases in leaf area and therefore biomass accumulation.

There is a relationship between water and nitrogen levels in Sandalwood that affects the Sandalwood water use efficiency (WUE). Sandalwood aligns its transpiration with that of its host to maintain lower leaf vapour pressure (ψ). A study with mistletoe by Schulze & Ehleringer (1984) found that increased transpiration was an adaptive response to situations where N levels in a host transpiration stream were low. Rado (1998) also found that *S. album* transpiration was found to be significantly lower when attached to the low N yielding host rather than when attached to hosts with low N. This suggests essentially little capacity to adapt to hosts with low N concentrations in their xylem stream or for that matter poorly hosting N fixing hosts.

Research has indicated at the ψ showed only slight differences when on different hosts or in the absence of a host. These differences proved to be insignificant, implying relatively little effect of a host on directing osmotic adjustment.

A better understanding of, if and how the current hosts ability to provide optimal requirements for best Sandalwood biomass growth is needed. For this, there is a requirement for a better grasp of the role water and nitrogen contribution effect Sandalwood growth and health.

As discussed in the introduction, nitrogen is the principal driver of biomass production. At present, a foliar nitrogen concentration of 2.5 to 3.5% in sandalwood is generally considered adequate for healthy growth. This benchmark is based largely on field observations, where higher levels are seldom recorded, and on comparisons with many commercial forest plantation species, which typically perform well within, or below, this ranger (Boardman, Cromer, Lambert , & Webb, 1997). The key question, however, is whether increasing foliar nitrogen beyond this range could further enhance biomass accumulation. Because sandalwood derives much of its nitrogen through its host, any increase in sandalwood nitrogen levels may require a corresponding increase in host nitrogen status. This, in turn, would likely depend on improving nitrogen availability within the shared root zone. Although research suggests that, once established on a host, sandalwood has limited capacity to absorb mineral nutrients independently (Bell & Adams, 2010), this assumption warrants further field investigation. A clearer understanding of these relationships could lead to significant improvements in plantation productivity

Key questions that require answers include:

- Studies show that 40% to 70% of Nitrogen requirements can be achieved from hosts, what ability does Sandalwood have to acquire Nitrogen by itself?
- How much Nitrogen are hosts providing? How much does this vary between species, sites and age?
- Phosphorous in Sandalwood is increasing but crashing in Host, what is going on?
- Other than Nitrogen, what other mineral nutrients does host supply?
- The literature says depending on the host, the amino acid composition in Sandalwood varies. Which is the most important for optimal growth/health?
- Once determined can how can this amino acid (or combination) be supplied or increased?
- What role does Abscisic acid (ABA) play? (increase leaf conc of ABA = increase total plant biomass). If this is the case, will application of ABA increase biomass?
- Do levels of ABA have an influence on heartwood development?

Root structure of both Sandalwood and key host

- To what level does the root interaction effect host efficiency?
- There can be competition for canopy space, is there similar potential for below ground competition?
- Does various species rooting architecture affect hosting efficiency?

Sandalwood plantation water requirements.

- What % is taken from host as opposed to picked up itself?
- How does water uptake from host affect host growth/health? (if water from the host is not available, can Sandalwood take up extra itself)
- Does host water use efficiency effect Sandalwood water uptake from host?

Transpiration rate

- The effect of this on solute transfer?
- Due to closed stomata, if the Sandalwood is not transpiring or has slowed transpiration, does solute transfer stop or slow and does this effect growth and health?
- If the host is not transpiring (has no leaves i.e Dalbergia) does solute transfer stop or slow?

Transition phase

- What happens to water uptake, mineral nutrients, amino acid, ABA levels during transition?
- Assuming levels have reduced, how can any negative effect on growth and health be reduced?
- A theory that early transition in plantation 'kick starts" HW development. This theory needs validation by experimentation..

d) Heartwood development

Physical evidence indicated that heartwood in Australian plantations is initiated during the transition phase from age three to five. By the rotation age of 15, the average heartwood percentage in harvested logs is approx. 30% (Quintis harvest data). However, this can vary to be as much as 60%. As the key commercial reason for these plantations is heartwood production, it is very much an economic imperative to try to increase the average percentage of heartwood across the stand.

Sandalwood initially produces what is known as transitional heartwood prior to oil-bearing heartwood forming. The transitional heartwood is heartwood without the essential volatile oil components. The average levels of volatile oils in Western Australian harvested oil-bearing heartwood 15 year old trees range between 1% to 5%. However in the transitional heartwood the volatile oils (alpha and beta santalols) are at less than 1%.

Figure 228: Cored Sandalwood tree. (plus tree 2-8), (Robson, 2022).

Almost all trees by age ten have both oil bearing and transitional heartwood. Although almost no research work has been undertaken to determine how to quicken the conversion from transitional to oil-bearing heartwood, it is known that the change will happen over time.

Research investigating how to increase the rate of change from transitional heartwood to oil bearing heartwood is required.

Another area of research that requires more work is the determination of heartwood in standing trees. The current method involves coring or felling trees.

Some studies have explored non-invasive methods to determine heartwood levels without damaging the tree. Acoustic Wave Tomography was studied on Sandalwood trees in mid-2010's. In this method, an acoustic signal (200 volts) was passed from a pulse generator to a transducer which converts the electrical signal into a sound wave and transmits it through the tree. At the time the equipment was too bulky and expensive for operational use but with advancements in smaller electronic equipment, further development could be beneficial.

Figure 229: The protype test acoustic belt, (Done, 2010).

Additionally, there was a theory that Ultrawide Broadband Radar could be used to determine heartwood. A project was initiated with the Commonwealth Scientific and Industrial Research Organisation (CSIRO) but did not proceed due to lack of funding.

In 2023, a project with University of Sunshine Coast's Forest Research Institute, led by Associate Professor David Lee, investigated the use of the RESI power drill instrument. Several Sandalwood cores were tested in the lab with encouraging results, and field testing was proposed.

e) Harvest

The Australian Sandalwood harvesting system operates much like a traditional cottage industry, characterised by a significant amount of manual labour. Trees are initially cut down using machinery, after which the stems are laid out in the field for further processing. This subsequent processing involves manually trimming stems with chainsaws and painstakingly loading them by hand to the plantation's edge. There is a pressing need for innovative systems to enhance the efficiency and reduce the costs associated with harvesting both stems and stumps, ensuring a more sustainable and economically viable operation.

2. PLANTATION PLAN

The plantation establishment system implemented from 2012 onwards across all three sites—ORIA, Northern Territory, and Queensland—proved to be highly successful, achieving over 95% seedling survival within the first two years post-planting. However, several issues emerged during the transition and post-transition period that can be addressed through some modifications.

This section describes the recommended modifications to ensure a more robust successful plantation through to rotation.

Although not specifically for dripper systems, this plantation recommendation would be more suited for those conditions.

a) Establishment

Plantation layout

The configuration will be STD with rows 3.6m apart and all species no closer than 3.0m apart.

Many current configurations have at least three LTH species. With the correct species selection this is not necessary. An optimal selection for the long-term host species would be *Dalbergia latifolia or Dalbergia lanceolaria* and *Casuarina cunninghamiana*.

The short-term host species would be *Sesbania formosa* and *Acacia trachycarpa*.

Sandalwood will preferably be an improved seed source.

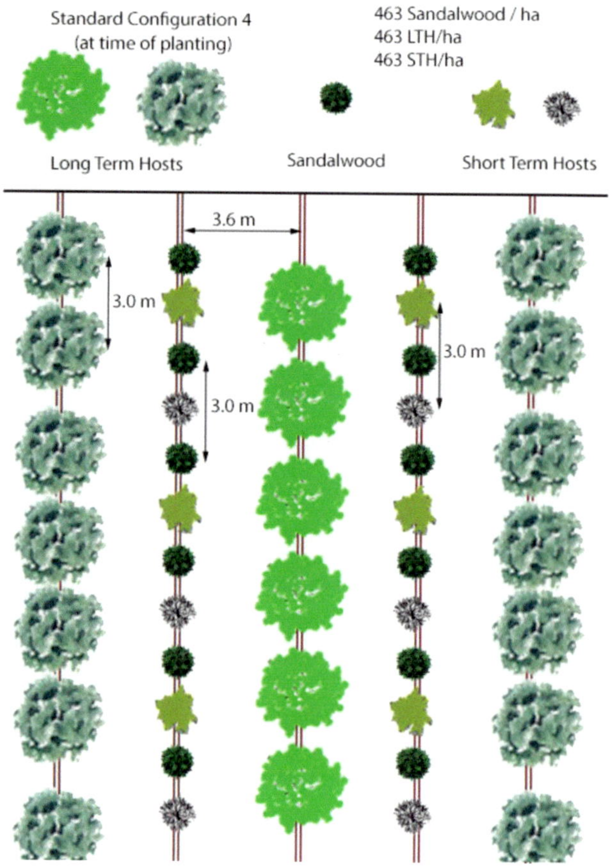

Figure 230: The STD configuration recommended for future plantations.

A cover crop of Wynn cassia (*Chamaecrista rotundifolia*) will be broad acre sown across the entire site during the cultivation stage of preparation.

On most sites, erosion due to slope is not an issue but in undulating sites the use of contour strips mounding must be utilised. However, dependant on the site the row direction must generally run in an east to westerly direction. Preferences is for row direction to run between 100° to 280°.

This will significantly reduce trunk expose to afternoon sun and reduce any change of sun scald.

Note that this is for plantations in northern Australia and this would vary for other geographic locations.

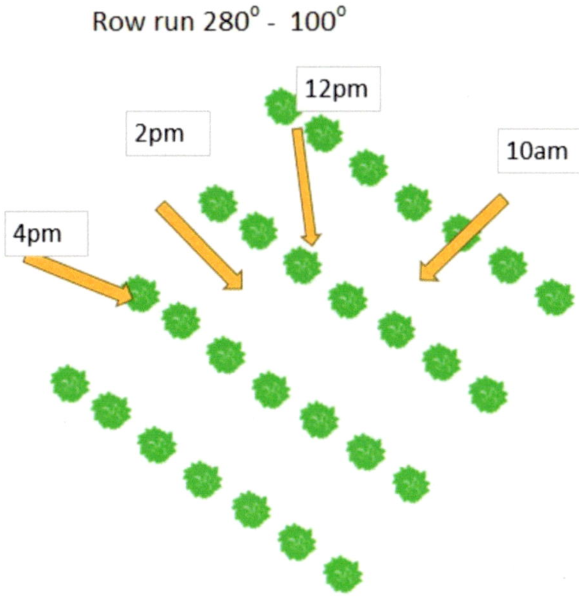

Figure 231: The row direction will determine how much afternoon sun can hit the side of the trees.

Site preparation

Cultivation should consist of ripping to a depth of at least 30 cm with small mounding over the rip line covering in the furrow. This cultivation should occur at least six to 12 months prior to proposed planting time. Site preparation should be completed by November the year prior to establishment. After cultivation the rows are left over the monsoonal wet season to settle and for weed and grass seed to germinate on the mound.

Any subterranean irrigation infrastructure should also be installed at this time.

Nursery

Sowing approximately 10% more than required for the species at the appropriate time is necessary. Good nursery protocols are required to produce seed to specifications by planting time.

Irrigation design

The current dripper system works well for all stages of the rotation; however, the optimal system would be to install a dual dripper system. A system for the Sandalwood and a system for the hosts. With a dual system in place the irrigation scheduling and water use efficiency could be better targeted to water requirements of the two groups of trees.

Suggested Timing

Year 1

Activity	Timing
Soil survey	March - June
Topography survey	March - June
Plantation design	July to Oct
Irrigation design	July to Oct
ripping	July to Oct
cultivation	July to Oct
irrigation infrastructure	Oct to Nov
cover crop sowing	Oct to Nov
mounding	Oct to Nov
Nursery sowing (Sandalwood & Dalbergia)	Oct

Year 2

Nursery sowing (Acacia)	January
Nursery sowing (Sesbania)	February
tape laying	Feb to March
Pre weed control spray	March to Apl
Planting	May to July

b) Management

The stand management of Sandalwood plantation is still an evolving methodology and most research has been aimed at managing poor establishment design. Optimally, by mid rotation, ages seven to nine, the stand should be balanced. This refers to the stand crown structure. Both Sandalwood and host should, by this stage, occupy the canopy evenly.

Figure 233: A seven year old unbalanced stand where the LTH dominate the site, (Midway A, Robson, 2019).

Figure 232: A nine year old balanced stand where both LTH and Sandalwood share space in the crown, (Midway B, Robson, 2021).

The approaches to how to obtain this are described in Chapter 5. A combination of stand management with timely thinning and appropriate crown manipulation.

Thinning

Research on appropriate thinning is still in its preliminary stages. However, initial results suggest that removing poor-quality individuals (thinning from below) effectively reduces within-row competition. The slightly wider spacing at establishment also delays the need for selective thinning.

The research shows that selective thinning is necessary for both Sandalwood and host species. In unimproved stands of Sandalwood, the selection process is simpler due to less uniform stem growth, making it easier to remove inferior quality or slower growing individuals. In more uniform stands with improved seed sources, selection for removal based on spacing may be required.

As discussed in Chapter 5, the selection of hosts for thinning is slightly more complicated than Sandalwood and in some cases removal of large dominate hosts may be the best option. This will leave smaller codominant hosts as the main hosts for Sandalwood and relieve some of the competition for space, in turn it would create the desirable balanced canopy that both Sandalwood and hosts share.

Nutrition

As discussed in Chapter 5, there has been no requirement for fertiliser application at establishment. However, for 2nd rotation or sites with no fertiliser history, it would be beneficial to apply a startup fertiliser to the land term hosts only. It is suggested that an individual tree application of 100grams/tree of a NPK fertiliser would be sufficient.

There may be a requirement for a broad acre application of a phosphorus fertiliser from age five. Foliar sampling for nutrient balance should be undertaken annually to determine when this application will be required.

c) Proposed Standard Silvicultural Regime (Sawyer, Robson, & Haberkorn, 2022)

At four to five years post-establishment, the Sandalwood stands will show notable development. Approximately 70% of the Sandalwood trees, which is about 280 stems per hectare, meet all specified growth criteria. These criteria include healthy canopy development, an upright posture, a clear single bole reaching two metres, and minimal bole sweep, kink, or wobble with less than an eight cm deflection in the two metre bole section. Additionally, 90% of the long-term host trees have a clear single upright bole of 2.4 metres and maintain healthy crown development.

Table 49: Suggested management prescription for various species groups up to age four.

Timeline	Tree type	Treatment	Requirement	Specifications
Years 0-1	Sandalwood	Tip prune	1-2 annually	As per Sandalwood pruning procedure
Years 0-4	Sandalwood	Form prune	1-2 annually	As per Sandalwood pruning procedure
Years 0-1	Long term hosts	Startup fertiliser	At planting	50kg/ha on host trees as an individual application.
Years 0-4	Long term hosts (LTH)	Form prune	1 annually	Prune up to 50% of crown annually Prune to select a single straight leader. Remove all other branches and leaders to achieve a single clear bole to 2.4m by age 5.
Year 4	Sandalwood	Thin		Thin 'from below' – cut to waste. Prioritise tree characteristics in the following order: 1. Healthy well-formed tree canopy 2. Tree upright 3. Tree straight – minimum kinks, wobble, sweep and branching 4. Tree height

By the eighth year, the expectation for the Sandalwood stands is approximately 300 trees per hectare meeting the specified criteria. The remaining trees will all show significant development, with diameters at 150 mm at DOB at 20cm above ground height. Each tree will have a clear, single bole of more than 2m with minimal wounds and no visible signs of rot. Additionally, the deflection of sweep, wobble, and kink in the 2m bole section will be under 100 mm, ensuring structural integrity. Canopy development will be notable, with tree canopies extending beyond 2m in diameter and overall tree heights exceeding 5m.

The overall stand will exhibit approximately 70% canopy closure, indicating advanced growth and a well-established ecosystem. The Sandalwood trees will transition effectively to the long-term hosts, emerging prominently within the canopy. This vertical development is encouraged by the tall canopy structure, which promotes the vertical growth of the Sandalwood trees. Meanwhile, host trees will also have a single bole reaching up to 2.4 metres, with the host basal area ranging between six to 8 m^2 per hectare. Notably, the basal area of the Sandalwood does not exceed 125% of the host basal area, maintaining a balanced and sustainable growth dynamic within the plantation.

Table 50: Suggested management prescription for various species groups between years five to eight.

Tree type	Treatment	Requirement	Specifications
Sandalwood	Assess for epicormic development	Prune as necessary	As per pruning procedure, pruning if 50% of the stand has vigorous epicormic shoots > 8mm.
Hosts	Options: 1. Thin hosts if basal area >8m²/ha 2. Side hedge maintaining stand height if canopy > 70% with host BA 6-8m²/ha.	As necessary	Maintain canopy cover at 70% and host Basal Area (BA) between 6-8m²/ha <u>Host thinning</u> Thin either medium- or long-term hosts 'from above and below'. Thin hosts with over dominant or underdeveloped canopies and hosts with poor form. The objective is to produce a more uniform stand of healthy – co-dominant host trees standing at a BA of between 6-8m²/ha.
Sandalwood	3. Thin Sandalwood if SW BA >10m2/ha. Thin to < 125% of host BA	As necessary	<u>Sandalwood thinning (if required)</u> Retained trees selected according to the following priority: 1. Diameter at 150mm height > 100mm 2. 2m clear single bole 3. Tree canopy > 2m in diameter 4. Wounds in bole <50mm² 5. Sweep, wobble, and kink deflection < large end diameter in 2m bole section 6. Tree height > 5m
All	Assess stand nutrition	Stand must exceed minimum nutrient thresholds	As per the procedure

By the 12 year, the plantation should maintain over 300 Sandalwood trees per hectare, with average diameters exceeding 17.5 cm at 20cm above ground height. Among these, 150 trees per hectare should have a clear bole of 2.4 metres. Wounds should be under 50 mm^2, without rot, and deflection for sweep, wobble, any kink in the bole section should be less than 10 cm. Host trees should have a basal area of six to 8 m^2 per hectare, with over 70% canopy coverage.

At 15 years, the stand should be thinned to 150 trees per hectare. Retained trees should have a diameter between 17.5 and 25 cm, a clear bole of 2.4 metres, wounds less than 5 cm^2, no rot, and deflection for sweep, wobble, and kinks under 8 cm. Host trees should have healthy crowns.

At 20 years, each Sandalwood tree should average 22 cm in diameter at 20 cm above ground height, with a clear bole of 2.4 metres. Trees must have minimal wounds under 5 cm^2, no rot, and deflection for sweep, wobble, and kinks under 8 cm, ensuring robust structural integrity. This growth ensures the plantation's sustainability and health for future rotations.

Table 51: Suggested management prescription for various species groups between years nine to harvest.

Tree type	Treatment	Requirement	Specifications
Sandalwood	Assess for epicormic development	Prune as necessary	As per pruning procedure, prune if 50% of the stand has vigorous epicormic shoots > 8mm.
Hosts	Options: 1. Thin hosts if basal area >8m^2/ha 2. Side hedge maintaining stand height if canopy > 70% with host BA 6-8m^2/ha	As necessary	Maintain canopy cover to at least 70% and host Basal Area (BA) between 6-8m^2/ha <u>Host thinning</u> Thin long-term hosts 'from above and below'. Thin hosts with over dominant, or underdeveloped canopies and hosts with poor form. The objective is to produce a more uniform stand of healthy – co-dominant host trees standing at a BA of between 6-8m^2/ha
Sandalwood	3. Thin Sandalwood if SW BA >10m2/ha. Thin to < 125% of host BA	As necessary	<u>Sandalwood thinning (if required)</u> Retained trees selected according to the following priority: 1. Diameter at 150mm height > 100mm 2. 2m clear single bole 3. Tree canopy > 2m in diameter 4. Wounds in bole <50mm^2 5. Tree height > 5m 6. Sweep, wobble, and kink deflection < large end diameter in 2m bole section
All	Assess stand nutrition	Stand must exceed minimum nutrient thresholds	As per the procedure

EPILOGUE

In April 2024, Quintis entered voluntary administration after receivers were unable to secure a buyer for the business. This followed a comprehensive review by KPMG in late 2023, which concluded that Quintis's managed investment schemes were no longer financially sustainable.

By the end of June 2024, much of the workforce had been made redundant. The company's difficulties were attributed primarily to a prolonged decline in global Sandalwood prices and mounting pressure from noteholders seeking repayment of outstanding debt. The administrator, KordaMentha, advised creditors that Quintis's financial position was untenable and recommended that the company be placed into liquidation.

Figure 234: Source: Forecast: Administrators Report to Creditors (May 2018, converted to AUD at forecast rates. Actual: Quintis MIS Investor update 1/2020, 6/2021, 5/2022, 9/2022 and 6/2023.

The collapse reflected a convergence of adverse market forces. Demand was significantly affected by China's prohibition on imports of raw Australian timber logs between 2020 and May 2023. At the same time, global supply increased following changes to Sandalwood harvesting regulations in India in 2022, contributing to a substantial oversupply. These pressures were compounded by rising operating costs, with inflation driving sharp increases in plantation maintenance, harvesting, and processing expenses.

Regrettably, much of the Quintis plantation estate may ultimately be converted to alternative agricultural uses, including sugar production in Queensland and cotton in Western Australia and the Northern Territory. Nevertheless, the outlook for the broader industry is not without hope. Several institutional investors, holding substantial Sandalwood assets, remain committed to the sector and continue to recognise the long-term value of this remarkable and enduring resource.

ACKNOWLEDGEMENTS

This work reflects the contributions of many colleagues and industry professionals over the past three decades. I am grateful to the managers, plantation teams, and collaborators who supported, challenged, and helped implement the ideas presented in this book. Their practical insight and willingness to engage with novel approaches have played a vital role in advancing the sandalwood plantation industry.

Although the list of individuals that have contributed is vast, I would like to give recognition and gratitude to the following individuals for their contributions: Frank Wilson (TFS), Bruce Matherson (ITC/ Santanol), Chris Done (TFS), Andy Wright (ITC), Mal Baker (TFS), Zoe Higgins (ITC), Andrew Brown (TFS/Quintis), Peter Jones (Elders), Wayne Davis (ITC/Elders/Santanol/Quintis), Matt Barnes (TFS/Quintis), Dannielle Wiseman (TFS), Charmaine Matherson (Santanol), Diane Robinson (Santanol), Dan Raymond (TFS), Michael Pattison (TFS/Quintis), Mitch Firth (TFS/Santanol), Ben Chesneau (TFS/Quintis), Dan Firth (TFS/Quintis/AFM), Dan Manson (TFS), Jess McAlister (TFS), Johan Nortier (TFS/Quintis), Emma Mende (TFS/Quintis), Bob Sandery (TFS/Quintis), Peter Van Beuzekom (TFS/Quintis), Micheal Hughes (Quintis), Kirk Hutchinson(TFS/Quintis), Loranne Walker (TFS/Quintis), Simon Ong (TFS/Quintis), Nur Saidah (Quintis), Precila Gonzales (Quintis), Felix Du (Quintis), Mitch Haberkorn (Quintis), Troy Sawyer (Quintis) and Steve Hindley (Quintis).

Many researchers outside the plantation companies have also made invaluable contributions to advancing Sandalwood science and narrowing critical knowledge gaps. In particular, Liz Barbour and John Brand from Western Australia, together with David Lee and Tony Page, deserve special recognition for their outstanding work and the significant advances they have made to our understanding of Sandalwood biology and plantation management.

I would like to particularly acknowledge Roger Underwood and Ray Fremlin, pioneering Western Australian foresters and steadfast mentors. Their wisdom, encouragement, and friendship over many years profoundly shaped my thinking and played a pivotal role in inspiring the writing of this book.

On a personal note, this book is as much a tribute to my family as it is a record of a lifetime's work. Over more than three decades devoted to building the Australian Sandalwood plantation industry, I was frequently away, and the responsibilities at home often fell to my remarkable wife, Kris. With unwavering strength, grace, and generosity, she carried our family through those years and remains my closest friend and constant support.

To Mitch and Ashlee, your willingness to listen, to question, and to challenge my thinking has influenced this journey more than you could ever know. You have both been a continual source of inspiration. Ashlee, your thoughtful editing and careful refinement of this manuscript brought clarity, polish, and vitality to these pages.

To all of you, this book belongs as much to you as it does to me.

Chapter 8. REFERENCES

Brazee, N. (2023). *Botryosphaeria Canker.* Retrieved from Centre for Agriculture, Food, and the Environment: https://ag.umass.edu/landscape

ALA. (2023). *Atlas of Living Australia.* Retrieved from https://www.ala.org.au/

Albrecht, W. A. (1938). Loss of Soil Organic Matter. *U.S. Dept. of Agric., Soils and Men, Yearbook of Agriculture* , 347 - 360.

Alcorn, P. J., Forrester, D. I., Thomas, D. S., James, R., Smith, R. G., Nicotra, A. B., & Bauhus, J. (2013). hanges in Whole-Tree Water Use Following Live-Crown Pruning in Young Plantation-Grown Eucalyptus pilularis and Eucalyptus cloeziana. *Forests* (4(1)), 106-121. Retrieved from https://doi.org/10.3390/f4010106

Allen, R. G., Pereira, L. S., & Smith, M. (1998). Crop evapotranspiration - Guidelines for computing crop water requirements. In *FAO Irrigation and drainage paper 56.* Retrieved from https://www.fao.org/3/X0490E/X0490E00.htm

Allen, R. G., Pereira, L. S., Raes, D., & Smith, M. (1998). Crop Evapotranspiration: guidelines for computing crop water requirements. *FAO Irrigation and Drainage paper No 56.*

Ananthapadmanabha, H. S. (2022). Indian Sandalwood Market Trend. In A. N. ArunKumar, G. Joshi, R. R. Warrier, & N. Nataraja Karaba (Eds.), *Indian Sandalwood A Compendium, Materials Horizons: From Nature to Nanomaterials* (pp. 479 - 489).

Angadi, V. G., Jain , S. H., & Shankaranarayana, K. H. (2003). Genetic diversity between sandal populations of different provenances in India. *Sandalwood research Newsletter*(17), 4 - 5.

Applegate , G. B., & McKinnell, F. H. (1993). The management and conservation status of Santalum species occurring in Australia. '. *Sandalwood in the Pacific region: proceedings of the XVII Pacific science congress,* . Honolulu, Hawaii .

Applegate, G. B., Davis, A. G., & Annable, P. A. (1990). Managing Sandalwood for conservation in north Queensland, Australia.'. In L. Hamilton, & C. E. Conrad (Ed.), *Proceedings of the symposium on Sandalwood in the Pacific,,* (pp. 12 - 18). Honolulu, Hawaii.

APVMA. (2024). Permits database.

Arumugasundaram, S., Ghosh, M., Veerasamy, S., & Ramasamy, Y. (2011). Species Discrimination, Population Structure and Linkage Disequilibrium in Eucalyptus camaldulensis and Eucalyptus tereticornis Using SSR Markers. *PLOS one*(6), 171 - 175.

Avery, T. E., & Burkhart, H. E. (2015). *Forest measurements.* Waveland Press.

Badami , V. K., & Venkata Rao, M. G. (1930). A preliminary note on varieties of Santalum album. L. *Mysore Sandal Spike Investigation Committee Bulletin*(1).

Baltunis, B. S., & Brawner, J. T. (2009). Clonal stability in Pinus radiata across New Zealand and Australia. I. Growth and form traits. *New Forests.* doi:DOI 10.1007/s11056-010-9201-4

Barbour, E. L., Moniodis, J., Finnegan , P., Page, T., Tungeon, J., & Tabi, M. (2020). *Enhancing the formation of heartwood in Sandalwood in Vanuatu.* ACIAR report FR2020-021 .

Barbour, L., Norris, L., & Burgess, T. (2010). Heartwood Rot Identification and Impact in Sandalwood (Santalum album),. *RIRDC Publication No. 10/179, Project No. PRJ-004677.*

Barbour, L., Plummer, J., & Norris, L. (2011). Flood Irrigated Tropical Timber Trials in the North of Western Australia. *RIRDC Publication No 11/RIRDC Project No. PRJ-002676.*

Barrett, D. R. (1998). Santalum album (Indian Sandalwood): Literature review . *Mulga Research Centre. Curtin University of Technology, Western Australia.*

Barrett, D. R., & Fox, J. E. (1997). Santalum album: kernel composition, morphological and nutrient characteristics of pre-parasitic seedlings under various nutrient regimes. *Annal of Botany, 79,* 59 - 66.

BASF Australia Ltd . (2024). BASTA Non-Selective Herbicide LABEL. *BASTA Non-Selective Herbicide LABEL.*

Bell, T. L., & Adams, M. A. (2010). Attack on all fronts: functional relationship between an aerial and root parasitic plants and their woody hosts and consequences for ecosystems. *Tree Physiology, 31,* 3-15.

Beltrame, R., Zanatta, P., de Avila Delucis, R., Lazarotto, M., Pereira AcostaA, Haselein, C., & Bisognin, D. (2023). Selection of eucalyptus hybrid clones planted in Brazil for growth and wood quality. *J Indian Acad Wood Sci, 20,* 191–196. doi:https://doi.org/10.1007/s13196-023-00318-x

Bhandari, M. S. (2023). Scientist, Forest research Institute. (K. Robson, Interviewer) Dehradun, India.

Binks , R., & Byrne, M. (2022). *Genetic analysis of Santalum album seedstock for provenance assessment.* WA Department of Biodiversity and Attractions. unpublished Quintis Internal Report.

Binks, R., & Byrne, M. (2017). *Provenance assessment of Santalum album.* Science and Conservation Division, WA Department of parks and Wildlife. unpublished Quintis Internal report.

Boardman, R., Cromer, R. N., Lambert, M. J., & Webb, M. J. (1997). Forest Plantations. In D. Reuter, & J. B. Robinson, *Plant Analysis: An Interpretation Manual* (2nd ed., pp. 505 - 559). CSIRO Publishing.

Brady, N. C., & Well, R. R. (2017). *The Nature and Properties of Soils.* (15 ed.). Pearson Education.

Brand, J. (1993). Preliminary observations on ecotypic variation in Santalum spicatum 2 Genetic variation. *Mulga Research Centre Journal, 11.*

Brand, J. (1994). Genotypic variation in Santalum album. *Sandalwood Research Newsletter*(2), 2 - 4.

Brown, A. (2020). *Interim Report on the Harvest of 20 Year Old Trees from PSR 2.* Quintis Internal Report.

Bui, D., Bush, D., Robson, K., & Lee, D. (2026). Host influence on survival and growth of two Sandalwood species during early development. *Forest Ecology and Management, 599.* doi:https://doi.org/10.1016/j.foreco.2025.123265

Burbidge, N. T. (1965). *Australian Tropical Rainforest Plants.* Retrieved from https://apps.lucidcentral.org/rainforest/text/entities/sesbania_formosa.htm

Burgess, T. (2014). *Disease associated with Sandalwood hosts in Kununurra .* Centre for Phyophthora Science and Management.

Burgess, T. I., Edwards, J., Drenth, A., Massenbauer, T., Cunnington, F., Mostowfizadeh-Ghalamfarsa, R., Tan, Y. P. (2021). Current status of Phytophthora in Australia. *Persoonia*, 151 - 177.

Burgess, T., White, D., Barbour, L., Norris, L., & Brand, J. (2014). *Indian Sandalwood pruning management to minimise fungal attack.* Rural Industries Research and Development Corporation. Publication No. 14/012.

Byrne, M. (2009). Genetic diversity: a basis for forestry breeding programs. *Australasian Forest Genetics Conference*, (pp. 20 - 22). Perth.

Byrne, M. (2014). *Genetic assessment of Santalum album.* Science and Conservation Division, Department of Parks and Wildlife. unpublished internal TFA Report.

Carter, J. L., Smith, A. J., Silberstein, R. P., Ali, S. M., Doody, T. T., Byrne, J,Palmer, D. (2010). *Interaction between trees and groundwater in the Ord River Irrigation Area.* CSIRO: Water for a Healthy Country National Research Flagship.

Celedon, J. M., Chiang, A., Yuen, M., Diaz-Chavez , M. M., Madilao, L. L., Finnegan , P. M., & Barbour, E. L. (2016). Heartwood specific transcriptome and metabolite signatures of tropical Sandalwood (Santalum album) reveal the final step of (Z)-santalol fragrance biosynthesis. *The Plant Journal, 86*, 289 - 299.

Chadrack , K., Lehnebach, R., Bourland, N., Beeckman, H., Van Acker, N., Luambua, N. K., & Van den Bulcke., J. (2024). Earlier onset and slower heartwood investment in faster-growing trees of African tropical species. *Annals of Botany* , 905 - 916 .

Chia, S. Y., & Lim, M. W. (2022). A critical review on the influence of humidity for plant growth forecasting. *IOP Conference Series: Materials Science and Engineering* . doi:10.1088/1757-899X/1257/1/012001

Clark, M. (2006). *Australia's Sandalwood industry : an overview and analysis of research needs; report for the Rural Industries Research and Development Corporation.* RIRDC publication, 1440-6845; no. 06/131. Retrieved from http://www.rirdc.gov.au/reports/EOI/06-131.pdf

Coder, K. (2018). *Drought, heat & trees: A learning manual.* manual. University of Georgia, Warnell School of Forestry & Natural Resources Outreach Publication WSFNR18-41.

Collins, S., Walker, S., & Haines, R. (2000). *SPRIG vegetative propagation completion report.* Internal, Queensland Forestry Research Institute., Gympie (Australia): .

Coops, N. C., Hermosilla, T., Hilker, T., & Black, T. A. (2017). Linking stand architecture with canopy reflectance to estimate vertical patterns of light-use efficiency. *Remote Sensing of Environment, 194*, 322 - 330. doi:https://doi.org/10.1016/j.rse.2017.03.025

Costantini, A., Podberscek, M., & Nester, M. R. (1995). Site preparation for Pinus establishment in south-eastern Queensland. 2. Effect of cultivation and cultivation width on growth. *Australian Journal of Experimental Agriculture, 38*(8), 1159 - 1164.

Cotterill , P. P., & Dean, C. A. (1990). *Successful tree breeding with index selection.* CSIRO, East Melbourne.

Craine, J. M., & Dybzinski, R. (2013). Mechanisms of plant competition for nutrients, water and light. *Functional Ecology, 27*(4), 833 - 840. doi: https://doi.org/10.1111/1365-2435.12081

Cuevas, M. V., Torres-Ruiz, J. M., Alvarez, R., Jimenez, M. D., Cuerva, J., & Fernandez, J. E. (2010). Assessment of trunk diameter variation derived indices as water stress indicators in mature olive trees. *Agricultural Water management, 97*(9), 1293 - 1302. doi:doi.org/10.1016/j.agwat.2010.03.011

Danial, K. (2022). What Happened to Surflan (oryzalin)? *Landscape report - Plant Doctor*. Retrieved from https://www.purduelandscapereport.org/article/what-happened-to-surflan-oryzalin/

Davidson, E., & Cunnington, J. (2005). Ord valley Briefing Note.

Delvalle, T. (2022, December). *Amines or Esters; Which Should You Use?* Retrieved from PennState Extension: https://extension.psu.edu/amines-or-esters-which-should-you-use#:~:text=2%2C%204%2DD%20and%20other,generally%20less%20volatile%20than%20esters.

Dhola Riya., C. A., Sukhadiya, M., Behera, L. K., & Mehta, A. (2019). *Dalbergia lanceolaria L.f.: Prospect of a lesser known tree species in India* (Vols. ISSN 2395-468X). Van Sangyan, Tropical Forestry Research Institute, Jabalpur, India.

Dodd, S. (2023). *Growhelp Job 6860.* Quintis External consultant.

Doddabasawa, B., Chittapur, M., & Lokesh, R. (2020). Parasitism ecology of Sandalwood (Santalum album L.) for commercial production in the semi-arid tropics. *Current Science, 119*, 699–703.

Doddabasawa, B., Lokesh , M., & Lokesh , R. (2020). Parasitism ecology of Sandalwood (Santalum album L.) for commercial production in the semi-arid tropics. . *Curr Sci 1, 19*(4), 699–703.

Done , C., Kimber, P., & Underwood, R. (2005). Development of the Indian Sandalwood Industry on the Ord River Irrigation Area. In L. Thomson (Ed.), *Proceedings of the Regional Workshop on Sandalwood Research, Development and Extension in the Pacific Islands and Asia.* Nadi, Fiji.

Doran, J. C., Thompson, L. A., & Brophy, J. J. (2002). Sandalwood, a global perspective. In Thompson, L. Bulai, S. and Sovea, L. 2005. Proceedings of a regional workshop on Sandalwood research and development and extension in the Pacific Islands and Asia Oc. *Proceedings of a regional workshop on Sandalwood research and development and extension in the Pacific Islands and Asia.* Noumea, New Caledonia: Secretariat of the Pacific community.

DSITI, & DNRM. (2015). *Guidelines for agricultural land evaluation in Queensland (2nd edn).* Brisbane, Queensland.: Queensland Government (Department of Science, Information Technology and Innovation and Department of Natural Resources and Mines).

Du, Q., Zhao, X., Jiang, X., Wang, X., Han, Y., & Yu, H. (2017). Effect of Potassium Deficiency on Root Growth and Nutrient Uptake in Maize (Zea mays L.). *Agricultural Sciences, 8*, 1263-1277. doi:https://doi.org/10.4236/as.2017.811091

Echo Community. (2024). Asia Seed facts. Retrieved from www.echocommunity.org

Eldridge, K., Davidson, J., Harwood, C., & van Wyk, G. (1993). Eucalypt Domestication and Breeding. . Oxford.: Clarendon Press, O.

El-Hage Scialabba, N. (2022). *Managing Healthy Livestock Production and Consumption.* Academic Press. doi:10.1016/C2019-0-05093-5

EPA. (1997). *RED facts: Paraquat Dichloride* . Retrieved from http://www.epa.gov/oppsrrd1/REDs/factsheets/0262fact.pdf

Fabiola, N., Giarola, B., Pires da Silva, A., Imhoff, S., & Dexter, A. R. (2003). Contribution of natural soil compaction on hard setting behaviour. *Geoderma*(113), 95–108.

Farm Forest Line. (2009). High volume production regimes. Retrieved from https://www.farmforestline.com.au/pages/5.9.1_high.html

Farrakh Nawaz, M., Bourrié, G., & Trolard , F. (2012). Soil compaction impact and modelling. A review. In j. Meynard, S. Recous, J. Le Bot, L. Hossard, P. Debaeka, & P. Hinsinger (Eds.), *Agronomy for Sustainable Development* (pp. 291 - 309). Retrieved from https://doi.org/10.1007/s13593-011-0071-8

Fathi, A., & Barari, D. (2016). Effect of Drought Stress and its Mechanism in Plants. *International Journal of Life Sciences, 10,* 1 - 6. doi:DOI: 10.3126/ijls.v10i1.14509

Fernandes, C., & Ravi, L. (2023). Screening of symbiotic ability of Rhizobium under hydroponic conditions. In D. Dharumadurai (Ed.), *Microbial Symbionts. Functions and Molecular Interactions on Host.* Academic Press.

Fonseca de Lima, C. F., Kleine-Vehn,, J., De Smet, I., & Feraru, E. (2021). Getting to the root of belowground high temperature responses in plants. *Journal of Experimental Botany, 72*(21), 7404–7413. doi:https://doi.org/10.1093/jxb/erab202

Fround-Williams, R. J. (2008). Chapter 2 Weed competition 16 - 71. In E. L. Naylor, & E. L. Naylor (Ed.), *Weed Management Handbook.* (pp. 1 - 71). Wiley.

Fuller, W. H. (1958). *Soil Compaction.* Agricultural Experiment Station, University of Arizona. Retrieved from http://hdl.handle.net/10150/637584

Gardner, C. A. (1929). A taxonomic study of the genus Santalum with special reference to the Sandalwoods of Australia. Forests Department Western Australia 44, 1–10. In *Forests Department Western Australia* (Vol. 44, pp. 1 - 10).

Ge, F., Zeng, W., Ma, W., & Meng, J. (2017). Does the Slope of the Self-thinning Line Remain a Constant Value across Different Site Qualities?—An Implication for Plantation Density Management. *Forests*(355). doi:https://doi.org/10.3390/f8100355

Gonzales, P. (2021). *Sustainable Biomass Waste Recovery and Value Adding Program.* Internal Quintis report.

Gourley, C. J. (1999). Potassium. In D. J. Reuter, K. I. Peverill, & L. A. Sparrow, *Soil Analysis, an interpretation manual* (pp. 229 - 245). CSIRO Publishing.

Gull , A., Lone, A. A., & Islam Wani, B. U. (2019). *Biotic and Abiotic Stresses in Plants.* (A. Bosco de Oliveira, Ed.) doi:DOI: 10.5772/intechopen.85832

Haberkorn, M. (2023). *2023 Inventory Report.* Quintis Internal.

Harbaugh, D. (2007). A taxonomic revision of Australian northern Sandalwood (Santalum lanceolatum, Santalaceae). *Australian Systematic Botany, 20,* 409 - 416. Retrieved from www.publish.csiro.au/journals/asb

Harbough, D. T., & Baldwin, D. T. (2007). Phylogeny and biogeography of the Sandalwoods (Santalum, Santalaceae): repeated dispersals throughout the Pacific:. *American Journal of Botany, 64,* 1028–1040.

Harrington, C. A., & Reukema, D. L. (1983). Initial shock and long-term stand development following thinning in a Douglas-fir plantation. *Forest Science, 29,* 33 - 46.

Harris, J. M. (1954). Heartwood formation in Pinus radiata (D. Don.). *New Phytologist, 53*, 517–524. doi:10.1111/j.1469-8137.1954.tb05258.x.

Hernandez, V. R., Comeau, P. G., & Bokalo, M. (2013). Static and dynamic maximum size-density relationships for mixed trembling aspen and white spruce in western Canada. *Forest Ecology and Management*(289), 300 - 311.

Holloway , R. E., Bertrand , I., frischke , A. J., Brace , D. M., McLaughlin, J. J., & Sheppard, w. (2001). Improving fertiliser efficiency on calcareous and alkaline soils with fluid sources of P, N and Zn. *Plant and Soil, 236*, 209 - 219. doi:DOI: 10.1023/A:1012720909293

Houghton, J. T., Ding, Y., & Griggs, J. D. (2001). Climate Change 2001: The Scientific Basis; Contribution of Working Group I to the Third Assessment Report of the Intergovernmental Panel on Climate Change. Cambridge and New.York: Cambridge Univ Press.

Huiqiong , L., Kunrong , H., ZhiQiang Zhang,, & Yanru , H. (2023). Molecular mechanism of phosphorous signalling inducing anthocyanin accumulation in Arabidopsis. *Plant Physiology and Biochemistry*. Retrieved from https://phys.org/news/2023-01-reveals-phosphorous-deficiency-anthocyanin-accumulation.html

Iba, K. (2002). Acclimative response to temperature stress in higher plants: approaches of gene engineering for temperature tolerance. *Annu. Rev. Plant Biology*(54), 225 - 245. doi:10.1146/annurev.arplant.53.100201.160729

Irwin, H. H., & Barneby, R. R. (1982). *Australian Tropical Rainforest Plants*. Retrieved from Memoirs New York Botanical Garden: https://apps.lucidcentral.org/rainforest/text/entities/senna_siamea.htm

Isbell, R. F. (1998). *The Australian Soil Classification.* CSIRO Publishing.

IUCN . (1998). *Santalum fernandezianum.* doi:http://dx.doi.org/10.2305/IUCN.UK.1998.RLTS.T30406A9544750.en

Iyengar, A. (1960). The relation of soil nutrients to the incidence of spike disease. *Indian Forester, 86*, 220–230.

Jain, S., Ranajan, A., & Hemany, K. (2007). Distribution of Sandal (Santalum album L.) current growth rates predicted yield of heartwood and oil content and future potential in semi-arid and arid regions of Rajasthan, India. *Forests, Trees and Livelihoods, 17*(3), 261 - 266.

Johnston, P., & Robinson, R. (2015). Forestry, SA History. *History Trust of South Australia, https://sahistoryhub.history.sa.gov.au/subjects/forestry.*

Jones , C. G., Plummer , J. A., Barbour, E. L., & Byrne, M. (2008). Genetic Diversity of an Australian Santalum album Collection – Implications for Tree Improvement Potential. *Silvae Genetica, 58*, 5 - 6.

Jones, C. G. (2001). *What Causes Insect and Disease Outbreaks on Trees?* Retrieved from TREE CARE INDUSTRY: (www.caryinstitute.org)

Jones, T. H. (2005). *Population genetics of native and domesticated Eucalyptus globulus.* University Of Tasmania Thesis. doi:https://doi.org/10.25959/23230604.v1

Karmakar. A, B. N. (2020). *Hemi-parasitic establishment in root zones of White Sandal (Santalum album)* (Vol. Vol. 12). International Journal of Current Research.

Keenan, R., Hambleton, A., Robson, K., & Webb, M. (1998). Growth response of rainforest cabinet timber species to fertiliser application in North Queensland plantations. In A. Schulte, & D. Ruhiyat (Ed.), *Soils of Tropical Forest Ecosystems.* Springer. doi:https://doi.org/10.1007/978-3-662-03649-5_11

Kew, G. (2010). *Readily available water holding capacity assessment for Sandalwood plantations in ORIA.* Internal Elders Report.

Kozlowski, T. T., & Pallardy, S. G. (1997). *Physiology of Woody Plants* (2 ed.). Academic Press.

Kumar, A. A., Srinivasa, Y. B., Joshi, G., & Seetharam., A. (2010). *Variability in and between tree growth, heartwood and oil content in Sandalwood (Santalum album L).* Institute of Wood Science and technology, Bangalore, India.

Lewis, F., Wildy, D., & Higgins, Z. (2006). *ORIA Irrigation Investigations.* ITC Internal Report.

Lines-Kelly, R. (1993). *Soil management guides: Cation exchange capacity.* Retrieved 2023, from https://www.dpi.nsw.gov.au/agriculture/soils/guides/soil-nutrients-and-fertilisers/cec

Lines-Kelly, R. (1994). *Protect your soil from compaction.* Retrieved from https://www.dpi.nsw.gov.au/agriculture/soils/more-information/structure/compaction#:~:text=What%20causes%20soil%20compaction%3F,and%20the%20type%20of%20soil.

Liu, X., Zhang, O., Hong, Z., & Xu, D. (2022). Induction of heartwood formation in young Indian Sandalwood (Santalum album L.) by gas elicitors. *Frontiers in Plant Science, 13.* doi:10.3389/fpls.2022.961391

Locke, S. (2012). Losing billions in agricultural MIS projects. *https://www.abc.net.au/news/rural/2012-12-21/losing-billions-in-agricultural-mis-projects/6151334.*

Lonergan, O. W. (1990). Historical review of Sandalwood (Santalum spicatum). *Research bulletin (Western Australia. Department of Conservation and Land Management); no. 4.*

Lue, A. (2012). Soil organic matter: Tips for responsible nitrogen management. *Acres USA.*

Mackey, A. P., Miller, E. N., & Palmer, W. (1997). *Sicklepod (Senna obtusifolia) in Queensland.* Department of Natural Resources and Mines, Qld.

Malinda, D. K. (1995). Factors in conservation farming that reduce erosion. *Australian Journal of Experimental Agriculture, 35*, 969-978.

Malinda, D. K., Fawcett, R. G., Little, D., & Bligh, K. (1998). The effect of grazing, surface cover and tillage on erosion and nutrient depletion. *Advances in GeoEcology, 31*, 1217-1224.

Manners, A. (2023). *Growhelp Report 6685.* Quintis External consultant.

Manson, D. (2013). *R028a Thinning trial establishment report.* TFS Internal Report.

Manson, D. (2014). *Host tree inoculation trials – Dalbergia inoculation.* Internal TFS Report.

Manson, D. (2016). *Use of Methyl jasmonate to induce heartwood in Sandalwood.* TFS Internal report.

McComb, J. A. (2009). Clonal Santalum album growth, oil content and composition on different hosts and at different locations. *Journal of the Royal Society of Western Australia*(92), 15 - 25.

McDowell, N., Pockman, W. T., Allen, C. D., Breshears, D. D., Cobb, N., Koib, T., . . . Yepez, E. (2008). Mechanisms of plant survival and mortality during drought: why do some plants survive while others succumb to drought? *New Phytologist Foundation, 178*(4), 719 - 739. doi:https://doi.org/10.1111/j.1469-8137.2008.02436.x

McKenzie, D. (2011). *Soil survey Taylors Park and Douglas Daly for Sandalwood plantations.* SoilMgt Consultancy report for TFS.

McKenzie, N. J., Grundy, M. J., Webster, R., & Ringrose-Voase, A. J. (2008). *Guidelines for surveying soil and land resources (2nd edn).* Collingwood, Victoria.: CSIRO Publishing.

McKenzie, N., Jacquie, D., Isbell, R., & Brown, K. (2004). *Australian Soils and landscapes , an illustrated compendium.* CSIRO Publishing.

McKinnell, F. A. (1990). Status of Management and Silviculture Research on Sandalwood in Western Australia and Indonesia. *Presented at the Symposium on Sandalwood in the Pacific, April 9-11,.* Honolulu, Hawai'i.: USDA Forest Service Gen. Tech. Rep. PSW-122.

Mende, E., Robson, K., & Llorente, G. (2019). *The degree of fig leaf beetle (FLB) feeding and their effect on the rate of growth of Sandalwood (Santalum album) trees in selected treated and untreated blocks.* unpublished Quintis internal Report.

Mertens, J., & Smolders, E. (2013). Zinc. In E. Alloway (Ed.), *Heavy Metals in Soils. Environmental Pollution* (Vol. 22). Springer, Dordrecht. doi:https://doi.org/10.1007/978-94-007-4470-7_17

Meteorology, A. B. (2023). *Climate Data online.* Retrieved from http://www.bom.gov.au/

Meteorology, A. B. (2023). *What are El Niño and La Niña events?* Retrieved from http://www.bom.gov.au/climate/enso/history/ln-2010-12/ENSO-what.shtml#:~:text=They%20occur%20when%20the%20Pacific,cooling%20of%20these%20same%20areas.

Miles, B., & McTainsh, G. (1994). Wind erosion and Land Management in the Mulga Lands of Queensland. *Australian Journal of Soil and Water Conservation, 7,* 41 - 42.

Moody, P. W. (2007). Interpretation of a single-point P buffering index for adjusting critical levels of the Colwell soil P test. *Australian Journal of Soil Research,, 45,* 55-62.

Moody, P. W., & Bolland, M. D. (1999). *12 Phosphorus. In Soil Analysis, an interpretation Manual.* (S. L. Peverill. K.I, Ed.) CSIRO Publishing.

Moore, G. (1998). *Soilguide: A Handbook for Understanding and Managing Agricultural Soils.* Department of Agriculture, Western Australia.

Moore, G. (2012). Flooding following Drought: A Swift and Subtle Killer of Stressed Trees. *TreeNet.* Retrieved from TreeNet.

Morris, H. (2013). Tree pruning: A modern approach. *International Dendrology Society , 209.*

Mosaic. (2023). *Essential role of Phosphorus in Plants.* Retrieved from Mosaic Crop Nutrition: https://www.cropnutrition.com/nutrient-management/phosphorus/

Muhairwe, C. K. (1994). Tree form and taper variation over time for interior lodgepole pine. *Can. J. For. Res, 24,* 1904–1913.

Neimisto, P. (1995). Influence of initial spacing and row-to-row distance on the crown and branch properties and taper of silver birch (Betula pendula). *Scand. J. For. Res, 10*, 235–244.

Nelder, J. A. (1962). New kinds of systematic designs for spacing experiments. *Biometrics, 43*, 283 - 307. doi:http://www.jstor.org/stable/2527473

Orwa, C., Mutua , A., Kind, R., Jamnadass, R., & Imons, A. (2009). *Agroforestry Database:a tree reference and selection guide version 4.0.* Retrieved from http://www.worldagroforestry.org/af/treedb/.

Page, T., Hanington, T., & Clement, B. (2012). Breeding behaviour of three Sandalwood species (Santalum album, S. austrocaledonicum and S. lanceolatum). *Regional Workshop on Sandalwood Development, Research and Trade in the Pacific and Asian Region*, (pp. 102 - 108).

Parker, P. G., Snow, A. A., Schug, M. D., & Booton, C. G. (1998). What molecules can tell us about populations: Choosing and using a molecular marker: Ecology. *Ecology*(79), 361–382.

Parthasarathi, K., Gupta, S. K., & Rao, P. S. (1974). Differential response in the cation exchange capacity of the host plants on parasitisation by sandal (Santalum album L). *Current Science, 43.*

Pinkard, E. A. (2003). Physiological and growth responses related to pattern and severity of green pruning in young Eucalyptus globulus. *Forest Ecology and Management, 183*(1-3), 231 - 245.

Pinkard, E. A., & Beadle, C. L. (1998). Effects of green pruning on growth and stem shape of Eucalyptus nitens (Deane and Maiden) Maiden. *New Forests, 18*, 107 - 126. doi:https://doi.org/10.1023/A:1006533423884

Pinkard, E. A., & Beadle, C. L. (2000). A physiological approach to pruning. *International Forestry review.*

Plantin. (2024). Siamese Cassia. Retrieved from https://plantiary.com/plant/senna-siamea_7018.html

Plonk, G. (2022). A History of Indian Sandalwood in Australia. In A. N. Arunkumar, G. Joshi, R. R. Warrier, & N. N. Karaba (Eds.), *Indian Sandalwood, A compendium. Materials Horizons: From Nature to Nanomaterials.* Springer Books.

Pretzch, H. (2014). Canopy space filling and tree crown morphology in mixed species stands compared with monocultures. *Forest Ecology and Management, 327*, 251 - 264. doi:https://doi.org/10.1016/j.foreco.2014.04.027

QLD Govt. (2024). *Termites.* Retrieved from Business Queensland: https://www.business.qld.gov.au/industries/farms-fishing-forestry/forests-wood/pests-diseases/trees-timber/termites

Quintis. (2019). *PSP Data.* Quintis Internal.

Quintis. (2024). Retrieved from https://quintis.com.au/

Radomiljac, A. (1993). Introduction and overview of Santalum research in Kununurra, Western Australia. *Sandalwood Research Newsletter.*

Radomiljac, A. (1998). Santalum Album L Plantations. A complex interaction between parasite and host. *Thesis presented for degree od Doctor of Philosophy of Murdoch University.*

Radomiljac, A. M., & Vernes, T. (2019). *Santalum album (Indian Sandalwood).* Retrieved from CABI Compendium: https://doi.org/10.1079/cabicompendium.50389

Radomiljac, A. M., McComb, J. A., & Pate, J. S. (1999). Gas exchange and water relations of the root hemi-parasite Santalum album L. in association with legume and non-legume hosts. *Annal of Botany*, 215 - 224.

Radomiljac, A. M., McComb, J. A., Pate , J. S., & Tennakoon, K. U. (1998). Xylem transfer of organic solutes in Santalum album L. (Indian Sandalwood) in association with legume and non-legume hosts. *Annals of Botany, 82*, 675–682.

Rai, R. K., Singh, V. P., & Upadhyay, A. (2017). *Planning and evaluation of irrigation projects. Chapter 17 - Soil analysis.* Academic Press. doi:doi.org/10.1016/B978-0-12-811748-4.00017-0

Rai, S. N. (1990). Status and Conservation of Sandalwood in India. *Proceeding of the Symposium on Sandalwood in the Pacific*, (pp. 66 - 77). Honolulu.

Ramaiah, P. K., Parthasarathi, K., & Rao, P. S. (1962). Studies on sandal spike: a histological study of sandal root haustoria in relation to mineral nutrition. *Journal of the Indian Academy of Wood Science, 56B*, 111–113.

Rangaswamy, C. R., Jain, J. H., & Parthasarthi, K. (1986). Soil properties of some Sandalwood bearing areas. *Van Vigyan, 24*, 61 - 68.

Rao, L. (1942). Parasitism in the Santalaceae. Ann Bot 6:131–151. *Annals of Botany, 6*, 131–151.

Rashkow, E. D. (2022). Indian Sandalwood: A History of Overexploitation and Endangerment. In G. J. A. N. Arunkumar (Ed.), *Indian Sandalwood, A compendium* (pp. 27 - 43). Springer Books.

Rathore, T. S., Rangaswamy, M., & Goyal, B. (2022). Micropropagation in Sandalwood (Santalum album L). In A. N. Arunkumar, G. Joshi, & R. R. Warrier (Eds.), *Indian Sandalwood, A compendium. Materials Horizons: From Nature to Nanomaterials.*

Reid, R. (2010). Thinning plantations for high quality sawlogs - How to grow sawlogs fast. *Agroforestry in Australia*. Retrieved from https://agroforestry.net.au/main.asp?_=Thinning

Reuter, D. J., Duncombe-Wall, D., & Judson, G. (1997). Potassium balance in Australian broad acre industries: a contemporary national and regional analysis. *Preceding of the 1st workshop on Potassium in Australian Agriculture*, (pp. 1 - 8). Geraldton, WA.

Reuter, D., & Robinson, J. B. (1997). *Plant Analysis: An Interpretation Manual* (2nd ed.). CSIRO Publishing.

Robson, A. D., Longnecker, N. E., & Osborne, L. D. (1992). Effects of heterogeneous nutrient supply on root growth and nutrient uptake in relation to nutrient supply on duplex soils. *Australian Journal of Experimental Agriculture, 32*, 879 - 886.

Robson, K. (1995). *Growth response of E. pellita to increasing levels of cultivation.* unpublished QFRI report.

Robson, K. (2011). *Development of efficient irrigation scheduling for flood systems of Sandalwood plantations at Kununurra.* Elders Internal report.

Robson, K. (2012). Variation in Sandalwood (Santalum album L) seed diameter and its effect on nursery and field growth. *Sandalwood research newsletter*(27).

Robson, K. (2019). *Sandalwood root dynamics study.* Internal Quintis report.

Robson, K. (2021). *Irrigation scheduling using Soil water balance calculations.* Quintis Internal Report.

Robson, K. (2021). *Sandalwood root dynamics in TP6.* Internal.

Robson, K., & Barnes, M. (2021). Challenges for Sandalwood plantation development in tropical Australia. In T. Page, J. Meadows, & T. Kalsakau (Ed.), *Sandalwood Regional Forum. Proceedings of a regional meeting November 2019,* (pp. 112 - 119). ort Vila, Vanuatu.

Robson, K., & Mende, E. (2016). *Timing of Sandalwood hosting with alternanthera.* Internal TFS Report.

Robson, K., & Tapscott, G. (2010). *Applications of phosphorus acid (Agri-Fos 600).* unpublished Elders Forestry internal Report.

Robson, K., McCallister, J., & Nortier, J. (2016). *Nursery germination testing for mechanisation.* Internal TFS Report.

Rugkhia, A., & McComb, J. (1993). Self and cross pollination in Santalum spicatum and S. album. *Sandalwood Research Newsletter.*

Sage, R. F., Sage, T. L., Pearcy, R. W., & Borsch, T. (2007). The taxonomic distribution of C4 photosynthesis in Amaranthaceae sensu stricto. *American Journal of Botany, 94 (12,* 1992–2003.

Sakalidis, M. L., Ray, J. D., Lanoiselet, V., Hardy, G. E., & Burgess, T. (2011). Pathogenic Botryosphaeriaceae ssociated with Mangifera indica in the Kimberley Region of W estern Australia. *European Journal of Forest Pathology*(130), 379 - 391.

Salih, T. K., Saied, M. Y., & Wali, S. (2019). Determination of Competition Factor for Tree Growth within A Stand. Journal of University of Duhok., Vol. 22, No.2 (), Pp 177-188,. *Agri. and Vet. Sciences.* doi:https://doi.org/10.26682/ajuod.2019.22.2.17

Sanjaya, Anathapadmanabha , H. S., & Rai, V. R. (2003). In vitro and in vivo micrografting of Santalum album shoot tips. *Journal of Tropical Forestry*(15), 234 - 236.

Sanquetta, C. R., Corte, A. P., & da Silva, F. (2011). Biomass expansion factor and root-to-shoot ratio for Pinus in Brazil. *Carbon Balance and Management, 6.* doi:doi.org/10.1186/1750-0680-6-6

Sawyer, T. (2024). *Cost-benefit Analysis – Thinning vs Hedging.* Quintis Internal, Quintis Forest Performance.

Sawyer, T., Robson, K., & Haberkorn, M. (2022). *Draft Standard Silvicultural Regime for 2nd Rotation Sandalwood Plantations.* Quintis.

Schulze , E. D., & Ehleringer, J. R. (1984). The effect of nitrogen supply on growth and water use efficiency of xylem tapping mistletoes. *Planta, 162,* 286 275.

Selvaraj, V., Suryanarayana, V., Venkataravanappa, V., Rao, G. P., & Mandal, B. (2022). Diagnosis of phytoplasma associated with the Sandalwood spike disease. *Indian Journal of Entomology.* doi:DOI: 10.55446/IJE.2022.456

Shain, L., Hillis, W., & Hillis, E. (1973). 'Ethylene production in xylem of Pinus radiata in relation to heartwood formation. *Canadian Journal of Botany-Revue Canadienne De Botanique, 51,* 1331-1335. doi:10.1139/b73-166

Shibata , T., Iwao, K., & Takano, T. (1995). Effect of Vertical Air Flowing on Lettuce Growing in a Plant Factory, . In *Greenhouse Environment Control and Automation* (Vol. 399, pp. 175–182). Retrieved from https://www.actahort.org/books/399/

Silveira Pontes, L., Rodrigues, D., Souza Carneiro, L., Chiezi Mendes, L., & Porfírio-da-Silva, v. (2024, February 7). *L. , D. et al. Impact of beef cattle on eucalyptus trees in a silvopastoral system,* doi:https://doi.org/10.21203/rs.3.rs-3914348/v1

Singh, U. V. (1995). Sandalwood - scope for commercial propagation on community lands in India. *Sandalwood Research Newsletter, 4.*

Skovsgaard, J. P., & Vanclay, J. K. (2008). Forest site productivity: a review of the evolution of dendrometric concepts for even-aged stands. Forestry 81:13–31. *Forestry, 81*, 13 - 31.

Soloman, D., & Zhang, L. (1998). Maximum size-density relationships for mixed-hardwood forest stands in New England. *Integrated Tools Proceedings.* Boise Idaho, USA,.

Spears, S. (2023). *What is Biochar?* Retrieved from Regeneration International: https://regenerationinternational.org/2018/05/16/what-is-biochar/

Srinivasan, V. V., Sivaramakrishnan, V. R., Ananthapadmanbha , H. S., & Shankaranarayana, K. H. (1992). Sandal - (Santalum album Linn.). *Indian Council of Forestry Research and Education,* 233.

Srinivasan, V. V., Sivaramakrisnan, V. R., Rangaswam, C. R., Ananthapadmanabha, H. S., & Shankarnarayana, K. H. (1992). *Sandal (Santalum album L).* ICFRE, Dehra Dun.

Streten, C., & Gibb, K. S. (2006). Phytoplasma diseases in sub-tropical and tropical Australia. *Australasian Plant Pathology, 35*, 129 - 146. doi: https://doi.org/10.1071/AP06004

Strong, W. M., & Mason, M. G. (1997). Nitrogen. In D. J. Reuter, & J. B. Robinson, *Soil Analysis. An interpretation manual* (2nd ed.). CSIRO Publishing.

Subasinghe, S. M. (2013). Sandalwood Research: A Global Perspective. *Journal of Tropical Forestry and Environment*, Vol. 3, No. 01 (2013) 1-8.

Summit Fertilisers. (2023). *The Phosphorus Buffering Index – What Does it Mean for Your Soil?* Retrieved from Summithttps://www.summitfertz.com.au/blog/the-phosphorus-buffering-index-what-does-it-mean-for-your-soil.

Swaminathan, M. H., Hosmath, B. J., & Mallesha , B. B. (1998). The status of Sandalwood in India- Karnataka. In A. M. Radomiljac, H. S. Ananthapadmanabha, R. M. Welbourn, & K. Salyanarayna Rao (Ed.), *) Sandal and Its Products. ACIAR Proceedings. 84.* Canberra, Australia,: ACIAR. Australian Centre for International Agricultural Research,.

Taylor , D., Swift , S., & Collins, S. (2000). Testing growth and survival of four Sandalwood species in Queensland. *Sandalwood Research Newsletter 10, 6–8., 10*, 6 - 8.

Taylor, A. M., Gartner , B. L., & Morrell , J. J. (2022). Heartwood formation and natural durability – A review. *Wood and Fiber Science, 34*, 587-611.

Teixeira da Silva, , J. A., Kher, M. M., Soner, D., Page, T., Zhang, X., Nataraj, M., & Ma, G. (2016). Sandalwood: basic biology, tissue culture, and genetic transformation. *Planta, 243*, 847–887. Retrieved from https://doi.org/10.1007/s00425-015-2452-8

Tennakoon, K. U., & Cameron, D. D. (2006). The anatomy of Santalum album (Sandalwood) haustoria. *Can. J. Bot 84*, 1608 – 1616.

Tewari, V. P., Naik, S. D., Veerapur, S. H., & Chandras. (2014). Distribution, Growth & Yield and Demand & Supply of Santalum album: Survey & Review. *International Seminar on 'Sandalwood: Current Trends and Future Prospects'.* Bangalore (India).

The Times of India. (2022). *The Times of India*. Retrieved from The Times of India: https://timesofindia.indiatimes.com/city/bengaluru/karnataka-approves-new-Sandalwood-policy-no-curbs-to-grow-and-sell/articleshow/95590280.cms

Thisleton, B. M., Neal, M. J., & Smith, S. C. (2013). *Management of Giant Northern termite (Mastotermes darwiniensis).* NT Government Ag Note.

Thompson, W. P., Cannon, M. G., Reid, R. E., & Baker, D. E. (1990). *Soils of the Lower Burdekin Valley, North Queensland.* Queensland Departments of Primary Industries.

TITAN AG Pty Ltd . (2024). PENDIMETHALIN 440 EC LABEL. *PENDIMETHALIN 440 EC LABEL.*

Underwood, R. (2023). Early forestry in WA. (K. Robson, Interviewer)

Underwood, R. (2024). Selection of suitable areas for growing Indian Sandalwood. (K. Robson, Interviewer)

Van Beuzekom, P. (2023). *Irrigation management plan.* Quintis Internal procedure.

Vanclay, J. K. (1994). Modelling forest growth and yield: applications to mixed tropical forests. School of Environmental Science and Management Papers, 537. *School of Environmental Science and Management Papers*(537).

Vanclay, J., & Sands, P. (2009). Calibrating the self-thinning frontier. *Forest Ecology and Management, 259*, 81 -85.

Verhaegen, D., Ofri, D., Fofana, I., Poitel, M., & Vaillant, A. (2005). Development and characterization of microsatellite markers in Tectona grandis (Linn. f). *Molecular Ecology Notes*(5), 945 - 947.

Walter, W. D., Garret, H. E., & Godsey, L. D. (2004). Use of herbicide to reduce stump-sprouting following thinning of an eastern black walnut agroforestry planting. *6th Walnut Council Research Symposium.*

Warburton , C. L., James, E. S., Fripp, Y. J., Trueman, S. J., & Wall, H. M. (2000). Clonality and sexual reproductive failure in remnant populations of Santalum lanceolatum (Santalaceae). . *Biological Conservation, 96*, 45 - 54. doi:doi: 10.1016/S0006-3207(00)00049-5

Weintraub, P. G., & Beanland, L. (2006). Insect vectors of phytoplasmas. *Annu. Rev. Entomol., 51*, 91 - 111. doi:DOI: 10.1146/annurev.ento.51.110104.151039

Weiss, E. A. (1997). Ch 13: Santalaceae. In *Essential Oil Crops* (pp. 516 - 555). CAB International.

West, P. W. (2013). Stand Density and Initial Spacing. In P. W. West, *Growing Plantation Forests* (pp. 99 - 114). Springer.

Wetherby, K. (2003). *Soil Description Book.*

Wheeler, W. D., Black, B., & Bugbee, B. (2023). Assessing water stress in a high-density apple orchard using trunk circumference variation, sap flow index and stem water potential. *Frontiers in Plant Science, 14.* doi:10.3389/fpls.2023.1214429

White, T. L. (1987). A conceptual framework for tree improvement programs. *New Forests, 1*, 325–342.

Wijayawardene , N. N., Hyde, K. D., Rajeshkumar, K. C., Hawksworth, D. L., Madris, H., Kirk, P. M., . . . Yurkov, A. (2017). Notes for genera: Ascomycota . *Fungal Diversity, 86*, 1-594. doi:https://doi.org/10.1007/s13225-017-0386-0

Wild, A. (1958). The phosphate content of Australian soils. *Australian Journal of Agricultural Science*, 193 - 204.

Williamson, L., Moore, S., & Warren, C. (2005). 1919 to 1935: A pivotal period for the forests of the southwest of Western Australia. *Proceedings 6th National Conference of the Australian Forest History Society Inc, Michael Calver et al. (eds).*

Winrock. (2024). Senna siamea – a widely used legume tree. Retrieved from https://winrock.org/

Xu, X., Du, X., Wang, F., Sha, J., Chen, Q., Tian, G., . . . Jiang, Y. (2020). Effects of Potassium Levels on Plant Growth, Accumulation and Distribution of Carbon, and Nitrate Metabolism in Apple Dwarf Rootstock Seedlings. *SEC Plant Nutrition, 11*. doi:https://doi.org/10.3389/fpls.2020.00904

Yaemphum, S., Unawong, W., & Tor-ngern, P. (2021). Sapwood area~DBH allometries for 14 common tree species in a successional tropical forest in Thailand. *Forestry: An International Journal of Forest Research, 95*(4), 562–571. doi:https://doi.org/10.1093/forestry/cpab054

Zemp, D. C., Ehbrecht, M., Seidel, D., Ammer, C., Craven, D., Erkelanz, J., . . . Kreft, H. (2019). Mixed-species tree plantings enhance structural complexity in oil palm plantations. *Agriculture, Ecosystems & Environment, 283*. doi:https://doi.org/10.1016/j.agee.2019.06.003

Made in the USA
Las Vegas, NV
03 November 2026

57420899R00176